Barocaloric Effects in the Solid State

Materials and methods

Online at: https://doi.org/10.1088/978-0-7503-4690-0

Barocaloric Effects in the Solid State

Materials and methods

Edited by
Pol Lloveras
Group of Characterization of Materials, Department of Physics, EEBE and Barcelona Research Center in Multiscale Science and Engineering, Universitat Politècnica de Catalunya, Barcelona, Spain

IOP Publishing, Bristol, UK

ISBN 978-0-7503-4690-0 (ebook)
ISBN 978-0-7503-4688-7 (print)
ISBN 978-0-7503-4691-7 (myPrint)
ISBN 978-0-7503-4689-4 (mobi)

DOI 10.1088/978-0-7503-4690-0

Version: 20231001

IOP ebooks

British Library Cataloguing-in-Publication Data: A catalogue record for this book is available from the British Library.

Published by IOP Publishing, wholly owned by The Institute of Physics, London

IOP Publishing, No.2 The Distillery, Glassfields, Avon Street, Bristol, BS2 0GR, UK

US Office: IOP Publishing, Inc., 190 North Independence Mall West, Suite 601, Philadelphia, PA 19106, USA

Contents

Preface

The term *barocaloric* was coined in 2000 by Strässle and Furrer in their paper *Cooling by adiabatic application of pressure—the barocaloric effect*, by analogy with the magnetocaloric effect as a potential new solid-state method for cooling. After a decade of modest research in this phenomenon, the paper by Mañosa, Tamarit and collaborators published in 2010 in *Nature Materials* reported a giant barocaloric effect in a Heusler alloy, which marked a turning point from which research in this field undertook a good push that still lasts today. Looking back just over a decade later, a giant leap is evident. Reported figures of merit of barocaloric materials have been improved dramatically: the maximum refrigerant capacity has increased by two orders of magnitude, greatly outperforming the best magnetocaloric, electrocaloric and elastocaloric values, whereas the required pressures have been reduced by one order of magnitude. In between, a sustained growth along the years with a continuous sequence of milestones is at the origin of this outstanding progress. An increasing annual number of peer reviewed publications, citations and research groups involved, its presence in international conferences, workshops and symposia and, most importantly, first steps towards the realization of barocaloric prototypes, demonstrate that this field enjoys a good state of health and promises a hopeful future. Definitely, the barocaloric effect has earned its own place as a standalone field in condensed matter physics and more specifically among caloric effects in the solid state. It is precisely due to the amount of accumulated knowledge and, more importantly, due to its relevance in terms of future potential, that we believe such a book is appropriate and timely to summarize the main advances as enumerated above.

Research on barocaloric effects establishes a common thread for a number of different approaches with multifaceted interests. On the one hand, the leitmotif of this field is the potential applicability as a more sustainable alternative for cooling (and heating) to the traditional vapor compression cycle. In this context, the investigation of figures of merit about cooling performance of materials are addressed within materials science research and a few engineering approaches have provided support as a feasible method from the technical point of view. Additionally, basic research on different areas of condensed matter physics has been conducted to unveil the physical mechanisms underlying the barocaloric response, generating knowledge with high scientific value beyond application. On the other hand, one of the keys for the success and diversity of this field relies on the rich chemical variety of potential barocaloric materials, which to date encompass magnetostructural alloys, pure ferroelastic materials, polymers, ferroelectric compounds, superionic conductors, spin-crossover complexes, plastic crystals, hybrid organic–inorganic perovskites and metalorganic frameworks. Each material family has shown particular pros and cons for a competitive device implementation and raises specific interest from the fundamental point of view. Last but not least, stimulated by the needs of this diversity of materials and approaches, a variety of techniques, both theoretical and experimental, are used, including standard commercial instruments and bespoke homemade equipment that have been developed specifically to face the particularities of the material families. This wealth of materials and methods is therefore inherent to the field of barocalorics. Although it is too early to anticipate whether barocaloric effects in the solid state will be used in future cooling devices or not, it can undoubtedly be stated that this field has already contributed to the progress of science and will continue doing so for a long time.

This book aims at presenting an extensive and multifaceted review on materials and methods and at giving engineering insights. It is conceived as a compendium of essential aspects along with cutting-edge research. The book is structured as follows: chapter 1 is devoted to a wide introduction presenting general aspects that are relevant for the field, ranging from fundamental thermodynamics and figures of merit, to measurement methods and some considerations for barocaloric applications. Chapters 2–7 deal with different barocaloric material families ordered according to the first publication for each type: polymers, magnetic alloys, fluorides and oxyfluorides, spin crossovers, hybrid compounds and plastic crystals. Chapter 8 presents the main advanced tools for theoretical modeling of barocaloric materials and effects, and chapter 9 addresses the engineering modeling of barocaloric devices. For the topics covered in this book, We hope it will raise interest for the wide community of condensed matter physicists, material scientists and engineers, both for those young and senior researchers interested in joining the field as well as for those already expert in barocalorics. Researchers will find here ideas for further investigations on already known materials and inspiration for the scrutiny of new or improved compounds.

Acknowledgements

I would like to thank all the contributors who have made it possible for this book to come to light.

Editor biography

Pol Lloveras

Courtesy of Jaume Solé.

Pol Lloveras is associate professor at the Group of Characterization of Materials of the Department of Physics of the Universitat Politècnica de Catalunya (Catalonia, Spain). He devoted his PhD (Universitat de Barcelona, Spain, 2010) to the theoretical study and modeling of structural and thermodynamic properties of ferroelastic and magnetoelastic systems, including the elastocaloric effect. After a first post-doctoral stage at the *Laboratoire Mchanique des Solides* of the École Polytechnique (Palaiseau, France), he joined his current Group in 2012. Since then, he has specialized in the analysis of barocaloric effects in the solid state for environmentally friendly cooling and heating methods. His works involve high-pressure techniques such as calorimetry, dilatometry and diffraction, the latter performed in a number of short stages in large-scale facilities. Over recent years, thanks to collaborations with other groups worldwide, he has been investigating the BCE of a vast range of materials including metallic alloys, ferroelectrics, hybrid perovskites, metalorganic frameworks, organic compounds and spin-crossover complexes. It is worth mentioning the colossal barocaloric effects unveiled in a number of plastic crystals and other compounds, which bring solid-state performances close to current harmful fluids. At present, his interests aim at exploring novel BC materials and addressing a more applied approach towards device design.

List of contributors

María Barrio
Group of Characterization of Materials, Department of Physics, EEBE and Barcelona Research Center in Multiscale Science and Engineering, Universitat Politècnica de Catalunya, Barcelona, Catalonia, Spain

Juan Manuel Bermúdez-García
Grupo Quimolmat, CICA-Centro Interdisciplinar de Química y Biología, Universidade da Coruña, A Coruña, Spain
and
Grupo Quimolmat, Departamento de Química, Facultade de Ciencias, Universidade da Coruña, A Coruña, Spain

Evgeniy V Bogdanov
Kirensky Institute of Physics, Federal Research Center KSC SB RAS, Krasnoyarsk, Russia
and
Institute of Engineering Systems and Energy, Krasnoyarsk State Agrarian University, Krasnoyarsk, Russia

Alexander Magnus G Carvalho
Departamento de Engenharia Mecânica, Universidade Estadual de Maringá, CEP, Maringá PR, Brazil

Claudio Cazorla
Department of Physics, Universitat Politècnica de Catalunya, Barcelona, Spain

Luca Cirillo
Department of Industrial Engineering, University of Naples Federico II, Naples, Italy

Ignacio Delgado-Ferreiro
Grupo Quimolmat, CICA-Centro Interdisciplinar de Química y Biología, Universidade da Coruña, A Coruña, Spain
and
Grupo Quimolmat, Departamento de Química, Facultade de Ciencias, Universidade da Coruña, A Coruña, Spain

Igor N Flerov
Kirensky Institute of Physics, Federal Research Center KSC SB RAS, Krasnoyarsk, Russia
and
Institute of Engineering Physics and Radioelectronics, Siberian Federal University, Krasnoyarsk, Russia

Javier García-Ben
Grupo Quimolmat, CICA-Centro Interdisciplinar de Química y Biología, Universidade da Coruña, A Coruña, Spain
and
Grupo Quimolmat, Departamento de Química, Facultade de Ciencias, Universidade da Coruña, A Coruña, Spain

Mikhail Gorev
Kirensky Institute of Physics, Federal Research Center KSC SB RAS, Krasnoyarsk, Russia
and
Institute of Engineering Physics and Radioelectronics, Siberian Federal University, Krasnoyarsk, Russia

Adriana Greco
Department of Industrial Engineering, University of Naples Federico II, Naples, Italy

Malcolm A Halcrow
School of Chemistry, University of Leeds, Woodhouse Lane, Leeds, United Kingdom

William Imamura
Departamento de Engenharia Mecânica, Universidade EM, CEP, Maringá PR, Brazil

Yukinobu Kawakita
J-PARC Center, Japan Atomic Energy Agency, Japan

Bing Li
Shenyang National Laboratory for Materials Science, Institute of Metal Research, Chinese Academy of Sciences, Shenyang, China
and
School of Materials Science and Engineering, University of Science and Technology of China

Kuo Li
Center for High Pressure, Science and Technology Advanced Research, China

Shangchao Lin
Key Laboratory for Power Machinery and Engineering of Ministry of Education, Institute of Engineering Thermophysics, School of Mechanical Engineering, Shanghai Jiao Tong University, China

Pol Lloveras
Group of Characterization of Materials, Department of Physics, EEBE and Barcelona Research Center in Multiscale Science and Engineering, Universitat Politècnica de Catalunya, Barcelona, Catalonia, Spain

Claudia Masselli
Department of Industrial Engineering, University of Naples Federico II, Naples, Italy

Xavier Moya
Department of Materials Science and Engineering, University of Cambridge, Cambridge, United Kingdom

Nilson Antunes de Oliveira
Instituto de Física Armando Dias Tavares, Universidade do Estado do Rio de Janeiro, Rio de Janeiro, RJ, Brazil

Lucas S Paixão
Departamento de Engenharia Mecânica, Universidade EM, CEP, Maringá PR, Brazil

Jorge Salgado-Beceiro
Grupo Quimolmat, CICA-Centro Interdisciplinar de Química y Biología, Universidade da Coruña, A Coruña, Spain
and
Grupo Quimolmat, Departamento de Química, Facultade de Ciencias, Universidade da Coruña, A Coruña, Spain

Karl G Sandeman
Department of Physics, Brooklyn College of The City University of New York, New York, USA
and
Physics Program, The Graduate Center, CUNY, New York, USA

Josep-Lluís Tamarit
Group of Characterization of Materials, Department of Physics, EEBE and Barcelona Research Center in Multiscale Science and Engineering, Universitat Politècnica de Catalunya, Barcelona, Catalonia, Spain

Dehong Yu
Australian Nuclear Science and Technology Organization, Australia

Ming Zeng
Group of Characterization of Materials, Department of Physics, EEBE and Barcelona Research Center in Multiscale Science and Engineering, Universitat Politècnica de Catalunya, Barcelona, Catalonia, Spain

Zhe Zhang
Shenyang National Laboratory for Materials Science, Institute of Metal Research, Chinese Academy of Sciences, Shenyang, China
and
School of Materials Science and Engineering, University of Science and Technology of China

Chapter 1

Introduction to the barocaloric effect

Pol Lloveras and Josep-Lluís Tamarit

1.1 Motivation

Cooling technologies are considered amongst the greatest transformational achievements of the 20th century [1]. Refrigeration and air conditioning have allowed the development of new, effective and efficient industrial processes, agricultural production and data centers, the storage and distribution of pharmaceutical drugs, vaccines and food, and thermal comfort of populated spaces, both in buildings and transport, thus leading to drastic changes in consumer habits and new human settlements in warm regions. Just how essential refrigeration is to the modern lifestyle is evident from the myriad of refrigeration appliances in use today, estimated at nearly 4 billion units worldwide, and that will continue to increase in the coming decades due to improvement of conditions in developing countries and global warming. According to one estimation [2], this number could rise to 14 billions by 2050 if all cooling needs are covered.

Around 80% of the total amount of current refrigeration devices employ the vapor compression cycle (VCC) [3, 4], a method that was developed in the 19th century concomitantly with the formulation of classical thermodynamics. It was a significant improvement over the passive evaporation methods existing since antiquity thanks to the active control of the exchange of the latent heat of evaporation by means of changes in the applied pressure. The ultimate aim of the cooling cycle is to bring heat from a cold end to a hot end and can be used as such also as a heat pump (see figures 1.1(a) and (b)). In the particular case of a reverse Brayton cycle (see figures 1.1(c) and (d)), this technique essentially achieves cooling through: (i) an adiabatic compression of the refrigerant gas leading to a temperature increase; (ii) thermalization of the refrigerant with the environment leading to its condensation; (iii) an adiabatic decompression leading to a temperature decrease; and (iv) thermalization with the target to be cooled leading to the evaporation of the refrigerant, which completes the cycle and so it is ready to start again.

doi:10.1088/978-0-7503-4690-0ch1

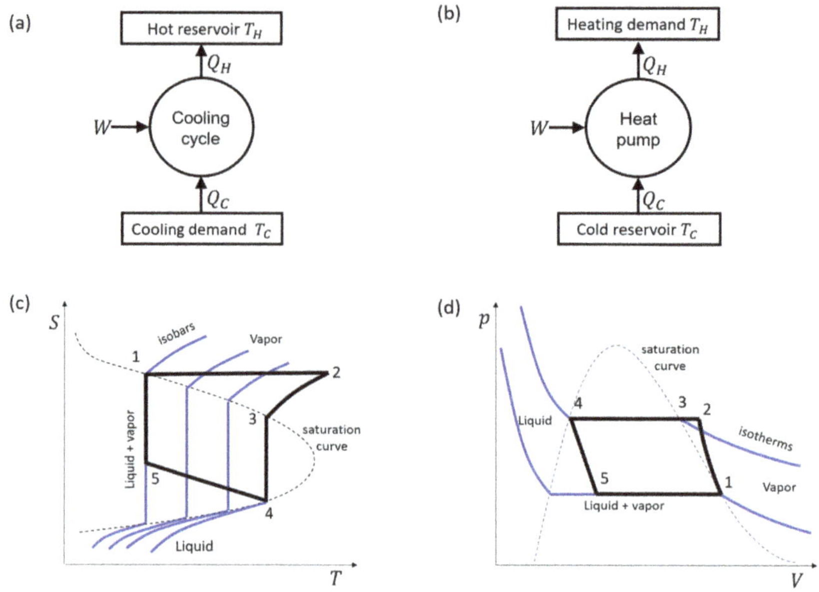

Figure 1.1. First principle of thermodynamics for (a) the refrigeration cycle and (b) the heat pump. VCC cycle in the (c) entropy-temperature and (d) pressure–volume diagrams. 1 → 2: compression of vapor. 2 → 3: release of excess vapor heat. 3 → 4: condensation of vapor. 4 → 5: decompression of liquid leading to partial evaporation. 5 → 1: evaporation of remaining liquid.

Early devices used the so-called *natural refrigerants*, like diethyl ether (R-610), dimethyl ether (R-170), carbon dioxide (R-744), ammonia (R-717), sulphur dioxide (R-764), methyl chloride (R-40), isobutane (R-600a) and propane (R-290) [5], but major safety drawbacks concerning high flammability, toxicity or low efficiency limited widespread use. In 1928, the synthesis by General Motors of a new family of compounds called halocarbons (containing carbon and halogens such as fluorine, bromine and chlorine atoms, as the well-known chlorofluorocarbons) marked a turning point because they were nonflammable and nontoxic, and rapidly dominated the domestic and industrial markets. However, the discovery of the ozone depleting potential (ODP) of these fluids in 1976 and, subsequently, the hole in the ozone layer in 1986 [6], led to their ban under the Montreal Protocol [7] following the Vienna Convention. These fluids were then replaced by hydrochlorofluorocarbons (HCFCs) with small ODP, and hydrofluorocarbons (HFCs) with zero ODP. However, HFCs and HCFCs have a huge heat-trapping ability, with a global warming potential (GWP) of typically more than a thousand times that of CO_2 when considering a time span of a hundred years. Regrettably, due to their accidental yet systematic release to the atmosphere due to the lack of proper maintenance and waste disposal, HFCs and HCFCs are deemed responsible for ~3.5% of total greenhouse gas emissions. On the other hand, the energy powering cooling devices, which present efficiencies in the range of 30%–60% and currently amount to 15% of global electricity consumption, are produced in part by fossil fuel plants, so that taking into account this contribution the overall environmental harm rises to ~8%

of total greenhouse gas emissions [4]. It is estimated that improvement of efficiency alongside the replacement of these fluids would save up to 210–460 gigatonnes of CO_2-equivalent in the next 40 years, depending on future rates of decarbonisation [8].

Consequently, the global warming emergency is now forcing the phase out of these refrigerants in the short to medium term, as agreed by the Kigali Amendment to the Montreal Protocol [9], which has been ratified in 2022 by 129 countries, including USA [10], EU [11], China [12] and India [13]. Thus, it is clearly urgent to develop more sustainable alternatives. As for manufacturers, most efforts are logically focused on finding new fluid refrigerants with much lower GWP due to the minimal associated costs of adapting a mature technology, as compared to the costs that would involve the engineering design of a completely new system. In this sense, formerly used natural refrigerants (isobutane, propane, ammonia, carbon dioxide among others) or a new family of low-GWP high-efficiency refrigerants called hydrofluoroolefins (HFOs), are already being employed in some new VCC devices [5], but again safety issues associated with toxicity and/or flammability, and other aspects such as limited availability or moderate efficiency [14, 15], leave room for improvement. As a particular yet significant example, the HFC with chemical formula CH_2FCF_3 (R134a), with a GWP of 1300 times that of CO_2 considering the time lapse of 100 years, and widely used in air conditioners of vehicles (a sector that particularly suffers from high leakage rates) contributes on average 1.3 tons of CO_2-equivalent over the lifetime of a vehicle [16]. For this reason, its use was already prohibited in Europe in all new cars sold from January 2017 due to the EU Regulation 2006/40/EC [17] so that it was first replaced by the HFO R1234yf. However, it was discovered that in case of accident, HFO ignition yields some very dangerous substances such as hydrogen fluoride, for which many important automobile manufacturers (such as Daimler, Toyota, Mazda and the Volkswagen group) have eliminated or will eliminate its use in their vehicles and in 2014, *Autobild*, the Europe's leading automobile magazine, stigmatized the HFO R1234yf as 'the killer refrigerant' [18]. This fluid is now being replaced by CO_2 which is less efficient and requires higher pressures of ~7 MPa.

It is worth mentioning here that current devices for heating also present critical issues in terms of environmental harm. Fuel-burning heaters, ranging from domestic gas-fired boilers to industrial appliances, show small efficiencies, whereas much more efficient heat pumps are also based on the VCC of harmful HFC fluids [19].

More disruptive technologies that avoid the use of harmful fluids are also being investigated but mainly by public research institutions due to the inescapable investment risk, which limits faster innovation and development. Examples are thermoacoustic, thermotunneling, thermoelectric or solid-state caloric methods, the latter being amongst the best positioned [20]. In general, caloric effects refer to changes in entropy or temperature undergone by a material when an externally applied field is modified isothermally or adiabatically, respectively, and can be used to generate a thermodynamic cooling cycle. While not limited to that, generally the magnitude of these quantities is enhanced close to first-order phase transitions (FOPTs). Essentially, most solid-state caloric methods are based on the same operation principle as VCC: the refrigerant is forced to undergo a FOPT forth and back assisted by an external field along each thermodynamic cycle so that at the exothermic transition the material expels heat to a thermal bath, whereas at the endothermic transition the material absorbs heat from the target to cool. The main difference relies on the substitution of the refrigerant fluid with a solid and the use of pressure with any external field (magnetic, electric, mechanical) that can drive the phase transition [21]. Liquid–vapor transitions used in the VCC are typically much more energetic than solid-to-solid FOPTs and require relatively small input work. Given the maturity of this technology, VCC is at present more favorable than solid-state caloric methods in terms of overall performance at the system level [22]. However, the aforementioned drawbacks exhibited by the fluid refrigerants are giving solid-state methods an opportunity to give proof of their competitivity. In this sense, there is evidence that solid-state caloric materials could feature improved properties compared to VCC, such as higher efficiency [23], low or negligible noise, low maintenance cost and easy handling, and downscalability [24, 25], allowing for portable battery-powered devices or even for micro- or nanoscale cooling, which can be a solution for the miniaturization of electronic devices critically restricted by failure due to overheating [26].

Solid-state caloric effects are classified according to the nature of the applied field. The barocaloric (hydrostatic-pressure driven, BC for short) effect was born as the fourth class of solid-state caloric effect, joining the existing elastocaloric (uniaxial stress-driven, first reported by Gough in 1805 [27]), magnetocaloric (magnetic field-driven, first observed by Warburg in 1881 [28] and reported by Weiss and Piccard in 1917 [29]) and electrocaloric effects (electric field-driven, first reported by Kobeko and Kurtschatov in 1930 [30]). Naturally, elastocaloric and barocaloric effects were then grouped within the so-called *mechanocaloric* effects to account for any caloric effect originated by the application of a mechanical field. Very recently, this group has been further extended to include other types of stress, like the torsiocaloric (or twistocaloric) effect (torsional stress) [31] and the flexocaloric effect (bending stress) [32]. To the present day, the magnetocaloric effect is by far the most investigated. Its technical viability has already been demonstrated by a number of prototypes and some devices are already commercialized. Main opportunities are found in both the commercial and industrial sectors and it is expected to have a market size of 4.5 million USD in 2022 and will reach 165.8 million USD by 2027. Bottlenecks limiting a widespread commercialization are the high initial investment, moderate magnetic field strengths created by permanent magnets and the lack of suitable magneto-caloric materials, which are expensive, toxic and little available [33]. While less investigated, elastocaloric and electrocaloric cooling have also demonstrated their technical viability [34, 35] but both suffer from fatigue upon cyclability along millions of cycles as required in real applications and large electrocaloric effects arise only in thin films due to electrical breakdown, which make necessary multilayer capacitors that hamper upscaling [36].

Solid-state BC effects were first reported in 1998 by Müller *et al* [37] but the research in this field was accelerated by the publication of Mañosa *et al* in 2010 [38]. At the beginning such studies were mainly focused on magnetostructural materials inspired by the volume changes displayed by magnetocaloric alloys, but soon the ubiquity of volume changes in matter (i.e. the basic requirement for BC effects to emerge) evidenced the large potential of this field and led the research on BC materials to experience great and fast progress and to an increasing number of research groups interested in this field [39]. This has resulted in the identification of colossal isothermal entropy changes, typically defined as those surpassing 100 J K^{-1} kg^{-1} in different material families including plastic crystals, polymers, hybrid organic–inorganic perovskites and spin crossover metal complexes, greatly outperforming caloric effects driven by other fields [39]. It is worth mentioning here several very recent outstanding achievements that report colossal or large reversible BC effects at low or very low pressures near room temperature so that the best solid-state BC materials are getting closer and closer to the performance of current harmful refrigerants. In particular, large caloric effects have been obtained in spin crossover complexes under 30 MPa [40] and 15 MPa [41] and colossal BC effects were reported in a layered perovskite [42] and in a plastic crystal [43] under 50 MPa pressure changes. These advances, along with the fact that pressure is easy to generate and that the possibility of using powder avoids any problem related to fatigue, are bringing good expectations and hopes to the field of barocalorics. It is worth mentioning here that pressure-induced gas adsorption in metal–organic frameworks have led to colossal caloric effects obtained under only 1.6 MPa [44]. Despite the external application of hydrostatic pressure, the physicochemical mechanism occurring in the system is completely different from canonical barocaloric effects, for which it has been named *breathing-caloric* effect.

The development of BC cooling devices is still very incipient [45–47] and there are limitations to overcome before any implementation of a BC technology can take place, for instance in terms of reducing the necessary pressures which now would require work recovery. However, some engineering numerical studies have suggested that, in addition to avoiding the use of harmful fluids, the BC technology could offer better efficiencies than VCC devices and more compact systems due to less compressor swept volume [23, 48] and also than other solid-state caloric methods [49]. Actually, it is still not very clear which would be the best materials to use in a future BC device as there is not a single figure of merit and some performance parameters may depend on the device design and application. As occurring in VCC devices, different materials could be the best candidates depending on the specific application. Naturally, most materials have been sought for domestic air conditioning or refrigeration.

On the other hand, in addition to solid-state cooling and heat pumping, caloric effects have also been proposed for other energy-related applications such as thermal energy management and power conversion [50, 51]. Beyond solar thermal energy, 70% of all energy produced by mankind is lost to the environment, posing a serious environmental thread in the form of not only a direct heating impact but with indirect consequences, as contributing to the hot island effect, which in turn increases energy consumption and affects air and water quality. However, this waste heat is viewed as a renewable energy source, thus offering a solution to the demand for improving process efficiencies. In the industrial sector, 20% to 50% of the consumed energy is dissipated, and it is deemed that up to 30% of it could be recovered for further reutilization [52–55]. Moreover, most of current heat recovery technologies require temperatures above 200 °C (high grade waste heat) but most waste heat is of low grade [56, 57]. Here, endothermic FOPTs in many BC materials (e.g. plastic crystals [58], layered hybrid perovskites [59]) were proposed many decades ago as *phase change materials* (PCMs) for passive thermal energy storage [60–62] because the very large latent heat is logically beneficial for this application. Furthermore, involving only solid phases allow the fabrication of functional composites, avoid leakages and facilitate their handling as compared to melting phases, that may require encapsulation [63, 64]. Actually, some solid BC materials have been proposed as active shape-stabilizers for molten PCMs [65].

Interestingly, pressure-assisted thermal management has recently been proposed [66–68] as an improved solution for waste heat management by actively controlling the absorption and storage of heat and its subsequent release when desired by the application of pressure. Therefore, it is conceivable that many BC materials could also offer this enhanced BC functionality, which in comparison to passive methods, takes advantage not only of the large latent heat but also of the pressure-sensitivity of the transition associated with typically large volume changes. This novel method could be viewed as an analogous advance to what the vapor compression cycle represented in the 19th century compared to ancient cooling methods based on passive evaporation. Interestingly, in contrast to cyclic cooling, harvesting and storage of heat does not require high operational frequencies, and transition hysteresis can be viewed as a positive feature as it enhances the actual temperature range of stability to the heat-storing phase. Therefore, for waste heat barocaloric management, solid-state phase change materials that transform above room temperature are required and a moderate hysteresis may be advantageous, which would otherwise rule them out for domestic refrigeration and air conditioning, as explained in section 1.2.4. Therefore, solid-state BC methods could be part of the solution to two of the *five thermal energy grand challenges for decarbonization* identified recently [69]. With respect to power conversion, it seems difficult to conceive the BC in solids as a direct means to produce electricity from waste heat as easily as can be done with other caloric methods. Theoretically, BC materials could be an intermediary agent to heat the boiling working fluids in the organic Rankine cycle [70] or as an active agent by using the high vapor pressure and/or relative easy sublimation ability of some of them [71].

The following sections are intended to provide a broad and basic scientific background that may be common to specific BC materials and methods covered in the subsequent chapters, but that could be omitted there for the sake of simplicity. This includes fundamental thermodynamics as well as an overview of the main methods and experimental techniques currently used in this field.

1.2 Thermodynamic background

Caloric effects are defined as isothermal entropy changes ΔS or adiabatic temperature changes ΔT that a system undergoes when subjected to changes in an external field. Therefore, the fundamental requirement for a caloric effect to emerge is that the caloric system undergoes a change in at least one physical quantity (generalized coordinate X) when the conjugated field y is changed, such that the energy in the form of work exchanged by the thermodynamic system is $\sim\int ydX$. Whether this work will transform to thermal energy or will result in an entropy change, or partially to both will depend on the specific conditions (isothermal, adiabatic, isofield, etc). In this section, the basic thermodynamics related to these quantities and phenomena will be reviewed. While the formalism can be carried out for any system yXT, here it will be particularized for the BC case, i.e. a pressure–volume–temperature pVT ($y = -p$, $X = V$) hydrostatic system [72].

1.2.1 Barocaloric effects in homogeneous systems

For a homogeneous system, let us consider[1] the entropy S as an exact differential in terms of T and p:

$$dS = \left(\frac{\partial S}{\partial T}\right)_p dT + \left(\frac{\partial S}{\partial p}\right)_T dp \tag{1.1}$$

Using the following thermodynamic relations

$$\left(\frac{\partial S}{\partial T}\right)_p = \frac{C_p}{T}, \tag{1.2}$$

$$\left(\frac{\partial S}{\partial p}\right)_T = -\left(\frac{\partial V}{\partial T}\right)_p, \tag{1.3}$$

equation (1.1) can be rewritten as

$$TdS = C_p\, dT - \left(\frac{\partial V}{\partial T}\right)_p dp \tag{1.4}$$

Given that S is a state function, by isolating dS and integrating the previous equation, entropy changes between any two states (T_0, p_0) and (T_1, p_1) can be obtained following any path linking these states:

$$S(T_1, p_1) - S(T_0, p_0) = \int_{T_0}^{T_1} \frac{C_p}{T} dT - \int_{p_0}^{p_1} \left(\frac{\partial V}{\partial T}\right)_p dp \tag{1.5}$$

Notice that taking $T_0 = 0$ K and $p_0 = 0$ Pa, and assuming that there is no residual entropy $[S(T = 0, p = 0) = 0]$, the absolute entropy can be calculated:

$$S(T, p) = \int_0^T \frac{C_p}{T'} dT' - \int_0^p \left(\frac{\partial V}{\partial T}\right)_{p'} dp' \tag{1.6}$$

[1] Notice that in the field of caloric effects. intensive quantities such as specific entropy, specific heat capacity, specific volume, etc are traditionally denoted by capital letters.

Using equation (1.5), isothermal entropy changes occurring upon a pressure change from p_0 to p_1 can be calculated by imposing $dT = 0$, obtaining:

$$\Delta S(T; p_1 \rightarrow p_0) = -\int_{p_0}^{p_1} \left(\frac{\partial V}{\partial T}\right)_p dp \tag{1.7}$$

Similarly, adiabatic temperature changes occurring upon a pressure change from p_0 to p_1, can be calculated by imposing $dS = 0$, obtaining:

$$\Delta T(S; p_1 \rightarrow p_0) = \int_{p_0}^{p_1} \frac{T}{C_p}\left(\frac{\partial V}{\partial T}\right)_p dp \tag{1.8}$$

Notice that the BC effects are directly correlated with the coefficient of isobaric thermal expansion $\alpha = (1/V)(\partial V/\partial T)_p$ and not with the isothermal compressibility $\chi_T = -(1/V)(\partial V/\partial p)_T$. From equations (1.7) and (1.8), it is clear that a large positive thermal expansion will give rise to large BC effects, whereas invar materials will exhibit null BC effects. In equation (1.8), two approximations are usually made: first, as C_p is difficult to measure at different pressures, it is assumed to be independent of pressure. Second, T is evaluated as a constant and equal to the initial temperature of the process. Accordingly, equation (1.8) results in the following expression:

$$\Delta T(Sp_0 \rightarrow p_1) \simeq \frac{T}{C_p}\int_{p_0}^{p_1} \left(\frac{\partial V}{\partial T}\right)_p dp = -\frac{T}{C_p}\Delta S(T, p_0 \rightarrow p_1) \tag{1.9}$$

The former expression indicates that, within this approximation, ΔT and ΔS are proportional to each other. Notice that equations (1.7) and (1.8) involve derivatives, so that it can strictly be used only if V is a continuous and differentiable function of temperature, as it does for a homogeneous system. As at FOPTs volume discontinuities appear, equations (1.7), (1.8) and (1.9) cannot be used and an alternative treatment is necessary (see section 1.2.2).

Despite the usual approximation $C_p(T, p) \simeq C_p(T, p_{\text{atm}})$ done in equation (1.9), the dependence of C_p on pressure as required in equation (1.8) can be evaluated by equating $\partial^2 S/\partial p \partial T = \partial^2 S/\partial T \partial p$, by taking derivatives with respect to p in equation (1.2) and with respect to T in equation (1.3), which leads to:

$$\left(\frac{\partial C_p}{\partial p}\right)_T = -T\left(\frac{\partial^2 V}{\partial T^2}\right)_p \tag{1.10}$$

Care must be taken not to misinterpret this equation. It entails that at temperatures where volume is linear with temperature, then C_p is independent of pressure. Notice, however, that this does not entail that S does not depend on pressure at these temperatures. Actually, equation (1.7) entails that S depends on pressure at any temperature where $(\partial V/\partial T)_p$ is just non-zero. In fact, as at $T = 0$ K, $S(T = 0\text{ K}, p) = 0$ for all p, then $\Delta S(T = 0\text{ K}; p_0 \rightarrow p_1) = 0$ under any pressure change $p_0 \rightarrow p_1$. Consequently, equation (1.7) entails that at $T = 0$ K, $(\partial V/\partial T)_p = 0$. As in general it can be stated that all materials show non-zero thermal expansion at some temperatures, $(\partial V/\partial T)_p \neq 0$, it can be easily deduced that necessarily at some temperature $(\partial^2 V/\partial T^2)_p \neq 0$. Therefore, equation (1.10) implies that C_p must depend on pressure at least at some temperature. Figure 1.2 shows a qualitative mathematical example aimed at illustrating the relationship existing between V, C_p and S. Let us assume that in the temperature interval below T_1, $V(T)$ is a second order polynomial concave function (see figure 1.2(a)). Therefore $(\partial^2 V/\partial T^2)_p = \text{const} > 0$. Therefore, according to equation (1.10), $\left(\partial C_p/\partial p\right)_T \propto -T$ so that C_p decreases with increasing pressure and the increasing rate is higher at higher temperatures (see figure 1.2(b)). In turn, $(\partial V/\partial T)_p$ increases when T increases

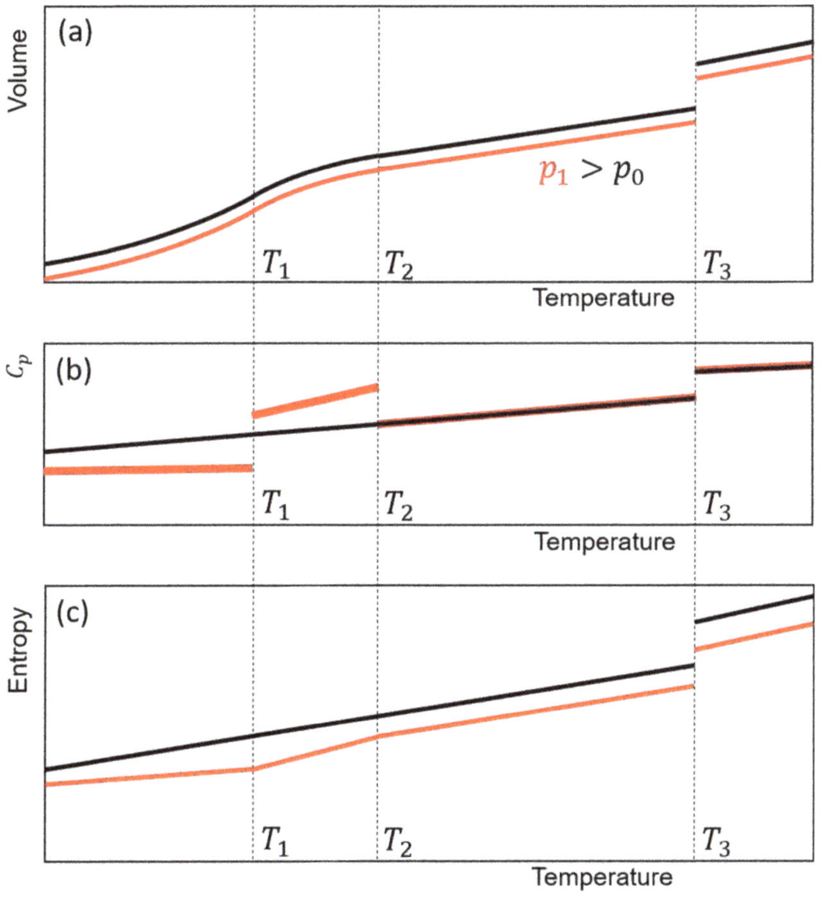

Figure 1.2. Qualitative example of relationship between isobaric (a) volume, (b) heat capacity and (c) entropy as a function of temperature for two different pressures. For more details, see the text.

and therefore $|\Delta S(T; p_0 \rightarrow p_1)|$ increases when increasing temperature (see figure 1.2(c)). Let us now assume that in the temperature interval $[T_1, T_2]$, $V(T)$ is a second order polynomial convex function (see figure 1.2(a)). Then, according to equation (1.10), C_p increases with increasing pressure and the increasing rate is higher at higher temperatures (see figure 1.2(b)). In turn, $(\partial V/\partial T)_p$ decreases when T increases and therefore $|\Delta S(T; p_0 \rightarrow p_1)|$ decreases when increasing temperature (see figure 1.2(c)). Finally, in the temperature interval $[T_2, T_3]$, $V(T)$ is linear (see figure 1.2(a)) so that $(\partial^2 V/\partial T^2)_p = 0$. Therefore, according to equation (1.10), C_p is independent of pressure (see figure 1.2(b)) but according to equation (1.7) S does depend on pressure (see figure 1.2(c)).

1.2.2 Barocaloric effects at first-order phase transitions

Caloric effects are usually sought near FOPTs because the transition latent heat promises a large heat exchange and the shift of the transition to the external field may give rise to a large temperature change. At ideal FOPTs, the equations presented in the former section cannot account for the discontinuities in temperature and pressure of some state functions such as V and S. Nonetheless, measurements across FOPTs in real materials often provide continuous changes in these variables, due to the nature of the measurements or to the athermal nature of materials themselves, or a combination of both. However, even in this case it has been shown that Maxwell relations may fail across FOPTs due to wrong protocols [73, 74] or strong non-equilibrium dynamics [75–77]. Furthermore, mathematical derivatives of measured quantities are critically affected by noise and sampling frequency, which makes the proper application of equations (1.7) and (1.8) very challenging [78]. Instead, to analyze the BC effects at FOPTs it may be more helpful to consider the (reverse) Brayton cycle, consisting of adiabatic and isobaric processes (see figure 1.3). It can be described as follows:

- (a) The BC material is at phase I, the high-temperature phase, at the lowest pressure (typically atmospheric pressure, p_{atm}) and at the temperature of a thermal bath, T_{B} (typically room temperature for air conditioning).
- (a)→(b) Pressure is increased adiabatically up to p so that the BC material increases temperature up to T_2.
- (b)→(c) The BC material is brought into thermal contact with a bath at $T_{\mathrm{B}} < T_2$ to provoke the material's thermalization at T_{B} through the release of heat. Notice that through this process the material typically transforms to phase II, the low-temperature phase, such that the amount of the heat transferred to the bath is of the order of the latent heat of the transformation.
- (c)→(d) Pressure is removed adiabatically down to p_{atm} so that the BC material cools down to $T_1 < T_{\mathrm{B}}$.
- (d)→(a) The BC material is brought into thermal contact with the target to be cooled such that the material absorbs heat from that target, thus transforming back to phase I and being ready to start the cycle again, whereas the target decreases its temperature, fulfilling the goal of the cycle.

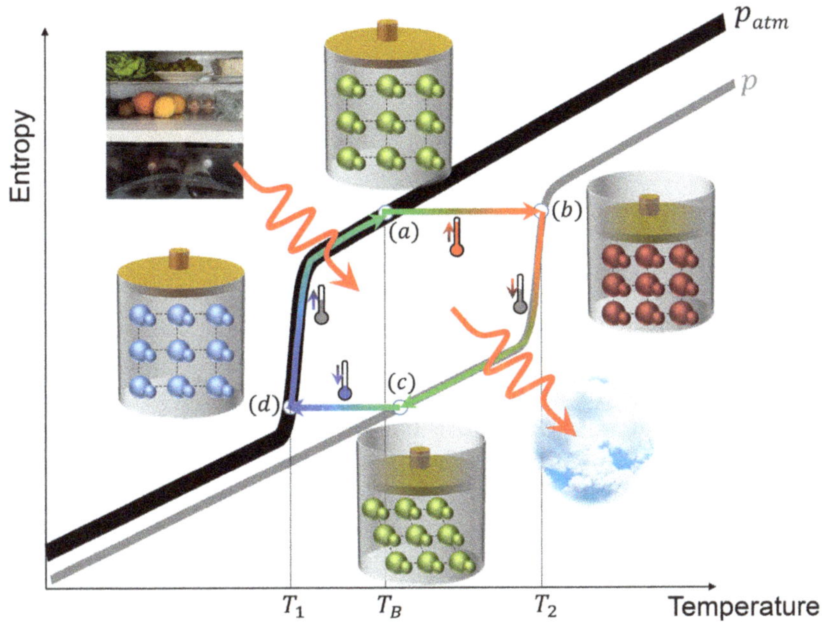

Figure 1.3. Schematic representation of the reverse Brayton cycle for cooling.

From the cycle above, two features can be deduced: (i) the amount of heat absorbed from the target to cool will be of the order of the latent heat of the FOPT; (ii) up to moderate pressures, the temperature change will be of the order of the temperature shift of the transition with the pressure change. Moreover, the temperature span achieved by the BC material must be such that the highest temperature (T_2) lies above the temperature of the hot end (i.e. T_B) and the lowest temperature (T_1) lies below the cold end (the lowest desired temperature of the target to be cooled). Evidently, the larger the difference between T_2 and the hot end, and between T_1 and the cold end, the faster the heat will be exchanged between the BC material and its surroundings. Interestingly, the two relevant quantities, the enthalpy change at the transition (latent heat) ΔH_t (or the entropy change at the transition ΔS_t) and the shift of the transition with the pressure change dT/dp are linked through the Clausius–Clapeyron (CC) equation. This equation can be derived from the Gibbs energy $G(p, T)$ as follows. Let us express the exact differential of $G(p,t)$:

$$dG = \left(\frac{\partial G}{\partial T}\right)_p dT + \left(\frac{\partial G}{\partial p}\right)_T dp \tag{1.11}$$

Taking into account that $(\partial G/\partial p)_T = V$ and $(\partial G/\partial T)_p = -S$ and writing the resulting expression for the two phases I and II, we obtain:

$$dG_{\mathrm{I}} = V_{\mathrm{I}}dp - S_{\mathrm{I}}dT; \quad dG_{\mathrm{II}} = V_{\mathrm{II}}dp - S_{\mathrm{II}}dT \tag{1.12}$$

Considering that through an infinitessimal path along the coexistence curve, infinitessimal changes of the Gibbs free energy of each phase must be the same, then $dG_{\text{I}} = dG_{\text{II}}$ which results in the following expression:

$$V_{\text{I}}dp - S_{\text{I}}dT = V_{\text{II}}dp - S_{\text{II}}dT \rightarrow (V_{\text{I}} - V_{\text{II}})dp = (S_{\text{I}} - S_{\text{II}})dT \tag{1.13}$$

$$\frac{dT}{dp} = \frac{\Delta V_{\text{t}}}{\Delta S_{\text{t}}} = \frac{T\Delta V_{\text{t}}}{\Delta H_{\text{t}}} \tag{1.14}$$

where dT/dp is the slope of the two-phase coexistence line $T(p)$ constituting the phase diagram, and thus represents the sensitivity of the transition temperature on the applied pressure, whereas $\Delta V_{\text{t}} \equiv V_{\text{I}} - V_{\text{II}}$ and $\Delta S_{\text{t}} \equiv S_{\text{I}} - S_{\text{II}}$ are the changes in volume and entropy at the transition, respectively. Equation (1.14) is the CC equation and, in fact it has been shown that in the discontinuity limit the Maxwell relation (see equation (1.3)) becomes the CC equation [79].

The CC equation shows that dT/dp and ΔS_{t} are inversely proportional and so the adiabatic temperature changes and isothermal entropy changes. This means that it may be difficult to find BC materials displaying ΔS and ΔT simultaneously large. As examples, plastic crystals with chemical formulas $(NH_2)C(CH_2OH)_3$ and $(NH_2)(CH_3)C(CH_2OH)_2$ display huge latent heats of ~280 J g^{-1} and ~224 J g^{-1} at solid transitions and very small or moderate shifts of dT/dp ~ 15 K GPa^{-1} and dT/dp ~ 85 K GPa^{-1}, respectively [80], whereas the shape memory alloy $Ni_{50}Mn_{31.5}Ti_{18.5}$ displays an unusually large ΔH_{t} ~ 20 J g^{-1} in this material family, with a very small dT/dp ~ 20 K GPa^{-1} [81]. With the opposite behavior we can find, for example, the hybrid perovskite [TPrA][$Mn(dca)_3$], which exhibits a moderate ΔH_{t} ~ 14 J g^{-1} simultaneously with a huge dT/dp ~ 231 K GPa^{-1} [82]. As indicated by the CC equation, a good tradeoff between the magnitude of these two quantities can be achieved by maximizing ΔV_{t} as much as possible.

1.2.3 Conventional and inverse barocaloric effects

Equations (1.7) and (1.8) indicate that homogeneous systems with positive thermal expansion experience a negative isothermal entropy change ($\Delta S < 0$) or an adiabatic temperature increase ($\Delta T > 0$) when the applied pressure increases (as occurs in the ideal gas). Conversely, a pressure decrease yields $\Delta S > 0$ or $\Delta T < 0$. BC effects with these signs are called *conventional* because they are the most usual behavior in nature (see figures 1.4(a) and (b)). Consistently, homogeneous systems with negative thermal expansion experience a positive isothermal entropy change ($\Delta S > 0$) or an adiabatic temperature decrease ($\Delta T < 0$) when the applied pressure increases (as occurs to liquid water below 4 °C). Conversely, a pressure decrease yields $\Delta S < 0$ or $\Delta T > 0$. BC effects with these signs are called *inverse* (see figures 1.4(c) and (d)). In turn, at FOPTs the CC equation indicates that endothermic (i.e. $\Delta S_{\text{t}} > 0$) transitions with $\Delta V_{\text{t}} > 0$ result in $dT/dp > 0$. In this case a pressure increase will yield a negative isothermal entropy change ($\Delta S < 0$) or an adiabatic temperature increase ($\Delta T > 0$), which are conventional BC effects (see figures 1.4(e) and (f)). Conversely,

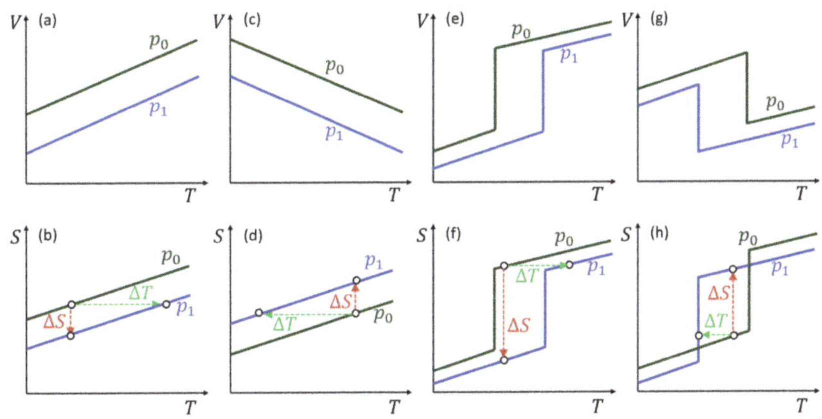

Figure 1.4. (a) Positive thermal expansion yields (b) conventional BC effects. (c) Negative thermal expansion yields (d) inverse BC effects. A first-order phase transition with (e) endothermic positive volume change yields (f) conventional BC effects. A first-order phase transition with (g) endothermic negative volume change yields (h) inverse BC effects.

endothermic transitions with $\Delta V_t < 0$ result in $dT/dp < 0$, and this behavior will yield inverse BC effects [83] (see figures 1.4(g) and (h)). A single material can exhibit both conventional and inverse BC effects at different temperature intervals [84, 85]. Figure 1.4 shows schematics of different volume–temperature behaviors and their relationship with conventional and inverse BC effects. Figure 1.5 shows representative reported examples in literature of conventional and inverse BC effects. Obviously, the conventional or inverse character of the BC effects has to be taken into account when designing a cooling device. Basically, if using a refrigerant exhibiting inverse BC effects, the steps in which the field is applied and removed in the cooling cycle must be exchanged.

1.2.4 Transition hysteresis and reversibility of the barocaloric effects

Hysteresis refers to the lag in the response of a system to an external stimulus when the latter is reversed. It is associated with out-of-equilibrium dynamics leading the system to remain for some time (that eventually can turn out to be extremely long compared to the human timescale) in a metastable state. In general, the occurrence of hysteresis may be very relevant for technological applications, either as a functional feature (for instance as a dissipative mechanism in seismic applications [89], for an enhanced stability of metastable states required in magnetic storage [90], for a desired delayed response in actuators, etc) or a drawback (usually due to associate losses with the subsequent decrease in performance in many devices as for instance in the core of transformers [91], etc).

At FOPTs, hysteresis appears as a difference in the (T, y) conditions between the endothermic and exothermic transitions, where y is the conjugate field, i.e. pressure in the case of BC materials [92]. From the point of view of hysteresis, FOPTs can be

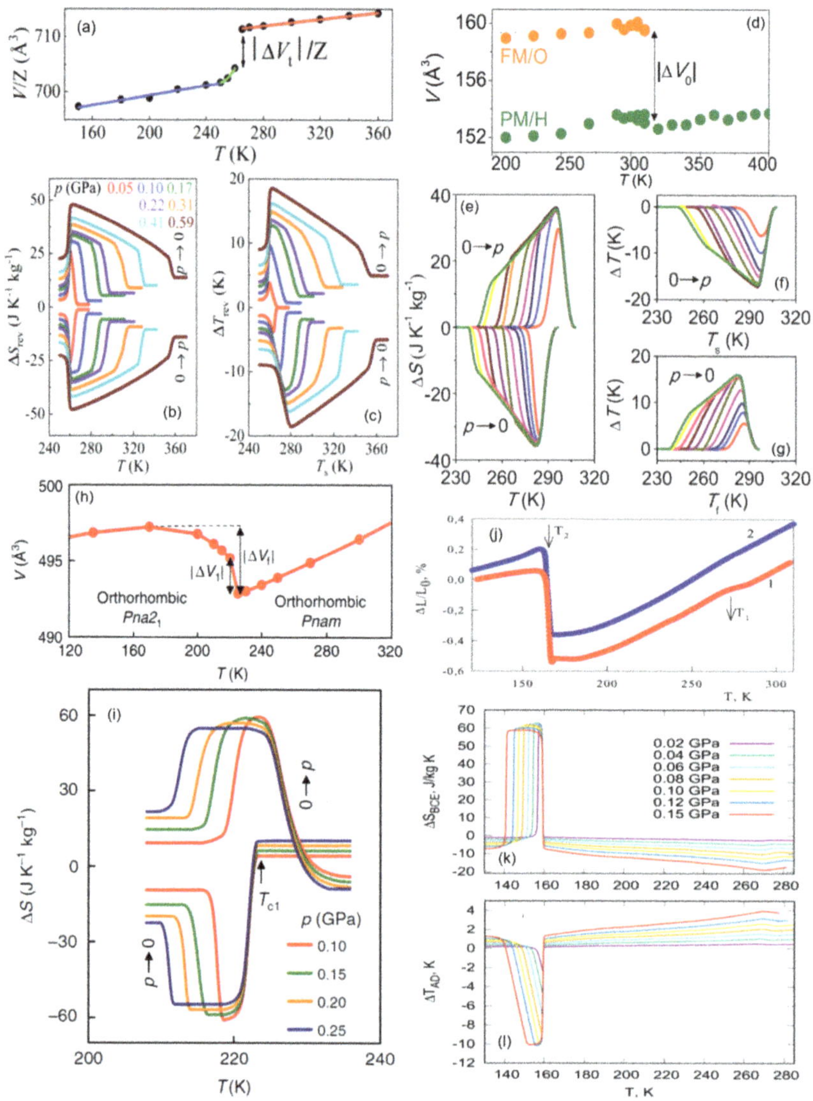

Figure 1.5. Conventional and inverse BC effects: (a) endothermic $\Delta V_t > 0$ and thermal expansion $\alpha > 0$ yields (b and c) conventional BC effects in fullerite C_{60} across and each side of the transition. Adapted and reproduced from [86] with permission from the Royal Society of Chemistry. (d) Endothermic $\Delta V_t < 0$ and $\alpha \sim 0$ yields (e–g) inverse BC effects at the transition and zero BC effects outside the transition in MnCoGe(B). Reprinted from [87] with permission from Wiley-VCH. (h) Endothermic $\Delta V_t < 0$, $\alpha < 0$ below the transition and $\alpha > 0$ above the transition yields (i) inverse BC effects at and below the transition and conventional BC above the transition in $(NH_4)_2SO_4$. Adapted and reproduced from [84] CC BY 4.0. (j) Endothermic $\Delta V_t < 0$ and $\alpha > 0$ in each phase (reproduced from [88], copyright (2017) with permission from Elsevier) yields (k and l) inverse BC effects across the transition and conventional BC effects in each phase in NH_4HSO_4 (reprinted from [85] with permission from Elsevier).

classified into *isothermal* and *athermal* transitions. In equilibrium, ideal isothermal FOPTs take place at constant temperature and field between two homogeneous phases with equal Gibbs energy, which define the two-phase coexistence line in the phase diagram. In these kinds of transition, which are triggered by thermal fluctuations, hysteresis arises as an extrinsic phenomenon of kinetic origin usually associated with the exothermic transition. There, the (out-of-equilibrium) (T, y) transition conditions depend on the rate of cooling or field change and, in general, on other external conditions such as grain size, thermomechanical history, etc. Moreover, hysteresis in thermally-driven transitions may be different from hysteresis in field-driven transitions. Consequently, the two-phase coexistence line for the exothermic transition is intrinsically not defined. As a matter of fact, experimental evidence has led to determine *equilibrium* phase diagrams of pure compounds using endothermic transition data [93]. On the other hand, in isothermal transitions the transition width in temperature or field is ideally zero by definition because transitions take place at a given constant temperature and field. However, typical differential scanning calorimetry experiments, that are performed commonly to determine the transition characteristics, yield a peak in the heat flow with a certain width that increases with the temperature rate. Therefore, the transition temperature is defined as the onset of the peak because this value should be temperature rate independent, as opposed to the temperature of the peak maximum. The onset transition temperature is strictly defined as the temperature at which the straight line representing the maximum slope of the peak curve crosses the baseline. These features can be visualized for a particular case in figure 1.6(a), that shows calorimetry measurements performed at different temperature rates across an isothermal transition in neopentyl glycol [94].

Contrary to the isothermal case, in athermal FOPTs, undergone by some alloys, thermal fluctuations do not play any role in the transition kinetics because higher intrinsic free energy barriers arise due to the presence of quenched disorder [96]. Then, in these systems the transition proceeds only if there is a change in the control parameter, either temperature or an external field, which yields a finite transition width. Therefore, hysteresis and transition width are intrinsic features of each particular sample (and not of a particular composition) and are roughly independent of the temperature and/or field rate within a reasonable range of rates. This type of material exhibits other characteristic behavior such as training on thermomechanical cycling [97, 98] and the virgin effect [87]. These features can be visualized for a particular case in figure 1.6(b), that shows calorimetry measurements performed at different temperature rates across an athermal transition in NiTi [95].

Both the transition hysteresis and calorimetric peak width are features that critically influence the BC response of the material in cooling applications working on cycles. In particular, the caloric material must undergo both the endothermic and exothermic phase transitions throughout each cycle in order to be able to absorb and release the associated latent heat but if at some point the transition hysteresis holds the material in a nontransformed metastable state, the desired transition does not occur and therefore the latent heat cannot be exchanged. Then, it is said that transition hysteresis may compromise the *reversibility* of the associated caloric

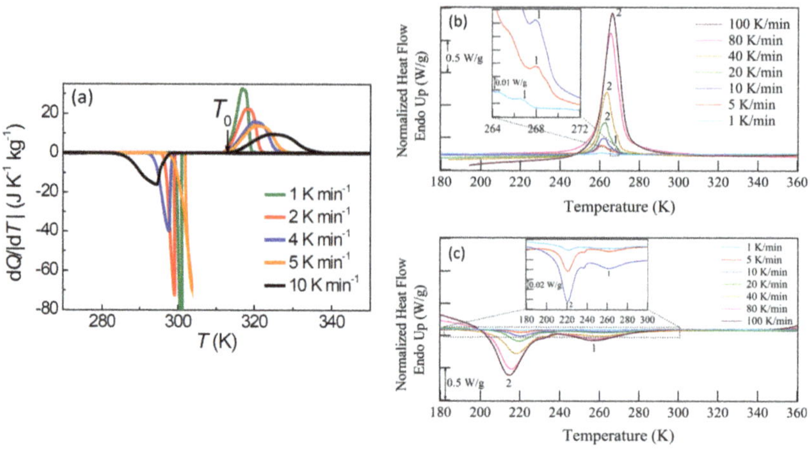

Figure 1.6. Differential scanning calorimetry measurements performed at different temperature rates in (a) neopentyl glycol (reproduced from [94] CC BY 4.0) and (b and c) NiTi (reproduced from [95] with permission from Springer). In (a), the onset transition temperature is independent of the temperature rate but the temperature of the endothermic peak maximum and the exothermic transition temperature (hence the hysteresis) do depend on the temperature rate, indicating that this is an isothermal transition. In (b), none of these features depend on the temperature rate, indicating that this is an athermal transition.

effects so that *reversible* effects are those that can be fully reproduced upon repetitive cycles, otherwise they are denoted as *irreversible*. As will be shown later, reversibility imposes a minimum magnitude for the external field y that must be applied to achieve reversible effects. Similarly, a large transition width is also detrimental for the reversibility of the BC effects because it also entails a larger driving field for the accomplishment of the full transition.

The relationship between the reversibility of caloric effects and the thermal hysteresis and transition width, was identified more than one decade ago [99, 100], by comparing temperature-variable isofield (i.e. *quasidirect*, see section 1.3) measurements with field-variable isothermal or adiabatic measurements (i.e. *direct*, see section 1.3) of magnetocaloric effects in magnetostructural materials. Then, a protocol to determine the reversible caloric effects from quasidirect measurements was well established [101]. These works were performed in magnetic shape memory alloys undergoing athermal FOPTs, and later the protocol was directly generalized to other types of materials. Currently this method is applied routinely for all types of caloric effects and to a variety of materials, many of them undergoing isothermal FOPTs. Then, the problem is that the essential differences between athermal and isothermal transitions raise doubts about the applicability of the protocol to isothermal FOPTs.

While many magnetocaloric and elastocaloric materials are considered to undergo athermal transitions, in electrocaloric materials and especially in BC materials, there are abundant isothermal transitions, so that this feature could particularly affect the unambiguous determination of BC effects. For instance, reference [102] reports direct electrocaloric measurements of ΔT in $Pb(Sc_{0.5}Ta_{0.5})O_3$

with very small hysteresis, that might be apparently inconsistent with quasidirect measurements. This could be due to effects arising in single phases or also to the fact that the aforementioned protocol could not be applicable in this case. In BC materials, this is aggravated because comparison between quasidirect and direct measurements are scarce [103] as the latter are difficult to perform in the BC field [43, 83] due to the impossibility of using infrared cameras, the difficulty in controlling accurately the pressure rate, the strongly nonadiabatic conditions when using liquid instead of gas as the pressure-transmitting fluid and the use of open systems where the mass of such fluids changes when pressure is modified. Therefore, it cannot be discarded that some of the reported reversible BC effects will be revisited in the future. Notwithstanding, in systems with large hysteresis with respect to the (measured) transition width, isothermal features may entail small relative errors in the determination of the reversible BC effects, and only in systems where the hysteresis is small, the relative uncertainties may become significant. Moreover on the one hand, most FOPTs are neither purely isothermal nor purely athermal [104–106]. On the other hand, it is often observed that within *non-extreme* conditions, the (T, y) values for exothermic isothermal transitions are in fact arranged in a certain orderly manner. Therefore, in many cases it is reasonable to use the protocol to obtain reversible BC effects in a wide range of isothermal FOPTs.

In the following, the standard protocol to determine the reversible barocaloric response from quasidirect measurements will be explained. For this purpose, we will disregard the consideration about the isothermal or athermal character of the transition and will consider that it is experimentally feasible to have a well-defined exothermic transition line which, along with the endothermic two-phase coexistence line, define an out-of-equilibrium phase diagram for a given material. According to this assumption, in figure 1.7 qualitative examples of out-of-equilibrium phase diagrams are plotted, where the red thick line indicates the endothermic (equilibrium) transition line, whereas the blue thick line corresponds to a hypothetical *exothermic (metastable) transition line*. Figure 1.7(a) evidences that for systems with $dT/dp > 0$, isothermal transitions driven on decompression are endothermic because

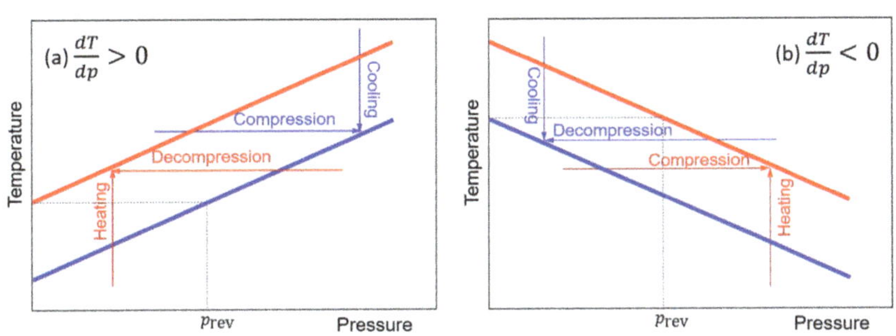

Figure 1.7. Schematic dimorphic phase diagram with (a) $\frac{dT}{dp} > 0$ and (b) $\frac{dT}{dp} < 0$. Red thick line indicates the endothermic (equilibrium) coexistence line, whereas blue line indicates the exothermic transition. Transitions on isobaric heating and cooling, and isothermal compression and decompression are indicated with arrows finishing at the corresponding line. The minimum pressure p_{rev} to obtain reversible BC effects is also indicated as explained in the text.

they lead to the same phase (I) as heating processes, whereas those driven on compression are exothermic because they lead to the same phase (II) as cooling processes. Therefore, the latter takes place with hysteresis, which means that the larger the hysteresis the larger the pressure increase that will be needed to induce the transition. Then, considering a cycle with the lowest pressure being atmospheric pressure, the minimum high pressure change to drive the cycle reversibly will end at the pressure $p_{rev} - p_{atm} \sim p_{rev}$ for which the exothermic transition temperature is equal to the endothermic transition temperature at atmospheric pressure, i.e., $T_{I\to II}(p_{rev}) = T_{II\to I}(p_{atm})$. This is shown schematically by the dashed lines in figure 1.7(a). Mathematically, p_{rev} can be computed as

$$p_{rev} = \int_{T_{I\to II}(p_{atm})}^{T_{II\to I}(p_{atm})} \left(\frac{dT_{I\to II}}{dp}\right)^{-1} dT \tag{1.15}$$

As for many materials $dT_{I\to II}/dp$ is constant within a large pressure domain, equation (1.15) can be simplified as $p_{rev} = (dT_{I\to II}/dp)^{-1}\left[T_{II\to I}(p_{atm}) - T_{I\to II}(p_{atm})\right]$. Conversely, for systems with $dT/dp < 0$, isothermal transitions driven on compression are endothermic, whereas those driven on decompression are exothermic. Following a similar argument to the conventional case, it can be shown that p_{rev} corresponds to the pressure value at which the endothermic transition temperature equals the exothermic transition temperature at atmospheric pressure, i.e., $T_{II\to I}(p_{rev}) = T_{I\to II}(p_{atm})$. This is shown schematically by the dashed lines in figure 1.7(b). In this case, p_{rev} can be expressed mathematically as

$$p_{rev} = \int_{T_{II\to I}(p_{atm})}^{T_{I\to II}(p_{atm})} \left(\frac{dT_{II\to I}}{dp}\right)^{-1} dT \tag{1.16}$$

Direct measurements, consisting of isothermal heat flow measurements or adiabatic temperature measurements, naturally yield reversible BC effects because they involve the field-driven transition hysteresis. However, direct measurements may be difficult to perform, especially for BC effects (see section 1.3). Instead, the quasidirect method, which is more usual, consists of isofield measurements and therefore it involves temperature-driven transition hysteresis that require additional data processing to determine the reversible effects. Following the argument explained above, in materials with conventional BC effects compression processes end at exothermic high-pressure entropy curves S_C, whereas decompression processes end at atmospheric-pressure endothermic entropy curves S_H. Notice that we define endothermic and exothermic entropies as those state functions associated with states reached at endothermic and exothermic processes, respectively. Consequently, adiabatic temperature changes obtained in compression–decompression cycles are obtained mathematically as

$$\Delta T_{rev}(T_s, p_{atm} \to p) = T(S_C, p) - T_s(S_H, p_{atm}) \tag{1.17}$$

$$\Delta T_{\mathrm{rev}}(T_{\mathrm{s}}, p \rightarrow p_{\mathrm{atm}}) = T(S_{\mathrm{H}}, p_{\mathrm{atm}}) - T_{\mathrm{s}}(S_{\mathrm{C}}, p) \tag{1.18}$$

Notice that in this case $\Delta T_{\mathrm{rev}}(T_{\mathrm{s}}, p \rightarrow p_{\mathrm{atm}}) \neq \Delta T_{\mathrm{rev}}(T_{\mathrm{s}}, p_{\mathrm{atm}} \rightarrow p)$. Here, ΔT is conveniently plotted as a function of the starting temperature T_{s} because the temperature changes along the adiabatic path. Moreover, it is straightforward to see that the reversibility of the process imposes that [107]

$$\Delta T_{\mathrm{rev}}(T_{\mathrm{s}}, p_{\mathrm{atm}} \rightarrow p) = -\Delta T_{\mathrm{rev}}(T_{\mathrm{s}} + \Delta T_{\mathrm{rev}}(T_{\mathrm{s}}, p_{\mathrm{atm}} \rightarrow p), p \rightarrow p_{\mathrm{atm}}). \tag{1.19}$$

Likewise, reversible isothermal entropy changes obtained in compression–decompression cycles can be obtained mathematically as:

$$\Delta S_{\mathrm{rev}}(T, p_{\mathrm{atm}} \rightarrow p) = S_{\mathrm{C}}(T, p) - S_{\mathrm{H}}(T, p_{\mathrm{atm}}) \tag{1.20}$$

$$\Delta S_{\mathrm{rev}}(T, p \rightarrow p_{\mathrm{atm}}) = S_{\mathrm{H}}(T, p_{\mathrm{atm}}) - S_{\mathrm{C}}(T, p) \tag{1.21}$$

and therefore $\Delta S_{\mathrm{rev}}(T, p_{\mathrm{atm}} \rightarrow p) = -\Delta S_{\mathrm{rev}}(T, p \rightarrow p_{\mathrm{atm}})$.

Notwithstanding, some studies have established an enhanced regime where reversible isothermal entropy changes ΔS_{rev} can be obtained, which should be determined from the overlapping between isothermal entropy changes ΔS derived from measurements performed on heating and on cooling [101, 108]. In this case the minimum reversible pressure for conventional BC effects can be expressed as

$$p_{\mathrm{rev}} = \int_{T^{\mathrm{f}}_{\mathrm{I}\rightarrow\mathrm{II}}(p_{\mathrm{atm}})}^{T_{\mathrm{II}\rightarrow\mathrm{I}}(p_{\mathrm{atm}})} \left(\frac{dT_{\mathrm{II}\rightarrow\mathrm{I}}}{dp}\right)^{-1} dT \tag{1.22}$$

where $T^{\mathrm{f}}_{\mathrm{I}\rightarrow\mathrm{II}}$ refers to the temperature at which the transition is completed. Notice that this quantity assumes that the transition occurs in a finite temperature range and therefore has a finite width. In athermal transitions for which an intrinsic transition width is well established, this enhanced reversible regime has been ascribed to the existence of minor transition loops. Instead, in the limit of a very narrow transition width (as ideal isothermal transitions are), the minor loops become negligible and equations (1.15) and (1.22) are equivalent. Then, the assumption of a finite transition width for isothermal transitions due to experimental measurements is safe because in the worst case it can result in an underestimation of the reversible BC effects. All the features concerning the reversibility of the caloric effects are illustrated qualitatively in figure 1.8.

1.2.5 Multicaloric effects

Some materials exhibit a significant coupling between different *ferroic* quantities (magnetic, polar, structural) which can manifest largely at FOPTs. Well-known examples are metamagnetic alloys, that display simultaneous changes in structural and magnetic degrees of freedom, or ferroelectric compounds, where changes in structure and polarization take place. These FOPTs are sensitive to the different conjugated fields of all generalized coordinates that are primary or secondary order parameters of the FOPTs. Therefore, caloric effects of different types can be driven

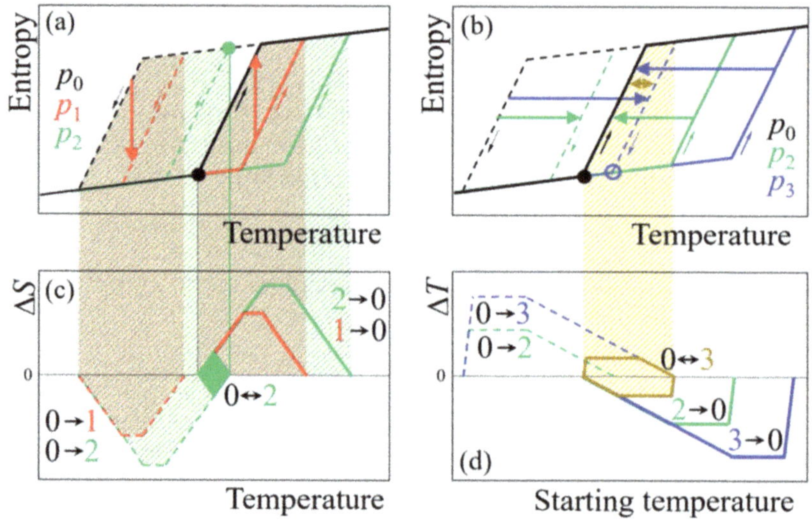

Figure 1.8. Hysteresis and reversibility for the particular case of a material displaying $\frac{dT}{dp} > 0$: (a) and (b) qualitative entropy functions $S(T, p)$ represented isobarically as a function of temperature at different values of pressure p_0 (black), p_1 (red, $p_1 > p_0$), p_2 (green, $p_2 > p_1$) and p_3 (blue, $p_3 > p_2$) on heating (continuous lines) and on cooling (dashed lines). Black filled symbols indicate the onset of the endothermic transition at pressure p_0, $T_{\mathrm{II}\to\mathrm{I}}(p_0)$, green filled symbol indicates the onset of the exothermic transition at p_2, $T_{\mathrm{I}\to\mathrm{II}}(p_2)$, and blue empty symbol indicates the finishing temperature of the transition on cooling at pressure p_3, $T^{\mathrm{f}}_{\mathrm{I}\to\mathrm{II}}(p_3)$. Vertical arrows in panel (a) and horizontal arrows in panel (b) indicate isothermal and adiabatic processes, respectively, under pressure changes. (c) and (d) Irreversible and reversible isothermal entropy changes (c) and adiabatic temperature changes (d) calculated from (a) and (b), respectively, following the quasidirect method, where continuous and dashed lines indicate irreversible changes on compression and decompression, respectively. The green surface in panel (c) stands for reversible ΔS obtained from overlapping of continuous and discontinuous lines and the dark yellow surface in panel (d) indicates the reversible ΔT as calculated from equations (1.17) and (1.18). Reproduced from [80] with permission from the Royal Society of Chemistry.

in a single material, which is then identified as *multicaloric*. Multicaloric effects are those effects that are obtained simultaneously or successively by application of more than one field type. The realization of simultaneous multicaloric effects may become technically difficult. As magnetic fields can be applied from a distance, only multicaloric effects involving a magnetic field, along with a mechanical stress or an electric field, have been reported so far. In fact, a commercial equipment is able to simultaneously apply magnetic field and hydrostatic pressure [109], which has led to several studies reporting the multicaloric effect under these two fields.

The thermodynamics of multicaloric effects was shown in references [110–112]. It establishes that for a system with generalized coordinates X_1 and X_2 with associated conjugate fields y_1 and y_2, respectively, the entropy changes occurring due to changes in both y_1 and y_2 can be determined as[2]:

[2] Here we assume X_i and y_i to be scalars for the sake of simplicity, but equivalent expressions can be obtained by replacing them by vectors or tensors.

$$\Delta S[(0,0)\rightarrow(y_1,y_2)] = \Delta S[(0,0)\rightarrow(y_1,0)] + \Delta S[(y_1,0)\rightarrow(y_1,y_2)] \tag{1.23}$$

$$= \Delta S[(0,0)\rightarrow(y_1,0)] + \Delta S[(0,0)\rightarrow(0,y_2)] + \int_0^{y_1}\int_0^{y_2}\left(\frac{\partial\chi_{12}}{\partial T}\right)dy_2'dy_1' \tag{1.24}$$

where

$$\Delta S[(0,0)\rightarrow(y_1,0)] = \int_0^{y_1}\left(\frac{\partial X_1}{\partial T}\right)_{y_1',y_2=0}dy_1' \tag{1.25}$$

$$\Delta S[(0,0)\rightarrow(y_1,0)] = \int_0^{y_1}\left(\frac{\partial X_1}{\partial T}\right)_{y_1',y_2=0}dy_1' \tag{1.26}$$

are the caloric effects associated with the application of each field individually and in the absence of the other. The last term in equation (1.23) contains the contribution of the off-diagonal term in the susceptibility symmetric tensor. It is calculated as

$$\chi_{12} = \left(\frac{\partial X_1}{\partial y_2}\right)_{T,y_1} = \left(\frac{\partial X_2}{\partial y_1}\right)_{T,y_2} \tag{1.27}$$

and represents the multiferroic cross-response. Interestingly, this term indicates that the multicaloric response of a material may be larger than the sum of the caloric responses obtained by the two fields individually. Furthermore, the combination of the two CC equations for each field individually yields the following equation valid at constant temperature:

$$\frac{dy_i}{dy_j} = -\frac{\Delta X_j}{\Delta X_i} \tag{1.28}$$

This set of equations shows that multicaloric effects could be used to shift the operational temperature regimes and/or optimize the overall caloric efficiency, either by enhancing the magnitude of the caloric effects or by reducing the required magnitude of the applied fields [113, 114]. Examples of these phenomena for the case of magneto-barocaloric effects are shown in figure 1.9. For instance, multicaloric effects have been proposed as a solution to overcome the transition hysteresis (see section 1.2.4) by designing smart thermodynamic paths in the T–y_1–y_2 phase diagram [115] (see figures 1.9(b)–(f)). Logically, the overall input work performed in a multicaloric cycle can be similar to that required in the single caloric cycle, because the lower applied fields is compensated by the sum over all the fields applied. However, the application of lower fields may be affordable in a more efficient way, as for instance, the use of permanent magnets to generate magnetic fields below 2 T, or thinner walls required at lower pressures, with the subsequent improvement of the heat exchange rate.

In contrast, in reference [116] a smart multicaloric cycle taking advantage of the transition hysteresis was proposed. There, the magnetic field has to be applied only for a little amount of time to achieve the high-magnetization phase because the thermal hysteresis is able to keep the system magnetized. A uniaxial stress is then used to drive the backward transition and thus recover the initial state.

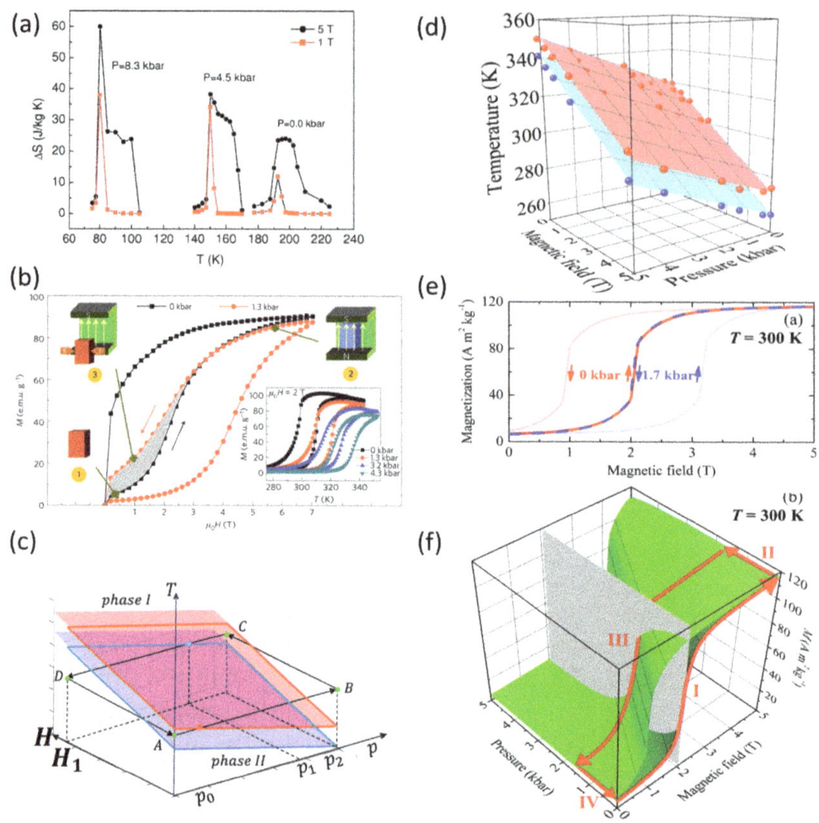

Figure 1.9. Multicaloric effects in magnetostructural alloys under simultaneous application of a magnetic field and pressure. (a) Pressure enhancement of the magnetocaloric effect in $LaFe_{11.6}Si_{1.4}$. Reproduced from [114] with permission from AIP Publishing. (b) Multicaloric strategy to overcome the thermal hysteresis using magnetic field and pressure in a Ni–Mn–In–(Co) Heusler alloy. Reprinted figures from [115] with permission from Springer Nature.(c) Representative temperature–pressure–magnetic field phase diagram where the red and blue surfaces correspond to the endothermic and exothermic transition surfaces, respectively. Along the path A → B under pressure sweep at constant magnetic field the endothermic transition II →I takes place but along the reverse path B → A the exothermic transition I → II does not take place and therefore, a reversible cycle A ↔ B cannot be established. Alternatively, if at state B the magnetic field is increased isobarically until reaching state C and then pressure is decreased under constant magnetic field until reaching state D, the exothermic transition can be forced at p_1 and H_1, as indicated by the blue point. Finally the magnetic field can be removed isobarically until reaching state A so that the cycle A → B → C → D → A is reversible. (d) $T(p, H)$ phase diagram and (e and f) multicaloric strategy to overcome the thermal hysteresis in $Fe_{49}Rh_{51}$. Reprinted figures with permission from [117] Copyright (2017) by the American Physical Society.

1.3 Methods and techniques

There exist the following three standardized methods to determine the BC response of a material (which can easily be extrapolated to other caloric effects), each involving specific experimental techniques and data processing:

- *Direct method.* It consists of measuring ΔS and ΔT under changes in the applied field. Therefore, the measurement outputs are the desired physical quantities so that neither additional data processing nor any thermodynamic equation are required. This is the most valuable method because it is closer to device operating conditions and upon the second application of the field, reversible effects are naturally obtained as it inherently identifies the hysteresis in pressure. In the case of BC effects, this method is scarcely used in literature because a good pressure control (as compared to controlling electric and magnetic fields) and adiabatic conditions may be difficult to achieve. A few examples of commercial equipment performing high-pressure calorimetry [118–121] are available, which may operate under pressure sweep in isothermal conditions, thus allowing direct measurements of isothermal entropy changes ΔS. They work with gas as a pressure-transmitting medium, which provides very good thermal insulation so that the heat flow recorded by the thermal sensors displays a low background noise and allows high-resolution measurements. As a counterpart, the use of gas compromises safety and the maximum working pressures are limited to barely 100 MPa or less. To date, direct measurements of ΔS are very scarce in literature of caloric effects in general [41, 103]. Direct measurements of adiabatic temperature changes have been achieved in bespoke experimental setups [43, 122], although adiabatic conditions are difficult to achieve because maximum experimental pressure rates are relatively low. On the other hand, infrared thermography is a technique commonly used to determine adiabatic temperature changes ΔT using the direct method in MC [123], EC [124] and eC materials [125] because they avoid heat leakage occurring in thermal sensors that must be attached to the samples. Unfortunately, relatively thick walls in pressure cells where BC samples must be placed, prevent the easy use of thermal cameras, although some commercial cells provided with sapphire windows could enable this technique in the future.

- *Quasi-direct method.* This method consists of performing calorimetry and thermometry measurements under isofield (isobaric for BC effects) while changing temperature. They are very common in literature because they are technically easier to perform than direct measurements. However, this method entails data processing, that transfers isofield conditions to isothermal and adiabatic conditions. For this purpose, the exothermic transition must be assumed to take place in the same conditions on cooling as on varying pressure, which means that pressure hysteresis is derived from the thermal hysteresis. Basically, it consists of subtracting isobaric entropy functions $S(T, p)$ isotermally to obtain ΔS or adiabatically to obtain ΔT. The mathematical procedure of this method is detailed in section 1.2.4. The commercial calorimeters mentioned for the direct method are often used in isobaric conditions, i.e., for the quasidirect method. A variety of commercial [126] or custom-built high-pressure calorimeters have been used for the determination of BC effects, based on a Paris–Edinburgh press [43], pressure cells using Bridgman pistons [84] or clamp cells [127].
- *Indirect method.* This method uses the Maxwell relations (see section 1.2.1 in general and equations (1.2) and (1.3) in particular). In caloric effects other than BC (MC, EC, eC), the indirect method has raised certain suspicions due

to the eventual misuse of this method associated with wrong protocols [73, 75] or non-equilibrium dynamics [74, 76, 77]. In the field of barocalorics, it is routinely applied to determine additional BC effects arising from continuous volume changes due to thermal expansion in single phases and far from FOPTs [84]. Therefore, this method complements the quasidirect method used across FOPTs. Instead, the reliability of the indirect method at FOPTs is compromised by the inadequacy of applying continuous mathematical quantities to physical discontinuous changes, although it has been shown that in the discontinuous limit equation (1.2) and the CC equation are equivalent [79]. Also, it is typically limited by the approximation $(\partial V/\partial T)_p \sim (\partial V/\partial T)_{p_{\mathrm{atm}}}$. In fact, at FOPTs this method is rarely applied and only $\Delta T \simeq -T/C_p\Delta S$ (equation (1.9)) has been frequently used to determine ΔT from ΔS (the latter being obtained by other methods). Alternatively, the CC equation (see section 1.2.2) is sometimes used to roughly estimate the BC effects through the approximations $\Delta S \sim \Delta S_{\mathrm{t}}$ and $\Delta T \sim \Delta V_{\mathrm{t}}/\Delta S_{\mathrm{t}}\Delta p$. However, for relatively large pressure changes this approximation can lead to significant uncertainties. As for the techniques to determine $V(T, p)$, temperature-dependent volume changes at atmospheric pressure are typically measured by means of conventional x-ray diffraction. Eventually, synchrotron x-ray [41, 94] or neutron [84] diffraction as well as dilatometry [94] can also be performed, to obtain pressure-dependence volume, which increases the reliability of indirect methods.

In addition to the determination of the barocaloric response of materials, many studies encompass the investigation of the underlying physics, which make them very interesting also for condensed matter physics. Here, a variety of techniques such as spectroscopy experiments (infrared, Raman, neutron scattering [42, 128–130]) are used to study the crystal structure dynamics. BC studies are often also supported by computer calculations which can provide predictions about good BC candidate materials, physical insights into the underlying physics, or information that is necessary to complement experimental data to fully characterize the BC response. They include density functional theory calculations [131, 132], molecular dynamics [133–135], a combination of both [136], mesoscopic modelling using Landau theory [137, 138] and finite element modelling to simulate heat transfer processes [48]. For instance, the origin of the large transition entropy change in ferrielectric ammonium sulphate has been unveiled in two recent studies by the same group [139, 140], each combining neutron scattering and theoretical modelling. Among other findings, the authors conclude that changes in the vibrational entropy may also be very significant in order–disorder transitions, in contrast to the simple picture mainly consisting of discrete changes in molecular configurations.

1.4 Barocaloric figures of merit and material requirements

Solids to be used as refrigerants in commercial devices must fulfill a number of requirements related not only to cooling performance parameters but also to safety concerns and cost [141], as occurs for fluid refrigerants. For this reason, it is difficult to carry out a simple yet comprehensive and fair comparison between BC materials with a single or even a few figures of merit. As a result, there is not a standard criterium in scientific articles to assess the competitivity of a material. Moreover, we can encounter different relevant quantities [142] about the material performance that can differ one or even two orders of magnitude from one material to another, and that sometimes are detrimental to each other. For instance, large isothermal entropy changes can go alongside low thermal conductivities and hinder large adiabatic temperature changes. Moreover, the fact that research on BC effects is still in an early stage, that BC materials are very diverse and that first BC devices are still under development, notably increases the difficulty to establish which families of materials will be more suitable for each market niche. Below we intend to provide a summary of the relevant quantities of BC materials for their applicability in cooling devices and heat pumps:

- *Isothermal entropy change* ΔS *and adiabatic temperature change* ΔT

  These quantities are chosen to parametrize and quantify the barocaloric effects because they define the fundamental Carnot cycle. The quantity $T\Delta S$ is the amount of heat exchanged with the environment when pressure is changed from p_0 to p_1 isothermally. In any cooling cycle (e.g. the reverse Brayton cycle depicted in figure 1.3), $T\Delta S$ will be an estimate of the amount of heat that the material can absorb from the target (at temperature T) to be cooled. Therefore, materials with large ΔS are sought. Although in condensed matter physics the most meaningful units for ΔS would be J K^{-1} mol^{-1}, this quantity is usually given in specific units J K^{-1} kg^{-1} because solid materials are purchased by weight, so it will mark the cost of the refrigerant. Alternatively, ΔS is sometimes given per unit volume J K^{-1} cm^{-3} because this will mark the energy density of the refrigerant and therefore the compactness of the cooling device. Therefore, a large mass density is also desirable.

  As for the adiabatic temperature change ΔT, in a simple cooling cycle $T_{bath} + \Delta T(p_0 \rightarrow p_1)$ and $T_{bath} - \Delta T(p_0 \rightarrow p_1)$ would be the highest and lowest temperature of the refrigerant, respectively, throughout the cooling (or heating) cycle, with T_{bath} being the temperature of the thermal bath which in a cooling cycle acts as the hot end. The lowest temperature of the refrigerant should be lower than the lowest temperature that the target to be cooled is intended to reach, i.e., the cold end (notice that in a heat pump the cold end is thermal bath, whereas the hot end is the target to be heated). Of course, to achieve a proper frequency of operation to reach the required cooling performance, the heat exchange between the refrigerant and both the hot and cold ends (i.e. thermalization) must be as fast as possible. For this purpose, in addition to high thermal conductivity, Fourier's law for heat conduction establishes that the temperature gradient between the two bodies exchanging heat should be maximized and so, BC ΔT. Notwithstanding, regeneration cycles [143, 144] or cascade systems [35, 145], widely used in prototypes based on other types of caloric effects, operate with several interconnected thermodynamic cycles in order to achieve much larger differences between the highest and the lowest temperatures than just that provided by ΔT.

It is worth recalling here that, as discussed in section 1.2.2, to obtain large ΔS and ΔT across a FOPT, ΔS_t and dT/dp should be as large as possible, but unfortunately the CC equation establishes an inverse proportional relationship between both quantities and a tradeoff is necessary.

- *Temperature span.* The temperature span is the interval between the highest and lowest temperatures within which the BC effects achieve at least a certain magnitude. As just discussed for ΔT, for cooling cycles these temperatures should be as high as possible over the temperature of the hot end to which the material expels the excess heat, and as low as possible over the temperature of the cold end from which the material absorbs heat. Therefore, the temperature span will mark the operational temperature of the cycle and can be enhanced using regeneration and/or cascading cycles.
- *Small hysteresis.* In section 1.2.4, it has been shown in great detail that for applications working in cycles, hysteresis is detrimental to the material performance as it increases the required input work to drive the caloric effects.
- *Refrigerant capacity (RC)* and *relative cooling power (RCP).* These quantities refer to the amount of heat that can be exchanged between the hot and the cold reservoirs throughout a cooling cycle and are related to an estimation of the area below the peak displayed by ΔS versus T. RC was defined as $\text{RC} = \Delta S \times \Delta T$ [146], although in reference [147] the following expression was used:

$$\text{RC} = \int_{T_1}^{T_2} \Delta S dT \tag{1.29}$$

where T_1 and T_2 are the temperatures of the cold and hot sinks, respectively. For caloric effects at second order phase transitions, the typical caret-like shape of the $\Delta S(T)$ suggests that T_1 and T_2 should be chosen as the temperatures corresponding to half of maximum value of ΔS on either side of the peak. In turn, RCP is defined as:

$$\text{RCP} = |\Delta S|_{\text{max}} \times \Delta T_{\text{FWHM}} \tag{1.30}$$

where ΔT_{FWHM} is the temperature width at half maximum of the $\Delta S(T)$ peak. It has been shown that in this context, RCP appears to be roughly 25% larger than RC. However, the widespread use of these quantities in other contexts has led to controversy or misuse of these quantities; for instance in reference [148] RC and RCP were considered to be equivalent. Notwithstanding, at first-order phase transitions and/or in materials exhibiting large additional caloric effects in each side of the transition, the use of RCP and the choice of T_1 and T_2 for the integral limits in the RC are highly disputable because $\Delta S(T)$ does not exhibit the caret-like shape, so in those cases adapted or more flexible definitions for RC have been used [80]. Of course, such flexibility hinders comparison between different materials.
- *Material performance parameters.* Previous quantities do not involve the required work to drive the cycles and therefore cannot determine the material

performance. Therefore, it may be convenient to use other quantities such as the coefficient of performance (COP or η), defined for cooling cycles as the ratio between the heat absorbed from the cold reservoir and the network required to drive the cycle. Alternatively, the Coefficient of Refrigerant Performance was also introduced [149] as

$$\text{CRP} \approx \frac{\Delta T_{\text{rev}} \Delta S}{W} \tag{1.31}$$

where W is the work needed to drive the BC effects reversibly which for BC effects becomes $W \simeq \frac{1}{2} p \Delta V_{\text{t}}$ [49, 150]. Also, the *barocaloric strengths* defined as the BC effects normalized by necessary pressure changes, have often been used to assess the material performance [43, 122, 151].

- *Low pressure*. Beyond the material's performance discussed above, the absolute value of pressure should be itself as low as possible because the required thickness of the walls holding pressure gradients is detrimental for the heat exchange rate and can result in too low operation frequencies, finally decreasing the device performance. Also, low pressures enhance safety and facilitate the use of off-the-shelf compressors.
- *Thermal conductivity*. Naturally, this property is key to assess the heat exchange between the refrigerant material and its surroundings. Unfortunately, the outstanding caloric properties of some BC materials rely on the emergence of strong structural disorder that provokes the shortening of the phonon mean free path, leading to low thermal conductivities of ~0.2 W m^{-1} K^{-1} [152, 153]. To improve this feature, within the context of phase change materials for thermal energy management, inclusion of nanoparticles or embedding the material in porous matrices have been proposed, leading to an increase in the thermal conductivity but at the expense of reducing the energy density [154–157].
- *Toxicity, flammability*. Safety issues, inherent in widespread marketing, have restricted the use of some fluid refrigerants with low or null GWP such as ammonia, hydrocarbons, or the newly synthesized hydrofluoroolefins [14, 15] and represent a main concern in the search for new fluid refrigerants, mainly those for domestic purposes.
- *Easy shaping*. The optimization of the heat exchange may demand specific structural features for the solid refrigerant, both at the material (grain size, microstructure, powder or non-powder form, etc) and device levels (macroscopic shape: tubes, plates, etc) [158].
- *Stability and cyclability*. Physical and chemical stability of the shaped material at the proper mass and time scale, and upon thermal and field cycling over millions of cycles, must be ensured. The solid refrigerant must not transform to a more stable state or to an uncontrolled metastable state, nor segregate (in the case of binary mixtures), nor degrade nor suffer corrosion (in the case of metals), nor alter the optimized shape, etc.

- *Cost, availability, easy large-scale production.* The risk of supply of raw materials for any mass market technology associated with geopolitical availability has to be taken into account. This is critical, for instance, for gadolinium needed in magnetocaloric coolers [159].

For other thermal management applications not involving high-frequency cycles, such as thermal energy harvesting, storage and reutilization, other requirements are demanded. For instance, transition hysteresis may become a positive feature because it enhances the actual temperature and/or field range of stability. On the other hand, adiabatic temperature changes may not be that relevant and RCP and RC parameters turn out to be useless.

1.5 Progress towards barocaloric devices

In recent years, several groups worldwide have announced projects aimed at constructing a BC prototype, one by the start-up *Barocal Ltd*, in UK [45, 160], one in USA [47] and one in China [161], and we are aware that at least three more research groups are addressing this task in the European Union. So far, the only reported setup closest to a BC proof-of-concept is limited to the realization of a nonautomatic toy device [46] driven by footsteps, with the purpose of demonstrating that required pressures can be easily generated and that the operation principle is simple. Thus, motor-driven devices working cyclically are still lacking.

Following the presentations by Stern-Taulats *et al* in the most important international conferences in this field (such as Thermag [160], MRS meeting [162], etc), the design of a BC prototype should pay particular attention to the following aspects:

- Heat exchange should be adapted to the temperature needs not only to the specific applications but also to the climatological environment. Thus, typical needs for the temperature span will range from ~20 K for refrigeration in indoor conditions, or air conditioning working in ambient conditions at subtropical, to ~70 K for a freezer working in tropical regions. Also, cooling capacities ranging from 100 W for domestic applications to 10 MW for the industrial sector must be borne in mind.
- A good heat exchange between the solid refrigerant and the thermal fluid is key to achieve reasonable operational frequencies but both must be separated by walls capable of holding high pressure gradients, as the thermal fluid will be at or nearly atmospheric pressure and the solid refrigerant will achieve pressures of the order of ~100 MPa inside pressure vessels. Unfortunately, thick walls are highly detrimental for the heat exchange, for which any reduction in the required pressures is very welcome, in addition to the consequent decrease in the necessary input work. Notwithstanding, as the wall thickness of pressure vessels rapidly increase as the inner volume increases, Stern-Taulats *et al* suggest a convenient strategy consisting of inserting the heat exchanger pipes inside the pressure vessel. As the pipes may have small inner diameters, then pipe thicknesses can be significantly reduced. As examples, for Cu–Be pipes under a hydrostatic pressure of ≃150 MPa,

inner diameters of 3 and 6 mm require wall thicknesses of $\simeq$0.5 and $\simeq$1.0 mm, respectively, whereas under $\simeq$200 MPa, wall thicknesses should be $\simeq$0.8 and $\simeq$1.5 mm, respectively. Notice that according to this configuration the heat exchange will occur inside the pressure vessel. Therefore, an optimal heat exchange will entail that the refrigerant should be isolated from the inner walls of the pressure vessel as they are made of highly-thermally conductive materials such as stainless steel or Cu–Be alloys. Notice that with this setup the wall thickness of the pressure vessel no longer plays a role. Nevertheless, the design of the pipes are of course critical to obtain a good heat exchange between all regions of the refrigerant.

- To achieve operational frequencies equal to or larger than 0.5 Hz, it is estimated that solid refrigerants with a minimum thermal conductivity of 1 W K^{-1} m^{-1} should be used.
- Pressure-transmitting fluids should be liquids because for volumes larger than 0.5 L the use of gas under the required pressures is unsafe due to the large gas volume changes, and rapid pressure changes are more challenging to perform.
- Wettability of the pressure-transmitting liquid is important for an efficient heat transfer.
- The material's mechanical response and distribution of internal stresses is crucial for the BC response. Poisson ratio is an important parameter such that its value should be as high as possible, of course the maximum value of 0.5 for an elastic material is desirable.
- For pressures below 200 MPa multiple off-the-shelf high-pressure components become available (pumps, vessels, hoses, tubing, valves, connectors, etc). A pressure of $\simeq$150 MPa is proposed as an upper limit for the BC device (fortunately, given the recent advances on BC materials, BC devices working at pressures much below 100 MPa are expectable).
- For the current required pressures, pressure recovery is crucial to obtain competitive efficiencies (in the field of caloric effects, work recovery consists of harvesting the work done by the system when removing the external field in order to be used in the next cycle to decrease the necessary input work).

Additionally, regeneration could help improve the efficiency and frequency of operation. First, BC prototypes, which are now eagerly awaited in this field, are reasonable to foresee in the next few years. After showing great promise due to excellent BC materials, the realization of a BC device should provide key insights into the strengths and weaknesses of BC applications and, therefore, a more realistic prospect about the future of this field.

References

[1] Constable G and Somerville B 2003 *A Century of Innovation: Twenty Engineering Achievements that Transformed Our Lives* (Washington, DC: The National Academies Press)

[2] University of Birmingham 2017 A cool world—defining the energy conundrum of cooling for all, https://www.birmingham.ac.uk/documents/college-eps/energy/publications/2018-clean-cold-report.pdf (accessed August 2022)

[3] She X, Cong L, Nie B, Leng G, Peng H, Chen Y, Zhang X, Wen T, Yang H and Luo Y 2018 *Appl. Energy* **232** 157–86
[4] McLinden M O, Seeton C J and Pearson A 2020 *Science* **370** 791–6
[5] Cavallini A 2020 *J. Phys.: Conf. Ser.* **1599** 012001
[6] Solomon S, Garcia R R, Rowland F S and Wuebbles D J 1986 *Nature* **321** 755–8
[7] UN Environment Program 1987 The montreal protocol on substances that deplete the ozone layer, https://ozone.unep.org/treaties/montreal-protocol/montreal-protocol-substances-deplete-ozone-layer (accessed July 2022)
[8] UN Environment Program 2020 Cooling emissions and policy synthesis report, https://www.unep.org/resources/report/cooling-emissions-and-policy-synthesis-report (accessed July 2022)
[9] UN Environment Program, Montreal Protocol Secretariat 2017 Ratification of the igali amendment information note, https://treaties.un.org/Pages/ViewDetails.aspx?src=TREATY&mtdsg_no=XXVII-2-f&chapter=27&clang=_en (accessed March 2022)
[10] US Environmental Protection Agency, https://www.epa.gov/climate-hfcs-reduction/final-rule-phasedown-hydrofluorocarbons-establishing-allowance-allocation (accessed June 2023)
[11] Regulation (EU) No 517/2014 of the European Parliament and of the Council of 16 April 2014 on fluorinated greenhouse gases and repealing Regulation (EC) No 842/2006 (2014), https://eur-lex.europa.eu/legal-content/EN/ALL/?uri=uriserv:OJ.L_.2014.150.01.0195.01.ENG (accessed July 2020)
[12] Ministry of Ecology and Environment, The People's Republic of China 2022 https://english.mee.gov.cn/News_service/media_news/202104/t20210428_831030.shtml (accessed March 2022)
[13] Press Information Bureau, Government of India 2022 https://pib.gov.in/PressReleasePage.aspx?PRID=1746946 (accessed August 2022)
[14] European Comission, Climate Action 2022 https://ec.europa.eu/clima/eu-action/fluorinated-greenhouse-gases/climate-friendly-alternatives-hfcs_en (accessed March 2022)
[15] Area: European organisation of refrigeration, air-conditioning and heat pump (RACHP) contractor 2022 https://area-eur.be (accessed July 2023)
[16] European Comission, Climate Action 2022 https://ec.europa.eu/clima/system/files/2016-11/leakage_rates_final_report_en.pdf (accessed July 2022)
[17] European Comission, Internal Market, Industry, Entrepreneurship and SMEs 2022 https://ec.europa.eu/growth/sectors/automotive-industry/environmental-protection/mobile-air-conditioning-systems-macs_en (accessed March 2022)
[18] Autobild 2022 https://www.autobild.de/artikel/killer-kaeltemittel-r1234yf-5180333.html (accessed July 2022)
[19] Chua K, Chou S and Yang W 2010 *Appl. Energy* **87** 3611–24
[20] US Department of Energy, Office of Energy Efficiency and Renewable Energy https://www.energy.gov/sites/prod/files/2017/12/f46/bto-DOE-Comm-HVAC-Report-12-21-17.pdf, Energy Savings Potential and RD&D Opportunities for Commercial Building HVAC Systems
[21] Moya X, Kar-Narayan S and Mathur N D 2014 *Nat. Mater.* **13** 439–50
[22] Moya X and Mathur N D 2020 *Science* **370** 797–803
[23] Greco A, Aprea C, Maiorino A and Masselli C 2019 *AIP Conf. Proc.* **2191** 020091
[24] Brück E, Tegus O, Thanh D and Buschow K 2007 *J. Magn. Magn. Mater.* **310** 2793–9
[25] Valant M 2012 *Prog. Mater. Sci.* **57** 980–1009
[26] Bianco V, De Rosa M and Vafai K 2022 *Appl. Therm. Eng.* **214** 118839

[27] Gough J 1805 *Mem. Lit. Phil. Soc. Manchester* **1** 288–95
[28] Warburg E 1881 *Ann. Phys.* **249** 141–64
[29] Weiss P and Piccard A 1917 *J. Phys. Theor. Appl.* **7** 103–9
[30] Kobeko P and Kurtschatov J 1930 *Z. Phys.* **66** 192–205
[31] Wang R, Zhou X, Wang W and Liu Z 2021 *J. Chem. Eng.* **417** 128060
[32] Porta M, Castán T, Saxena A and Planes A 2021 *Phys. Rev.* B **104** 094108
[33] Research and Markets, https://www.researchandmarkets.com/reports/5180906/magnetic-refrigeration-market-by-product?utm_source=BW&utm_medium=PressRelease&utm_code=9j8ngg&utm_campaign=1457997+-+Global+Magnetic+Refrigeration+Market+(2020+to+2025)+-+High+Potential+in+Industrial+Sector+Presents+Opportunities&utm_exec=jamu273prd, Magnetic Refrigeration Market by Product (Refrigeration Systems (Beverage Cooler, Cabinet Display, Refrigerator), Air Conditioning Systems), Application (Domestic, Commercial, Transportation, and Industrial), and Geography—Global Forecast to 2027
[34] Bruederlin F, Bumke L, Chluba C, Ossmer H, Quandt E and Kohl M 2018 *Energy Technol.* **6** 1588–604
[35] Meng Y, Zhang Z, Wu H, Wu R, Wu J, Wang H and Pei Q 2020 *Nat. Energy* **5** 996–1002
[36] Kar-Narayan S and Mathur N D 2009 *Appl. Phys. Lett.* **95** 242903
[37] Müller K A, Fauth F, Fischer S, Koch M, Furrer A and Lacorre P 1998 *Appl. Phys. Lett.* **73** 1056–8
[38] Mañosa L, González-Alonso D, Planes A, Bonnot E, Barrio M, Tamarit J-L, Aksoy S and Acet M 2010 *Nat. Mater.* **9** 478–81
[39] Lloveras P and Tamarit J-L 2021 *MRS Energy Sustain.* **8** 3–15
[40] Romanini M *et al* 2021 *Adv. Mater.* **33** 2008076
[41] Seo J, Braun J D, Dev V M and Mason J A 2022 *J. Am. Chem. Soc.* **144** 6493–503
[42] Li J, Barrio M, Dunstan D J, Dixey R, Lou X, Tamarit J-L, Phillips A E and Lloveras P 2021 *Adv. Funct. Mater.* **31** 2105154
[43] Zhang K, Song R, Qi J, Zhang Z, Zhang Z, Yu C, Li K, Zhang Z and Li B 2022 *Adv. Funct. Mater.* **32** 2112622
[44] García-Ben J, López-Beceiro J, Artiaga R, Salgado-Beceiro J, Delgado-Ferreiro I, Kolen'ko Y V, Castro-García S, Señarís-Rodríguez M A, Sánchez-Andújar M and Bermúdez-García J M 2022 *Chem. Mater.* **34** 3323–32
[45] Barocal 2022 https://barocal.com (accessed December 2022)
[46] Garcia-Ben J, Delgado-Ferreiro I, Salgado-Beceiro J and Bermudez-Garcia J M 2021 *Materials* **14** 5947
[47] American Chemical Society 2022 https://phys.org/news/2022-08-environmentally-friendly-air-conditioner.html (accessed December 2022)
[48] Aprea C, Greco A, Maiorino A and Masselli C 2020 *Energy* **190** 116404
[49] Crossley S, Mathur N D and Moya X 2015 *AIP Adv.* **5** 067153
[50] Kitanovski A 2020 *Adv. Energy Mater.* **10** 1903741
[51] ExtractEnergy, https://www.extractenergy.com/our-technology/ (accessed July 2022)
[52] Waste Heat 2022 https://www.waste-heat.eu/ (accessed May 2022)
[53] Arpa-e Project, Yale University 2022 https://arpa-e.energy.gov/technologies/projects/closed-loop-system-using-waste-heat-electricity (accessed May 2022)
[54] Arpa-e Project, Yale University 2022 https://arpa-e.energy.gov/technologies/projects/waste-heat-recovery-system (accessed May 2022)

[55] Jones, N, Yale Environment 360 2022 https://e360.yale.edu/features/waste-heat-innovators-turn-to-an-overlooked-renewable-resource (accessed May 2022)
[56] Jouhara H, Khordehgah N, Almahmoud S, Delpech B, Chauhan A and Tassou S A 2018 *Therm. Sci. Eng. Prog.* **6** 268–89
[57] Loibl W, Stollnberger R and Österreicher D 2017 *Sustainability* **9** 250
[58] Benson D, Burrows R and Webb J 1986 *Sol. Energy Mater.* **13** 133–52
[59] Busico V, Carfagna C, Salerno V, Vacatello M and Fittipaldi F 1980 *Sol. Energy* **24** 575–9
[60] Fallahi A, Guldentops G, Tao M, Granados-Focil S and Van Dessel S 2017 *Appl. Therm. Eng.* **127** 1427–41
[61] Moldgy A and Parameshwaran R 2018 *Mater. Today: Proc.* **5** 16840–8 *Int. Conf. Advanced Materials (SCICON'16), December 19–21, 2016*
[62] Usman A, Xiong F, Aftab W, Qin M and Zou R 2022 *Adv. Mater.* **34** 2202457
[63] Bentz D P and Turpin R 2007 *Cem. Concr. Compos.* **29** 527–32
[64] Jamekhorshid A, Sadrameli S and Farid M 2014 *Renew. Sustain. Energy Rev.* **31** 531–42
[65] Serrano A, Dauvergne J-L, Doppiu S and Palomo Del Barrio E 2019 *Materials* **12** 3169
[66] Tokoro H *et al* 2015 *Nat. Commun.* **6** 7037
[67] Ohkoshi S-i, Tokoro H, Nakagawa K, Yoshikiyo M, Jia F and Namai A 2019 *Sci. Rep.* **9** 13203
[68] Salvatori A, Negrier P, Aznar A, Barrio M, Tamarit J L and Lloveras P 2022 *APL Mater.* **10** 111117
[69] Henry A, Prasher R and Majumdar A 2020 *Nat. Energy* **5** 635–7
[70] Alvi J Z, Feng Y, Wang Q, Imran M and Pei G 2021 *Energy* **223** 120006
[71] Font J T and Muntasell J 1994 *Mater. Res. Bull.* **29** 1091–100
[72] Zemansky M and Dittman R 1997 *Heat and Thermodynamics* 7th edn (New York: McGraw-Hill)
[73] Caron L, Ou Z, Nguyen T, Thanh D C, Tegus O and Brück E 2009 *J. Magn. Magn. Mater.* **321** 3559–66
[74] Young J S 2012 Indirect measurement of the electrocaloric effect *PhD thesis* University of Cambridge, Cambridge, UK
[75] Niemann R, Heczko O, Schultz L and Fähler S 2014 *Int. J. Refrig.* **37** 281–8
[76] Liu Y, Scott J F and Dkhil B 2016 *Appl. Phys. Rev.* **3** 031102
[77] Li J, Li J, Qin S, Su X, Qiao L, Wang Y, Lookman T and Bai Y 2019 *Phys. Rev. Appl.* **11** 044032
[78] Tušek J, Engelbrecht K, Mañosa L, Vives E and Pryds N 2016 *Shape Mem. Superelasticity* **2** 317–29
[79] Planes A, Mañosa L and Acet M 2009 *J. Condens. Matter Phys.* **21** 233201
[80] Aznar A, Lloveras P, Barrio M, Negrier P, Planes A, Mañosa L, Mathur N D, Moya X and Tamarit J-L 2020 *J. Mater. Chem.* A **8** 639–47
[81] Aznar A, Gràcia-Condal A, Planes A, Lloveras P, Barrio M, Tamarit J-L, Xiong W, Cong D, Popescu C and Mañosa L 2019 *Phys. Rev. Mater.* **3** 044406
[82] Bermúdez-García J M, Sánchez-Andújar M, Castro-García S, López-Beceiro J, Artiaga R and Señarís-Rodríguez M A 2017 *Nat. Commun.* **8** 1–8
[83] Mañosa L, Gonzalez-Alonso D, Planes A, Barrio M, Tamarit J-L, Titov I S, Acet M, Bhattacharyya A and Majumdar S 2011 *Nat. Commun.* **2** 1–5
[84] Lloveras P *et al* 2015 *Nat. Commun.* **6** 8801

[85] Gorev M V, Mikhaleva E A, Flerov I N and Bogdanov E V 2019 *J. Alloys Compd.* **806** 1047–51
[86] Li J, Dunstan D, Lou X, Planes A, Mañosa L, Barrio M, Tamarit J-L and Lloveras P 2020 *J. Mater. Chem.* A **8** 20354–62
[87] Aznar A, Lloveras P, Kim J-Y, Stern-Taulats E, Barrio M, Tamarit J L, Sánchez-Valdés C F, Sánchez Llamazares J L, Mathur N D and Moya X 2019 *Adv. Mater.* **31** 1903577
[88] Mikhaleva E, Flerov I, Kartashev A, Gorev M, Bogdanov E and Bondarev V 2017 *Solid State Sci.* **67** 1–7
[89] Symans M, Charney F, Whittaker A, Constantinou M, Kircher C, Johnson M and McNamara R 2008 *J. Struct. Eng.* **134** 3–21
[90] Wäckerlin C *et al* 2016 *Adv. Mater.* **28** 5195–9
[91] Roshen W A 2007 *IEEE Trans. Emerg. Sel.* **22** 30–40
[92] Brokate M and Sprekels J 1996 Introduction *Hysteresis and Phase Transitions* (New York: Springer) pp 1–9
[93] Rietveld I B, Céolin R and Tamarit J L 2021 *Polymorphism* (Cham: Springer International Publishing) pp 41–76
[94] Lloveras P *et al* 2019 *Nat. Commun.* **10** 1803
[95] Lanba A and Hamilton R F 2014 *Metall. Mater. Trans.* A **45** 2732–40
[96] Pérez-Reche F J, Vives E, Mañosa L and Planes A 2001 *Phys. Rev. Lett.* **87** 195701
[97] Carrillo L and Ortín J 1997 *Phys. Rev.* B **56** 11508–17
[98] Pérez-Reche F-J, Stipcich M, Vives E, Mañosa L, Planes A and Morin M 2004 *Phys. Rev.* B **69** 064101
[99] Shamberger P J and Ohuchi F 2009 *Phys. Rev.* B **79** 144407
[100] Krenke T, Aksoy S, Duman E, Acet M, Moya X, Mañosa L and Planes A 2010 *J. Appl. Phys.* **108** 043914
[101] Emre B, Yüce S, Stern-Taulats E, Planes A, Fabbrici S, Albertini F and Mañosa L 2013 *J. Appl. Phys.* **113** 213905
[102] Nouchokgwe Y, Lheritier P, Hong C-H, Torelló A, Faye R, Jo W, Bahl C R H and Defay E 2021 *Nat. Commun.* **12** 3298
[103] Salgado-Beceiro J, Nonato A, Silva R X, García-Fernández A, Sánchez-Andújar M, Castro-García S, Stern-Taulats E, Señarís Rodríguez M A, Moya X and Bermúdez-García J M 2020 *Mater. Adv.* **1** 3167–70
[104] Kakeshita T, Fukuda T and Saburi T 1996 *Scr. Mater.* **34** 147–50
[105] Kustov S, Golovin I, Corró M L and Cesari E 2010 *J. Appl. Phys.* **107** 053525
[106] Sajkiewicz P 2003 *J. Polym. Sci. B Polym. Phys.* **41** 68–79
[107] Nielsen K K, Bahl C R H and Smith A 2010 *Phys. Rev.* B **81** 054423
[108] Gottschall T, Stern-Taulats E, Mañosa L, Planes A, Skokov K P and Gutfleisch O 2017 *Appl. Phys. Lett.* **110** 223904
[109] Physical Property Measurement System, Quantum Design North America 2022 https://www.qdusa.com/products/ppms.html (accessed November 2022)
[110] Vopson M M 2013 *J. Phys.* D **46** 345304
[111] Planes A, Castan T and Saxena A 2014 *Philos. Mag. Lett.* **94** 1893–908
[112] Stern-Taulats E, Castán T, Mañosa L, Planes A, Mathur N D and Moya X 2018 *MRS Bull.* **43** 295–9
[113] Morellon L, Arnold Z, Magen C, Ritter C, Prokhnenko O, Skorokhod Y, Algarabel P, Ibarra M and Kamarad J 2004 *Phys. Rev. Lett.* **93** 137201

[114] Sun Y, Arnold Z, Kamarad J, Wang G-J, Shen B-G and Cheng Z-H 2006 *Appl. Phys. Lett.* **89** 172513
[115] Liu J, Gottschall T, Skokov K P, Moore J D and Gutfleisch O 2012 *Nat. Mater.* **11** 620–6
[116] Gottschall T, Gracia-Condal A, Fries M, Taubel A, Pfeuffer L, Manosa L, Planes A, Skokov K P and Gutfleisch O 2018 *Nat. Mater.* **17** 929–34
[117] Stern-Taulats E, Castán T, Planes A, Lewis L H, Barua R, Pramanick S, Majumdar S and Manosa L 2017 *Phys. Rev.* B **95** 104424
[118] Lievois J, Perkins R, Martin R and Kobayashi R 1990 *Fluid Phase Equilib.* **59** 73–97
[119] Micro DSC 7 evo 2022 Evo, https://intercovamex.com/wp-content/uploads/2019/06/
[120] TA Instruments 2022 DSC25P, https://www.tainstruments.com/discovery-dsc-25p/ (accessed July 2022)
[121] DSC 204 HP Phoenix 2022 Netzsch, https://analyzing-testing.netzsch.com/en/products/differential-scanning-calorimeter-dsc-differential-thermal-analyzer-dta/dsc-204-hp-phoenix (accessed November 2022)
[122] Imamura W, Usuda É O, Paixão L S, Bom N M, Gomes A M and Carvalho A M G 2020 *Chin. J. Polym. Sci.* **38** 999–1005
[123] Hirayama Y, Iguchi R, Miao X-F, Hono K and Uchida K I 2017 *Appl. Phys. Lett.* **111** 163901
[124] Sebald G, Seveyrat L, Capsal J-F, Cottinet P-J and Guyomar D 2012 *Appl. Phys. Lett.* **101** 022907
[125] Wei Z, Sun W, Shen Q, Shen Y, Zhang Y, Liu E and Liu J 2019 *Appl. Phys. Lett.* **114** 101903
[126] Lin J *et al* 2020 *Mater. Horiz.* **7** 2690–5
[127] Vallone S P, Tantillo A N, Dos Santos A M, Molaison J J, Kulmaczewski R, Chapoy A, Ahmadi P, Halcrow M A and Sandeman K G 2019 *Adv. Mater.* **31** 1807334
[128] Gao Y *et al* 2022 *NPG Asia Mater.* **14** 34
[129] Li F B, Li M, Xu X, Yang Z C, Xu H, Jia C K, Li K, He J, Li B and Wang H 2020 *Nat. Commun.* **11** 4190
[130] Ren Q *et al* 2022 *Nat. Commun.* **13** 2293
[131] Aznar A *et al* 2017 *Nat. Commun.* **8** 1–6
[132] Hao J *et al* 2020 *Chem. Mater.* **32** 1807–18
[133] Li B *et al* 2019 *Nature* **567** 506–10
[134] Li F, Li M, Niu C and Wang H 2022 *Appl. Phys. Lett.* **120** 073902
[135] Miliante C M, Christmann A M, Usuda E O, Imamura W, Paixão L S, Carvalho A M and Muniz A R 2020 *Macromolecules* **53** 2606–15
[136] Sau K, Ikeshoji T, Takagi S, Orimo S i, Errandonea D, Chu D and Cazorla C 2021 *Sci. Rep.* **11** 11915
[137] Mendive-Tapia E and Castán T 2015 *Phys. Rev.* B **91** 224421
[138] Corrales-Salazar A, Brierley R T, Littlewood P B and Guzmán-Verri G G 2017 *Phys. Rev. Mater.* **1** 053601
[139] Meijer B E, Cai G, Demmel F, Walker H C and Phillips A E 2022 *Phys. Rev.* B **106** 064302
[140] Yuan S, Meijer B E, Cai G, Dixey R J C, Demmel F, Dove M T, Liu J, Playford H Y, Walker H C and Phillips A E 2022 *Adv. Funct. Mater.* **32** 2207717
[141] Gottschall T, Skokov K P, Fries M, Taubel A, Radulov I, Scheibel F, Benke D, Riegg S and Gutfleisch O 2019 *Adv. Energy Mater.* **9** 1901322
[142] Griffith L D, Mudryk Y, Slaughter J and Pecharsky V K 2018 *J. Appl. Phys.* **123** 034902

[143] Zhang M, Abdelaziz O, Momen A M and Abu-Heiba A 2017 *Sci. Rep.* **7** 13962
[144] Torelló A, Lheritier P, Usui T, Nouchokgwe Y, Gérard M, Bouton O, Hirose S and Defay E 2020 *Science* **370** 125–9
[145] Kitanovski A and Egolf P W 2006 *Int. J. Refrig.* **29** 3–21
[146] Wood M and Potter W 1985 *Cryogenics* **25** 667–83
[147] Gschneidner K A J and Pecharsky V K 2000 *Annu. Rev. Mater. Sci.* **30** 387–429
[148] Chaturvedi A, Stefanoski S, Phan M-H, Nolas G S and Srikanth H 2011 *Appl. Phys. Lett.* **99** 162513
[149] Brück E, Yibole H and Zhang L 2016 *Phil. Trans. R. Soc.* A **374** 20150303
[150] Mañosa L and Planes A 2020 *Appl. Phys. Lett.* **116** 050501
[151] Stern-Taulats E, Planes A, Lloveras P, Barrio M, Tamarit J-L, Pramanick S, Majumdar S, Frontera C and Mañosa L 2014 *Phys. Rev.* B **89** 214105
[152] Akbulut S, Ocak Y, Keşlioğlu K and Maraşlı N 2009 *J. Phys. Chem. Solids* **70** 72–8
[153] Krivchikov A I, Vdovichenko G A, Korolyuk O, Bermejo F J, Pardo L, Tamarit J L, Jezowski A and Szewczyk D 2015 *J. Non-Cryst. Solids* **407** 141–8
[154] Ibrahim N I, Al-Sulaiman F A, Rahman S, Yilbas B S and Sahin A Z 2017 *Renew. Sustain. Energy Rev.* **74** 26–50
[155] Praveen B and Suresh S 2018 *Eng. Sci. Technol. Int. J.* **21** 1086–94
[156] Zhang N, Song Y, Du Y, Yuan Y, Xiao G and Gui Y 2018 *Adv. Eng. Mater.* **20** 1800237
[157] Liu Y, Zhou H, Xu Z, Liu D, Li J, Hu F and Ma T 2022 *Mater. Res. Lett.* **10** 675–81
[158] Serrano A, Garrido I, Santos S, Duran M, Dauvergne J-L, Carmona M and Palomo Del Barrio E 2022 *J. Energy Storage* **55** 105677
[160] Stern-Taulats E, D'Ammaro A, Avramenko A, Robinson A and Moya X 2018 First steps towards barocaloric refrigeration, Thermag VIII *8th IIR, IIF Int. Conf. Caloric Cooling Int.* (Institute of Refrigeration)
[159] Gauß R, Homm G and Gutfleisch O 2017 *J. Ind. Ecol.* **21** 1291–300
[161] Bing L 2021 Large-scale facilities: grand opportunities for caloric materials research *EuroTherm Seminar on Caloric Heating and Cooling* (online)
[162] Stern-Taulats E, Bermúdez J and Moya X 2020 Practical considerations for barocaloric cooling *MRS Virtual Spring/Fall Meeting*

Chapter 2

Barocaloric effects in polymers

Alexander Magnus G Carvalho, Lucas S Paixão and William Imamura

2.1 Introduction

If applying a mechanical stress in a material we observe temperature and entropy changes, and removing that mechanical stress we observe opposite changes in temperature and entropy, we say the material presents mechanocaloric effect. *Barocaloric effect* is a particular case of mechanocaloric effect, in which the mechanical stress is an isotropic stress (called *hydrostatic pressure* if it is compressive).

The first study reporting mechanocaloric effect in a polymer dates back to the beginning of the 19th century. In 1805, the blind scientist John Gough published his sensorial experience when stretching and releasing an Indian-rubber slip in slight contact with his lips. He mentioned that 'the increase of temperature, which is perceived upon extending a piece of caoutchouc (Indian rubber) may be destroyed in an instant, by permitting the slip to contract again' [1]. In fact, what Gough reported is another particular case of mechanocaloric effect called elastocaloric effect. In 1859, J P Joule published the first systematic study on tractive elastocaloric effect in Indian rubber, vulcanized Indian rubber and gutta-percha, and compressive elastocaloric effect in vulcanized Indian rubber [2].

The first systematic studies on the barocaloric effect in polymers were only published in the 1980s. In 1982, Rodriguez and Filisko reported a detailed study on the barocaloric effect in poly(methyl methacrylate) (PMMA) [3]. To the best of our knowledge, it was the first work dedicated to the barocaloric effect in a polymer (in this case, a rigid polymer). Afterwards, two other rigid polymers were investigated in detail in terms of the barocaloric effect: polytetrafluoroethylene (PTFE) [4] and polyethylene (PE) [5]. In the case of polyethylene, the authors investigated the barocaloric effect in a low-density PE and in a high-density PE.

In 1986, Rodriguez and Filisko also reported the barocaloric effect in non-vulcanized styrene-butadiene rubber (SBR) [6]. It was the first systematic investigation on the barocaloric effect in an elastomer. In spite of the promising results for

doi:10.1088/978-0-7503-4690-0ch2

an elastomer, this class of polymers was neglected over the next several years. Only 31 years later, and almost 160 years after Joule's work on the elastocaloric effect in vulcanized Indian rubber (or vulcanized natural rubber), Bom *et al* [7] and Usuda *et al* [8] investigated the barocaloric effect in vulcanized natural rubber (VNR), sparking interest in studying the barocaloric effect in elastomers. In the following year, Bom *et al* showed the deleterious influence of the glass transition on the barocaloric effect directly measured in VNR [9]. This influence was also observed in nitrile butadiene rubber (NBR) [10]. In addition to VNR and NBR, it was recently reported giant barocaloric effects in other two elastomers: polydimethylsiloxane (PDMS) [11] and polyurethane (PU) [12]. Moreover, supergiant (or colossal) barocaloric effects were reported for acetoxy silicone rubber (ASR) [13].

Polymeric composites were also investigated recently in terms of barocaloric effect. Bom *et al* reported a systematic study on the barocaloric effect in waste tire rubber/natural rubber composites [14]. After that, Imamura *et al* published an extensive investigation on the barocaloric effect in natural graphite/PDMS composites [15]. These works were the first ones to report systematic studies on barocaloric effects in polymeric composites.

2.2 Polymers

The term *polymeric* was first introduced in the early 1830s, by the Swedish chemist Jöns Jacob Berzelius, to describe any compound with molar mass that was a multiple of the molar mass of another compound with the same elemental composition (e.g., ethylene C_2H_4 and isobutylene C_4H_8) [16, 17]. According to the Merriam-Webster Dictionary, the first known use of *polymer*, as a chemical compound or mixture of compounds formed by polymerization and consisting essentially of repeating structural units, was in 1866 [18]. The modern concept of polymers as substances composed of macromolecules (polymer molecules) was proposed in 1920 by the German chemist and Nobel Prize laureate Hermann Staudinger [19]. As stated in the *Purple Book* of the IUPAC, a polymer molecule is a molecule of high relative molecular mass, the structure of which essentially comprises the multiple repetition of units derived, actually or conceptually, from molecules of a prepolymer with low relative molecular mass, known as monomers, which have one repeating unit, or oligomers, which have a small number of repeating units [20].

Polymers can be classified in various ways because they are inherently complex multicomponent materials having several simple and distributed properties [21, 22]. To begin with, the simplest way to classify a polymer is based on its source availability. A *natural polymer* can be found in plants (e.g., cellulose, agar, starch, rubber, etc) and animals (e.g., chitin, gelatin, alginates, xanthum gum, etc). A *semisynthetic polymer* is derived from a natural polymer that undergoes some chemical modification; common examples include the vulcanization of cis-1,4-polyisoprene (natural rubber) to afford an elastomer, and the chemical transformations of cellulose, which can yield cellulose ethers, cellulose esters, cellulose nitrates, cellulose acetates, etc. In contrast, a *synthetic polymer* is synthesized chemically from synthetic monomers or oligomers, usually derived from gases and oils in petrochemistry; examples include polyethylenes, polyamides, polysiloxanes, epoxy resins, etc.

Another way to classify a polymer is by looking at its backbone, also known as main chain — defined as the linear chain to which all other chains, long or short or both, may be regarded as being pendant [20]. An *organic polymer* comprises essentially carbon atoms in the backbone (the repeating unit may contain covalent bonds between carbon and other elements); e.g., polyethylenes, polypropylenes, polyvinylchlorides, polyamides, etc. An *inorganic polymer* does not include carbon atoms in the backbone; e.g., polysilanes, polysiloxanes, polygermanes, polysulfides, etc. And an *inorganic–organic* polymer includes inorganic and organic repeating units; e.g., polycarbosilanes, polyphenylenesulfides, etc. Alternatively, a polymer can be classified according to its molecular architecture (which refers to the shape of the polymer molecule) [23]; a schematic representation of polymers with different molecular architectures is shown in figure 2.1.

A very useful and widespread classification for polymers is based on their properties as follows [20]:

(a) Thermoset: a cross-linked polymer not able to melt (infusible), due to its high concentration of crosslinks, formed by curing a soft solid or viscous prepolymer; it remains in a permanent solid after being cured by an irreversible hardening; e.g.: bakelite, epoxy resins;

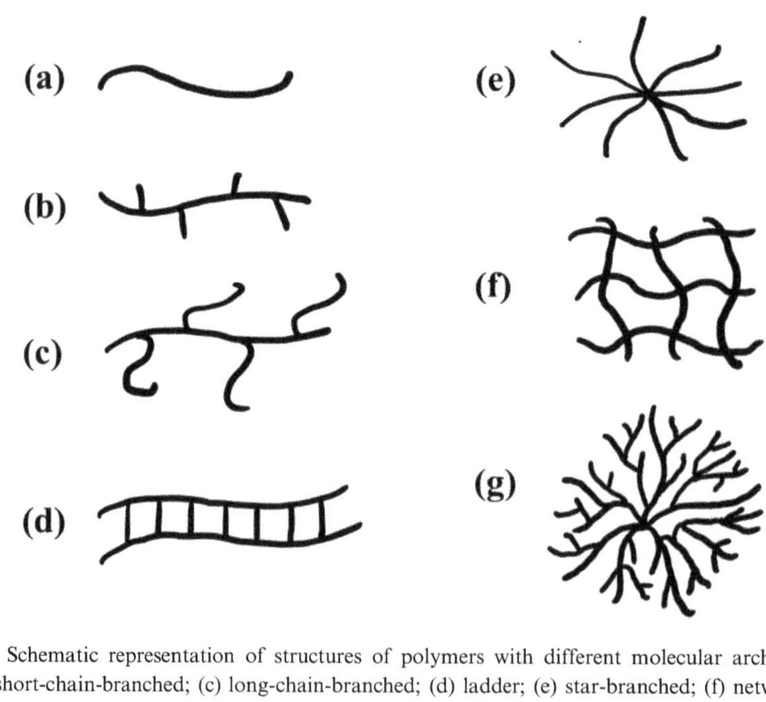

Figure 2.1. Schematic representation of structures of polymers with different molecular architectures: (a) linear; (b) short-chain-branched; (c) long-chain-branched; (d) ladder; (e) star-branched; (f) network; and (g) hyper-branched [23].

(b) Thermoplastic: a polymer that can melt, due to its low concentration or absence of crosslinks; it is flexible and moldable at elevated temperature and solidifies upon cooling; e.g.: polypropylene, polyethylene, polyethylene terephthalate;
(c) Elastomer: a cross-linked polymer that can be stretched several times its original length and returns very nearly to its original length when released (rubberlike elasticity); the concentration of crosslinks is generally smaller than that in a thermoset; e.g.: vulcanized natural rubber, silicone rubbers;
(d) Thermoplastic elastomer (thermoplastic rubber): a polymer blend or copolymer that is melt-processable, in which a continuous elastomeric phase is reinforced by a dispersed hard phase that acts as junction points over a limited range of temperature; or an elastomer that comprises a thermoreversible network; e.g.: styrenic block copolymers, thermoplastic polyolefins, thermoplastic polyurethanes, thermoplastic polyamides.

Depending on the application and related geometries, polymers can also be classified into: absorbent, additive, adhesive, container, extrudate, fiber, filler, film, foam, membrane, molding, monofilament, photoresist, rope, sealant, textile, etc [21–23]. It is also possible to find classifications based on step-growth and chain polymerization mechanisms, biodegradability, production volume concerning commodities, etc [24].

2.3 Theoretical approaches

2.3.1 Basic thermodynamics

Let us gather and briefly discuss important thermodynamic relations for later use in the present chapter. The adiabatic temperature change (ΔT_S) and the isothermal entropy change (ΔS_T) are the quantities of interest from the experimental point of view. Base equations to assess the performance of materials can be obtained from integration as below

$$\Delta T_S = \int_{p_1}^{p_2} \frac{\partial T}{\partial p}\bigg|_S dp \tag{2.1}$$

$$\Delta S_T = \int_{p_1}^{p_2} \frac{\partial S}{\partial p}\bigg|_T dp \tag{2.2}$$

From the first law of thermodynamics, $dU = TdS - pdV$, we can promptly derive

$$\frac{\partial T}{\partial p}\bigg|_S = -\frac{1}{c_p}\frac{\partial U}{\partial p}\bigg|_T - \frac{p}{c_p}\frac{\partial V}{\partial p}\bigg|_T \tag{2.3}$$

$$\left.\frac{\partial S}{\partial p}\right|_T = \frac{1}{T}\left.\frac{\partial U}{\partial p}\right|_T + \frac{p}{T}\left.\frac{\partial V}{\partial p}\right|_T \tag{2.4}$$

where $c_p = T\left.\frac{\partial S}{\partial T}\right|_p$ is the specific heat at constant pressure. The equations above are useful when we are interested in a microscopic description of the barocaloric effect. Internal energy and volume are quantities available via molecular dynamics simulations, as will be discussed later.

From a macroscopic point of view, deformations can be described by the Helmholtz free energy per unit volume [25]

$$F\left(T, \varepsilon_{ij}\right) = F_0(T) - B\alpha(T - T_0)\varepsilon_{kk} + \frac{B}{2}\varepsilon_{kk}^2 + G\left(\varepsilon_{ij} - \frac{1}{3}\varepsilon_{kk}\delta_{ij}\right)^2 \tag{2.5}$$

Above, F_0 is the energy of the underformed body, B and G are the bulk and shear moduli, α is the thermal expansion coefficient, and δ_{ij} is the Kronecker delta. The notation adopted here implies summation over repeated indexes; therefore, $\varepsilon_{kk} = \varepsilon_{xx} + \varepsilon_{yy} + \varepsilon_{zz}$. The stress tensor

$$\sigma_{ij} = \left.\frac{\partial F}{\partial \varepsilon_{ij}}\right|_T \tag{2.6}$$

and the entropy

$$S = -\left.\frac{\partial F}{\partial T}\right|_{\varepsilon_{ij}} \tag{2.7}$$

can be found with the equations above, and will be used in the following discussions.

From the theory of elasticity, we have that the sum of the diagonal elements of the strain tensor equals the relative volume change of a material, i.e., $\varepsilon_{kk} = \Delta V/V$. As a result, considering a hydrostatic pressure p, the first law can be written in terms of the strain tensor as $dU = TdS - pd\varepsilon_{kk}$, where all the extensive quantities refer to a unit volume. From this expression, the following Maxwell relation is derived

$$\left.\frac{\partial S}{\partial p}\right|_T = -\left.\frac{\partial \varepsilon_{kk}}{\partial T}\right|_p \tag{2.8}$$

Directly integrating the equation above, we obtain

$$\Delta S = -\int_{p_i}^{p_f} \left.\frac{\partial \varepsilon_{kk}}{\partial T}\right|_p dp \tag{2.9}$$

which represents the entropy change per unit volume, or alternatively

$$\Delta S = -\frac{1}{\rho_0}\int_{p_i}^{p_f} \left.\frac{\partial \varepsilon_{kk}}{\partial T}\right|_p dp \tag{2.10}$$

representing the entropy change per unit mass, where ρ_0 is the sample density. Here it is important to recall that $\varepsilon_{kk} = Tr(\varepsilon) = \varepsilon_{xx} + \varepsilon_{yy} + \varepsilon_{zz}$.

2.3.2 Consequences of uniaxial confined compression

In what follows, we are going to consider deformations in which the strain tensor is constant through the volume of the body, i.e., it will not be a function of the position inside the body. It does not mean, however, that deformations along different axes will be necessarily the same. Our basic relation to describe such deformations is the Helmholtz free energy, equation (2.5), which is valid while the deformations are small.

To derive thermodynamic relations from equation (2.5), it is important to obtain its differential form

$$dF = \left(\frac{dF_0(T)}{dT} - B\alpha\varepsilon_{kk}\right)dT + [-B\alpha(T - T_0) + B\varepsilon_{kk}]d\varepsilon_{kk} + 2G\left(\varepsilon_{ij} - \frac{1}{3}\varepsilon_{kk}\delta_{ij}\right)d\left(\varepsilon_{ij} - \frac{1}{3}\varepsilon_{kk}\delta_{ij}\right) \quad (2.11)$$

We can work out the first term using $d\varepsilon_{kk} = \delta_{ij}d\varepsilon_{ij}$, while for the second term we have[1]

$$2G\left(\varepsilon_{ij} - \frac{1}{3}\varepsilon_{kk}\delta_{ij}\right)d\left(\varepsilon_{ij} - \frac{1}{3}\varepsilon_{kk}\delta_{ij}\right) = 2G\left(\varepsilon_{ij} - \frac{1}{3}\varepsilon_{kk}\delta_{ij}\right)d\varepsilon_{ij} \quad (2.12)$$

Equation (2.11) then becomes

$$dF = \left(\frac{dF_0(T)}{dT} - B\alpha\varepsilon_{kk}\right)dT + \left[-B\alpha(T - T_0)\delta_{ij} + B\varepsilon_{kk}\delta_{ij}\right]d\varepsilon_{ij} + 2G\left(\varepsilon_{ij} - \frac{1}{3}\varepsilon_{kk}\delta_{ij}\right)d\varepsilon_{ij} \quad (2.13)$$

Finally, from the differential above we get the stress tensor

$$\sigma_{ij} = \left.\frac{\partial F}{\partial \varepsilon_{ij}}\right|_T = -B\alpha(T - T_0)\delta_{ij} + B\varepsilon_{kk}\delta_{ij} + 2G\left(\varepsilon_{ij} - \frac{1}{3}\varepsilon_{kk}\delta_{ij}\right) \quad (2.14)$$

Now we let us consider the particular case of uniaxial confined compression. By definition, the strain tensor is

$$\varepsilon_{ij} = \begin{cases} \varepsilon_{zz}, & i = j = z \\ 0, & \text{otherwise} \end{cases} \quad (2.15)$$

where we are assuming that the body is deformed along the z-axis. Applying this definition to equation (2.14) we obtain the diagonal components of the stress tensor as

[1] $2G\left(\varepsilon_{ij} - \frac{1}{3}\varepsilon_{kk}\delta_{ij}\right)d\left(\varepsilon_{ij} - \frac{1}{3}\varepsilon_{kk}\delta_{ij}\right) = 2G\left\{\left(\varepsilon_{ij} - \frac{1}{3}\varepsilon_{kk}\delta_{ij}\right)d\varepsilon_{ij} - \frac{1}{3}\left(\varepsilon_{ij}\delta_{ij} - \frac{1}{3}\varepsilon_{kk}\delta_{ij}\delta_{ij}\right)d\varepsilon_{kk}\right\}$
Since $\varepsilon_{ij}\delta_{ij} = \varepsilon_{kk}$ and $\delta_{ij}\delta_{ij} = 3$,
$\left(\varepsilon_{ij}\delta_{ij} - \frac{1}{3}\varepsilon_{kk}\delta_{ij}\delta_{ij}\right) = 0$. Therefore,
$2G\left(\varepsilon_{ij} - \frac{1}{3}\varepsilon_{kk}\delta_{ij}\right)d\left(\varepsilon_{ij} - \frac{1}{3}\varepsilon_{kk}\delta_{ij}\right) = 2G\left(\varepsilon_{ij} - \frac{1}{3}\varepsilon_{kk}\delta_{ij}\right)d\varepsilon_{ij}$.

$$\sigma_{zz} = -B\alpha(T - T_0) + B\varepsilon_{zz} + \frac{4}{3}G\varepsilon_{zz} \tag{2.16}$$

$$\sigma_{xx} = \sigma_{yy} = -B\alpha(T - T_0) + B\varepsilon_{zz} - \frac{2}{3}G\varepsilon_{zz} \tag{2.17}$$

while the off-diagonal components vanish, as easily verified. For the sake of simplicity, let us consider the temperature $T = T_0$. From equations (2.16) and (2.17), the ratio of those components of the strain tensor is

$$\frac{\sigma_{xx}}{\sigma_{zz}} = \frac{3B - 2G}{3B + 4G} \tag{2.18}$$

This relation takes a convenient form considering

$$B = \frac{Y}{3(1 - 2\nu)} \qquad G = \frac{Y}{2(1 + \nu)} \tag{2.19}$$

where Y is the Young modulus and ν is the Poisson's ratio. From equations (2.18) and (2.19), we arrive at a simple relation between the stress components parallel and perpendicular to the applied tension

$$\frac{\sigma_{xx}}{\sigma_{zz}} = \frac{\nu}{1 - \nu} \tag{2.20}$$

We can draw a straightforward interpretation from the relation above. Materials with $\nu = 0.5$ (elastomers in general present such a property) exhibit an interesting behavior: under uniaxial compression, they observe $\sigma_{xx} = \sigma_{yy} = \sigma_{zz}$. The restriction of the Poisson's ratio is the basic condition a material must fulfil in order to experience isostatic compression when subject to uniaxial loading and uniaxial deformation. In this situation, the material itself is soft enough to isotropically transmit all applied stress when confined (subject to the constraint of equation (2.15)). This property is the basis for the proper working of an experimental setup to be discussed later, in section 2.4.

To push forward the discussion about the barocaloric effect, the entropy is certainly necessary. From equation (2.5), the entropy is

$$S(T) = -\frac{\partial F}{\partial T}\bigg|_{\varepsilon_{ij}} = S_0(T) + B\alpha\varepsilon_{kk} \tag{2.21}$$

Let us continue to consider a uniaxial deformation, where $\varepsilon_{kk} = \varepsilon_{zz}$. Now, using equation (2.16), the relation between entropy and the uniaxial stress can be expressed as

$$S(T) = S_0(T) + \frac{3B\alpha}{3B - 4G}\sigma_{zz} \tag{2.22}$$

Again, we can use equation (2.19) to obtain an equivalent equation for the entropy in terms of the Poisson ratio

$$S(T) = S_0(T) + \frac{1 + \nu}{-1 + 5\nu}\alpha\sigma_{zz} \tag{2.23}$$

In the limit of soft materials ($\nu = 0.5$) the entropy takes the simpler form

$$S(T) = S_0(T) + \alpha\sigma_{zz} \tag{2.24}$$

Such linear dependence is only true for low pressures, as the theory described so far is valid for small deformations, i.e., within the limit of validity of the expansion in equation (2.5). Many polymers easily surpass this limit of small deformations, revealing the need for a better theoretical description.

2.3.3 Modified thermodynamic elasticity

Experimentally we know that entropy is not a linear function of stress. At least a quadratic dependence between entropy and pressure, which fits experimental data well, should appear in the equations; we should have at least $\frac{\partial F}{\partial T} \propto p^2$. Besides, the specific heat is a function of pressure, i.e., $\frac{\partial^2 F}{\partial T^2} = f(T, p)$. Equation (2.5) (the Helmholtz free energy) is a second-order expansion suitable to describe elastic behavior under small deformations. However, it does not cover the barocaloric phenomenology of polymers. One simple way to obtain such dependence is to assume a quadratic temperature dependence on the term proportional to ε_{kk}^2 of the Helmholtz energy. The modified expression then becomes

$$F(T, \varepsilon_{ij}) = F_0(T) - B\alpha(T - T_0)\varepsilon_{kk} + \frac{B}{2}\left[1 - \beta(T - T_0)^2\right]\varepsilon_{kk}^2 + G\left(\varepsilon_{ij} - \frac{1}{3}\varepsilon_{kk}\delta_{ij}\right)^2 \tag{2.25}$$

Let us discuss the consequences of the expression above. First, we can think of temperature $T = T_0$, where the extra term vanishes and equations (2.5) and (2.25) become identical. Therefore, equation (2.20), which states the condition for isostatic uniaxial compression, remains valid.

Now we need to evaluate the entropy, which motivated equation (2.25). But since we are interested in the entropy as a function of the applied stress, it is important to obtain the stress tensor first. From the Helmholtz energy, we have

$$\sigma_{ij} = \left.\frac{\partial F}{\partial \varepsilon_{ij}}\right|_T = -B\alpha(T - T_0)\delta_{ij} + B[1 - \beta(T - T_0)^2]\varepsilon_{kk}\delta_{ij} + 2G\left(\varepsilon_{ij} - \frac{1}{3}\varepsilon_{kk}\delta_{ij}\right) \tag{2.26}$$

Let us work out the sum of its diagonal components

$$\sigma_{kk} = -3B\alpha(T - T_0) + 3B[1 - \beta(T - T_0)^2]\varepsilon_{kk} \tag{2.27}$$

from which the sum of the diagonal strain components can easily be found

$$\varepsilon_{kk} = \frac{\sigma_{kk} + 3B\alpha(T - T_0)}{3B[1 - \beta(T - T_0)^2]} \tag{2.28}$$

Above, the temperature dependence of the strain is not linear, but it is what we observe experimentally. Now we are finally in a position to evaluate the entropy:

$$S(T) = -\frac{\partial F}{\partial T}\bigg|_{\varepsilon_{ij}} = S_0(T) + B\alpha\varepsilon_{kk} + B\beta(T - T_0)\varepsilon_{kk}^2 \tag{2.29}$$

Given the linear dependence between stress and strain (equation (2.28)), the entropy above can be rewritten as

$$S(T) = S_0(T) + a(T)\sigma_{kk} + b(T)\sigma_{kk}^2 \tag{2.30}$$

The exact equation depends on the parameter β, for which we did not describe the physical meaning. Therefore, equation (2.30) is as valid as the exact equation, with the advantage of being simpler; it also demonstrates the quadratic dependence between entropy change and pressure.

2.3.4 Molecular dynamics

Molecular dynamics (MD) simulations of polymers are not simple to perform due to limitations in simulated time and length scales when compared to the scales involved in physical processes. First, typical relaxation times for polymers are 10^{-6}–10^{2} s [26, 27]; on the other hand, time scales attainable in MD simulations lie within 10^{-9}–10^{-6} s. Likewise, the number of atoms included in a typical atomistic model is 10^4 atoms, which is a small number compared to a macroscopic sample.

Classical MD simulations successfully created a model of non-vulcanized natural rubber (*cis*–1,4-polyisoprene) [28]. The simulations used CHARMM general force field (CGenFF) [29]. In this force field, the intramolecular portion of the potential energy includes terms for the bonds, valence angles, torsion/dihedral angles (proper and improper dihedrals); the intermolecular terms include electrostatic and van der Waals interactions. Therefore, the energy contributions are

$$E_{\text{pot}} = E_{\text{intra}} + E_{\text{inter}} \tag{2.31}$$

$$E_{\text{intra}} = E_{\text{bond}} + E_{\text{angle}} + E_{\text{dihed}} \tag{2.32}$$

$$E_{\text{inter}} = E_{\text{coulomb}} + E_{\text{vdw}} \tag{2.33}$$

The total energy is therefore the sum of potential and kinetic contributions.

$$E_{\text{tot}} = E_{\text{pot}} + E_{\text{kin}} \tag{2.34}$$

where the latter represents temperature.

Barocaloric quantities ΔT_S and ΔS_T are characteristic of adiabatic and isothermal processes, respectively. In order to mimic such conditions, simulations were carried out in both NPH and NPT ensembles. Adiabatic processes, where enthalpy is constant ($dq = 0 \rightarrow d(u + pV) \equiv dh = 0$), were simulated in the NPH ensemble. From such simulations, the temperature change can be obtained directly. Isothermal processes were simulated in the NPT ensemble. In such a case, temperature and entropy changes can be evaluated from equations (2.1) and (2.2), respectively, which rely on the knowledge of the total internal energy.

Simulated cycles of adiabatic compression–decompression revealed the mechanism behind the barocaloric effect. A range of pressures and temperatures were simulated, and to circumvent the limitations highlighted before, simulations must be carried out at higher temperatures than the temperatures where the corresponding experiments are performed. The higher simulation temperatures ensure that relevant vibrational and torsional modes of the chains are excited. Such strategy allows the simulated system to visit a substantial number of microstates in the configuration phase space within the short time scale of simulations. To confirm this hypothesis, reference [28] presents a thorough comparison between experimental data and simulated data at several temperatures. Here, figure 2.2 compares simulated data at 545 K and experimental data at 300 K, presenting a good agreement.

The following discussion will focus on the results for 545 K. Under an adiabatic process, the work performed on the system by the compressing force must be equal to the variation of the internal energy. Such energy comprises the terms indicated in equations (2.31)–(2.34). Figure 2.3(a) shows the evolution of each energy contribution during a compression–decompression cycle. The rise in kinetic energy is responsible for the temperature rise, observed experimentally. An interesting behavior is observed in the potential energy; as pressure is applied, the potential energy drops. Since the total energy must be equal to the compressive work, this drop in potential energy boosts the kinetic energy rise.

Looking closer at the potential energy, we see that, because of compression, all three components of the intramolecular energy rise, while the intermolecular energy sharply drops. The compressive force reduces the polymer free volume, reducing the distance between the chains. This extra confinement of the chains limits their possibility to translate, vibrate and rotate, ultimately increasing the intramolecular energy. However, the interactions between the chains are predominantly attractive, leading to the reduction of the intermolecular energy when pressure increases.

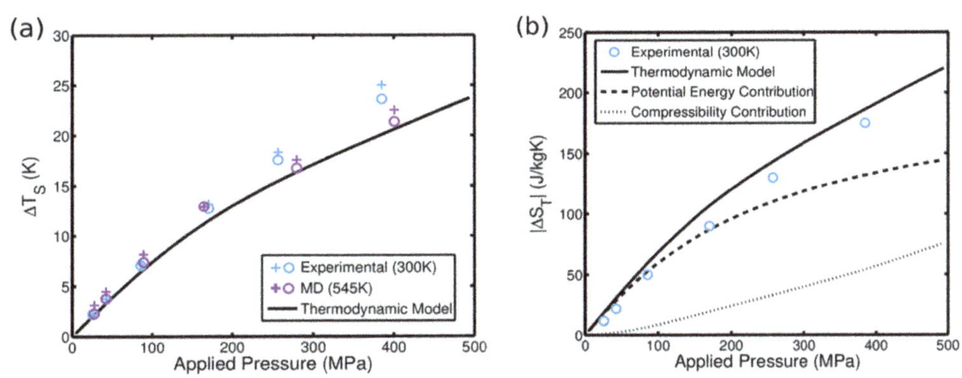

Figure 2.2. Temperature and entropy changes from MD simulations (at 545 K) compared with experimental data (at 300 K). (a) Temperature change obtained from NPH simulations (symbols) and from NPT simulations (line); the latter uses simulated data and equations (2.1) and (2.3). (b) Entropy change obtained from NPT simulations and equations (2.2) and (2.4). Reproduced with permission from reference [28], copyright (2020) American Chemical Society.

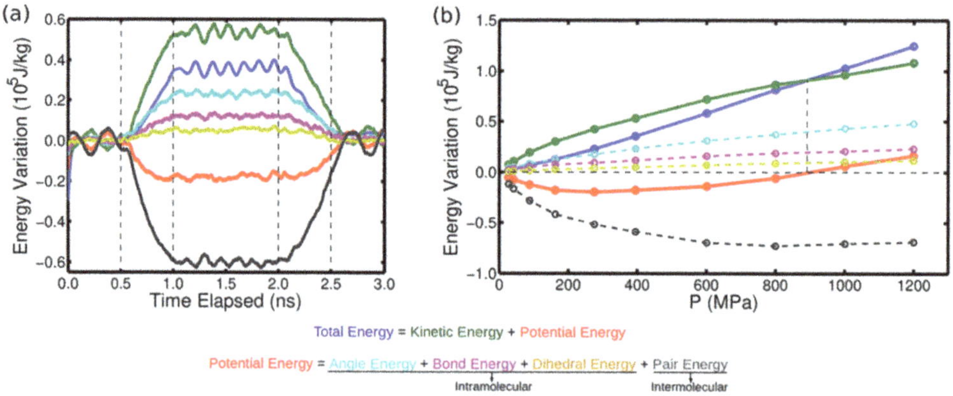

Figure 2.3. Energy components in adiabatic (NPH) simulations at a temperature of 545 K. The reference energy is the energy of the uncompressed initial state. (a) Energy as a function of time during a compression–decompression cycle (applied pressure is 390 MPa). (b) Energy variation during compression as a function of the applied pressure. Reproduced with permission from reference [28], copyright (2020) American Chemical Society.

The energy variation is temperature- and pressure-dependent, the pressure dependence is shown in figure 2.3(b). The attractive character of the intermolecular interaction persists under low pressures. At around 900 MPa the proximity between the chains makes their interaction repulsive, the intermolecular energy starts increasing. Around the same pressure, the total potential energy variation becomes positive and the gain in kinetic energy becomes lower than the total energy variation, hence lower than the external work.

2.4 Experimental methods

2.4.1 Direct adiabatic temperature change

Direct adiabatic temperature changes in polymers or polymer-based materials were reported using different pressure-cell setups [3, 7]. In these experiments, the sample is enclosed in a cylindrical pressure cell with a thermocouple placed inside the sample in direct thermal contact with it, as illustrated in two examples shown in figure 2.4.

In figure 2.4(a), pressurized water is used as the pressure-transmitting fluid in the system, where the sample is subject to a hydrostatic pressure using mercury as pressure transmission medium. The initial temperature is controlled by an oven, which can be set to different temperatures above room temperature.

In figure 2.4(b), a hydraulic press or a universal testing machine applies a load on the piston. The initial temperature is set by a fluid (such as water, ethylene glycol with water, or liquid nitrogen) that flows through a copper serpentine coil simultaneously with heat elements for fine temperature control. When the sample is in a liquid state, the pressure can be calculated by dividing the applied force on the piston by the cross-sectional area. For solids materials, a true hydrostatic pressure provided by this setup can only be considered for isotropic solid materials that have a value of ~0.5 for the Poisson's ratio (ν) [25, 30]. Since the vast majority of soft materials and elastomers have $\nu \approx 0.5$, they can effectively act as their own pressure transmission medium.

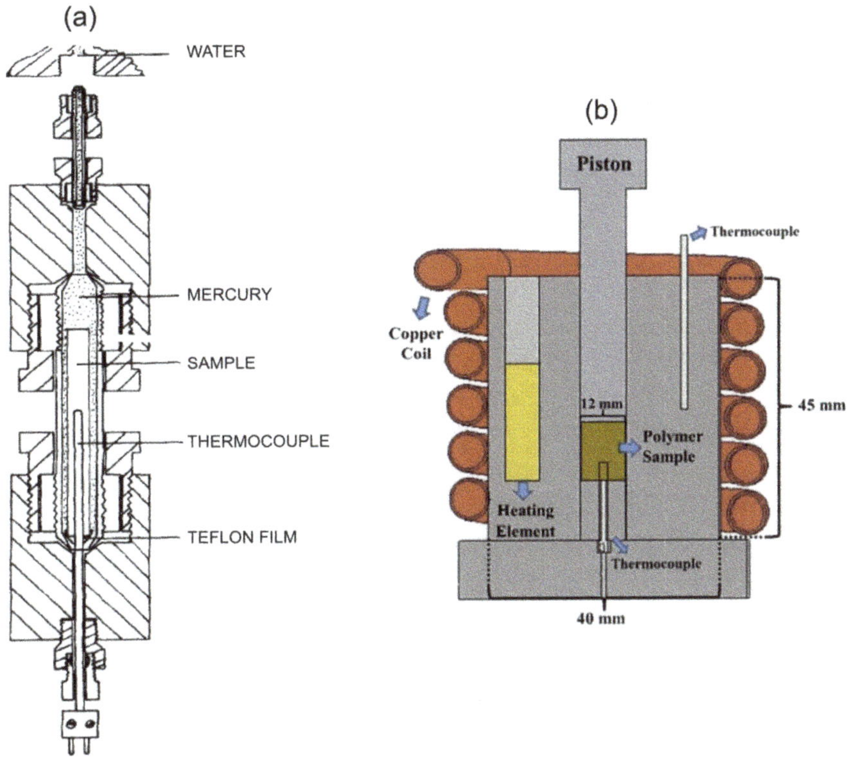

Figure 2.4. Pressure-cell setup developed by (a) Rodriguez and Filisko [3], and (b) Bom *et al* [7]. Figure 2.4(a) was reproduced from reference [3] with permission from AIP Publishing; figure 2.4(b) was reproduced from reference [7] with permission from AIP Publishing.

The mathematical formulation states that the transverse stress ($\sigma_T = \sigma_{xx} = \sigma_{yy}$) experienced by the confined isotropic material is related to the axial stress applied by the piston (σ_{zz}) follows the equation (2.20). Naturally, if $\nu \neq 0.5$ in equation (2.20), the setup measures a confined mechanical caloric effect, which is not a barocaloric effect. We can propose a minimum limit for the Poisson's ratio of 0.490 to denote a confined compression as a barocaloric effect. In this case, $\nu \geqslant 0.490$ leads to $\sigma_T \geqslant 0.961\ \sigma_{zz}$.

Both devices depicted in figure 2.4 consider that the compression and decompression processes are quasi-adiabatic processes, because the pressure varies quickly enough to preclude significant heat exchange between the sample and its surroundings. Once the initial temperature is stabilized, ΔT_S data for compression and decompression processes can be obtained in two different ways: (1) the pressure is quickly applied, and then it is quickly removed once the maximum temperature change is recorded (figure 2.5(a)); (2) the pressure is quickly applied, then there is enough time for thermal relaxation to the initial temperature while the sample is still under pressure, then the pressure is quickly removed, and finally there is enough time again for thermal relaxation to the initial temperature while the sample is not under pressure (figure 2.5(b)).

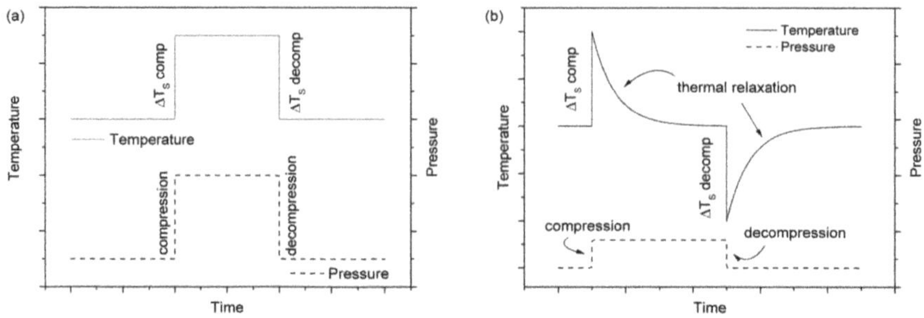

Figure 2.5. Schemes of temperature and pressure curves as a function of time during direct adiabatic temperature change measurements: (a) without thermal relaxation, and (b) with thermal relaxation.

2.4.2 Indirect isothermal entropy change

The setup illustrated in figure 2.4(b) can also be used as a dilatometer in different isobaric conditions. Once the pressure is set, the temperature is varied at a constant rate over the desired temperature range, and the sample's length is recorded. The observed expansion must be corrected by subtracting the mechanical and thermal contributions of the system; these contributions can be evaluated from a blank run or from two runs with two samples of the same material with different lengths.

Recalling volumetric strain is defined as $\varepsilon_{kk} \equiv \Delta V / V_0$, where ΔV is the volume change of the sample and V_0 is its initial volume, ε_{kk} for a confined sample can be expressed as

$$\varepsilon_{kk} = \Delta l / l_0 \tag{2.35}$$

due to the constant section area, where Δl is the length change of the sample, and l_0 is its initial length.

Isothermal entropy changes (ΔS_T) are indirectly obtained by taking the numerical derivative from different ε_{kk} versus temperature curves for heating or cooling process according to equation (2.10) (reproduced below for the reader's convenience):

$$\Delta S_T(T,\, \Delta p) = -\frac{1}{\rho_0} \int_{p_1}^{p_2} \left. \frac{\partial \varepsilon_{kk}}{\partial T} \right|_p dp \tag{2.36}$$

where ρ_0 is the density of the sample at atmospheric pressure and room temperature, and $\Delta p = p_2 - p_1$ is the pressure change from p_1 to p_2.

ΔS_T from the equation above is evaluated using techniques of numerical integration, as the Midpoint Rule or Trapezoidal Rule. Owing to this, the quality of the indirect ΔS_T is directly affected by the quantity of the isobaric ε_{kk} versus temperature curves considered during the calculation.

Another approach to obtain ΔS_T indirectly is through the following approximation:

$$\Delta S_T(T,\, \Delta p) \cong -\frac{c_p(T)\Delta T_S(T,\, \Delta p)}{T} \tag{2.37}$$

where $c_p(T)$ data can be obtained, for instance, by differential scanning calorimetry (DSC), and $\Delta T_S(T,\, \Delta p)$ data is obtained directly as described in section 2.1. It is important to mention that equation (2.37) is valid only far from transitions and for moderate pressures, otherwise considering c_p independent of pressure might not be a good approximation.

2.5 Barocaloric effects in rigid polymers

As mentioned before, Rodriguez and Filisko, to the best of our knowledge, published the first systematic study on the barocaloric effect in a polymer [3]. They studied the poly(methyl methacrylate), also known by its acronym, PMMA, which is a rigid polymer. The authors analyzed a PMMA sample available commercially. The sample was confirmed to be pure and had a glass transition temperature (T_g) of 378 K.

Figure 2.6(a) shows the adiabatic temperature change (ΔT) as a function of temperature for PMMA reported by Rodriguez and Filisko [3]. As expected, ΔT increases with pressure. Less obvious is the observed ΔT increase as the temperature increases. This increase gets more and more accentuated as the temperature approaches T_g. The authors mention that this behavior is 'probably associated with the onset of the glass transition'. For the maximum reported pressure change (Δp) of 193.1 MPa, PMMA presents a ΔT of 4.9 K at room temperature, and reaches 8.9 K at 368 K, near T_g.

The barocaloric effect in a commercial sample of polytetrafluoroethylene (PTFE) was studied by Rodriguez [4]. PTFE has a T_g around 173 K [31] and two crystal–crystal transitions around room temperature (292 K and 303 K) [4, 32].

Figure 2.6(b) shows the adiabatic temperature change (ΔT) as a function of temperature for PTFE reported by Rodriguez [4]. It is easy to observe a maximum in ΔT around 300 K and larger pressure changes. At lower pressures, the maximum seems to be shifted to lower temperatures. From calorimetry, we know the transition at 292 K has enthalpy change much higher than at 303 K [4]. Therefore, the observed maxima in ΔT are due to the first-order transition at 292 K (at atmospheric pressure), and this transition shifts to higher temperature with pressure.

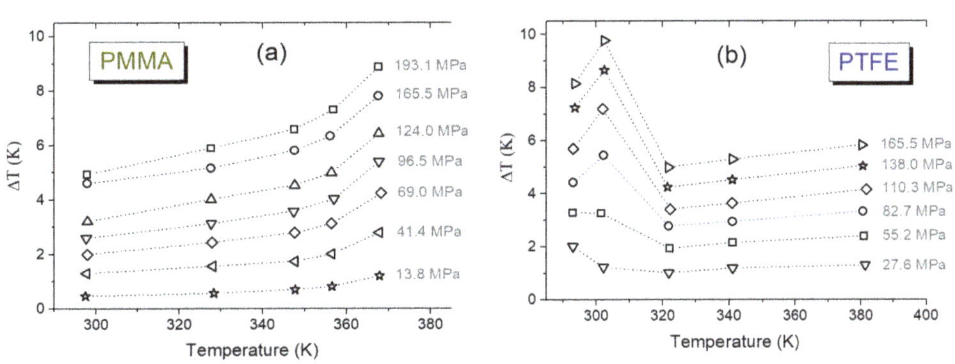

Figure 2.6. Adiabatic temperature change (directly measured) as a function of temperature for (a) poly(methyl methacrylate) (PMMA) (reprinted from [3] with the permission of AIP Publishing) and (b) polytetrafluoroethylene (PTFE) (reprinted from [4], copyright 1986 John Wiley & Sons, Inc.).

Following the same method used for PMMA and PTFE, Rodrigues and Filisko studied the barocaloric effect (only the adiabatic temperature change) in commercial low-density polyethylene (LDPE) and high-density polyethylene (HDPE) [5]. The glass transition temperature (T_g) of polyethylene is, typically, lower than 253 K at atmospheric pressure [33]. The melting point (T_m) of polyethylene can vary within a typical temperature range (from ~382 K to ~410 K), at atmospheric pressure, depending on its density and degree of crystallinity [34].

Figure 2.7(a) shows, for three selected pressure changes, the adiabatic temperature change (ΔT) as a function of temperature for LDPE and HDPE reported by Rodriguez and Filisko [5]. Clearly, ΔT increases as the temperature increases. Besides, this increase gets more accentuated at higher pressure changes and approaching T_m. At the lowest pressure change (13.8 MPa), there is no difference in ΔT between LDPE and HDPE. As the pressure change increases, ΔT for HDPE becomes higher and higher than for LDPE.

A few years before Rodriguez and Filisko's work about polyethylene, Ainbinder *et al* reported temperature changes in high-density polyethylene due to compression and decompression processes at different temperatures [35]. Although Ainbinder and coworkers claimed they measured HDPE in isostatic conditions, it seems not true. They used a high-pressure chamber similar to that described in figure 2.4(b), but HDPE has typical values of Poisson's ration lower than 0.46 [36]. Even so, it is interesting to compare the adiabatic temperature change for HDPE reported in both works. This comparison is shown in figure 2.7(b). We observed that, for low pressure changes ($\Delta p <$ 60 MPa), at 334 K, LDPE and HDPE reported by Rodriguez and Filisko [5] and HPDE reported by Ainbinder *et al* [29] present very similar values of adiabatic temperature change. Nevertheless, above 90 MPa, Ainbinder's HDPE presents higher and higher ΔT when comparing with Rodriguez and Filisko's HDPE, while HDPE and LDPE from Rodriguez and Filisko do not differ significantly from each other.

The barocaloric effect of PVDF–TrFE–CTFE (polyvinylidene di-fluoride tri-fluoroethylene-chlorotrifluoroethylene) is often referred to as giant [37], due to the

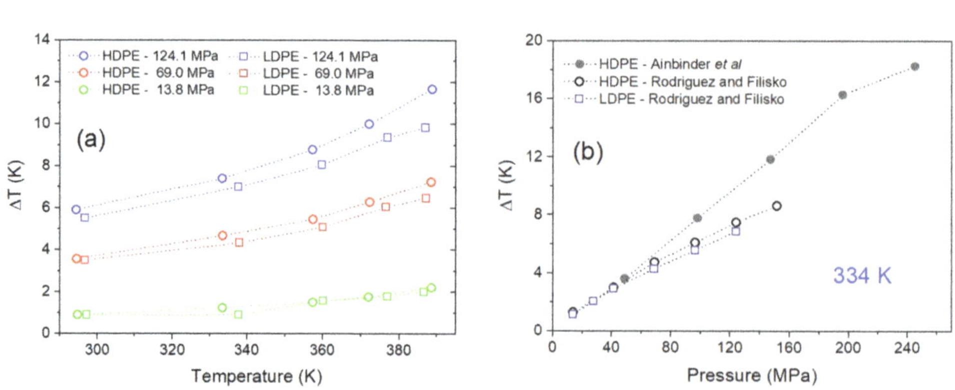

Figure 2.7. (a) Adiabatic temperature change (directly measured) as a function of temperature for low-density polyethylene (LDPE) and high-density polyethylene (HDPE), for selected pressure changes. (b) Comparison among HDPE and LDPE (reprinted from reference [5], copyright (1987), with permission from Springer) and HPDE from reference [35].

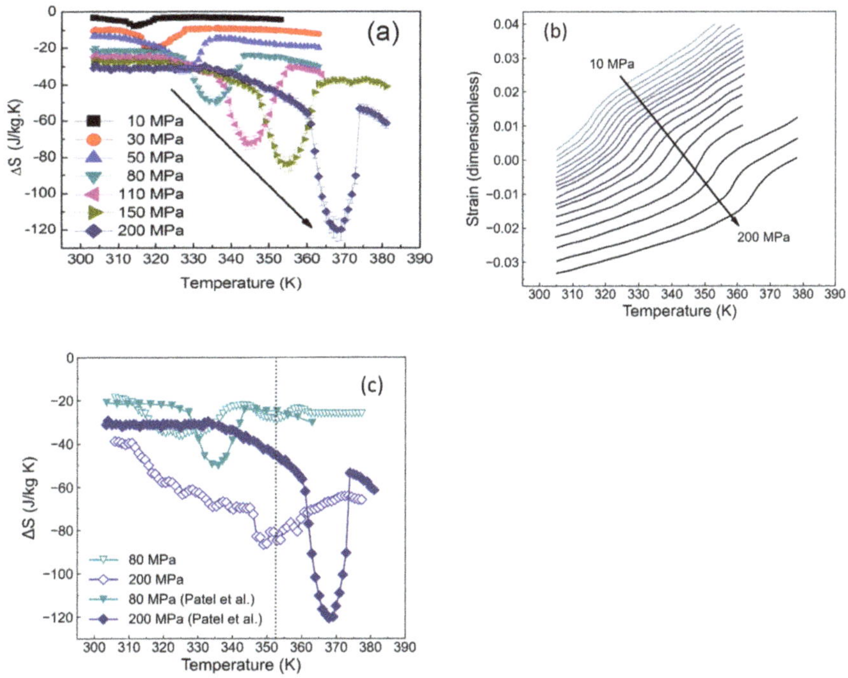

Figure 2.8. Barocaloric effect of PVDF. (a) Entropy change (reprinted from [38], with the permission of AIP Publishing). (b) Strain as a function of temperature for pressures of 10, 15, 20, 25, 30, 35, 40, 45, 50, 60, 70, 80, 95, 110, 130, 150, 175, and 200 MPa (reprinted with permission from [39], copyright (2004) American Chemical Society. (c) Entropy change evaluated from data of (b) using equation (2.36) and density 1828 kg m^{-3}.

entropy change [38] shown in figure 2.8(a). Although such effect is indeed giant, the reported maximum values have been overestimated. Apparently, the authors considered an approximation of equation (2.10) regarding the derivative of the strain as independent of pressure, retrieving an entropy change proportional to such derivative. After using strain data [39] (figure 2.8(b)), to evaluate equation (2.10), we obtain the entropy change of PVDF–TrFE–CTFE, shown in figure 2.8(c) for pressures 80 MPa and 200 MPa. Since the common temperature range for the strain is restricted to the upper limit of 353 K, we performed a linear extrapolation on the strain curves up to 380 K before evaluating the entropy change. The dotted line in figure 2.8(c) indicates the temperature above which the entropy change has been calculated from extrapolated strain data. We see that the peaks of $\frac{\partial \varepsilon}{\partial T}$ at each pressure sum up resulting in broad smaller peaks rather than sharp peaks near the transition temperatures.

2.6 Barocaloric effects in elastomers

The first systematic investigation on the barocaloric effect in an elastomer was also performed by Rodriguez and Filisko [6]. They studied a commercial non-vulcanized styrene-butadiene rubber (SBR), following the same method used for the rigid polymers mentioned in the last section. Typically, the glass transition temperature for SBR is in the temperature range of 218–238 K, at atmospheric pressure, depending on the styrene and butadiene ratio [40].

In figure 2.9, we reproduce the adiabatic temperature change (ΔT) as a function of temperature (and different pressure changes) for non-vulcanized SBR reported by Rodriguez and Filisko [6]. As expected, ΔT enhances with pressure, reaching ~11.7 K at about 404 K for Δp of 151.7 MPa. Besides, ΔT increases linearly with temperature in the reported temperature range. We are able to notice the rate $\frac{d\Delta T}{dT}$ slightly increases when the pressure change increases. The linear behavior observed in figure 2.8 is an indication that there is no significant phase transition in the reported pressure and temperature ranges.

The barocaloric effect in vulcanized natural rubber (VNR) was first systematically investigated by Bom *et al* [7] and Usuda *et al* [8] in 2017. Both works analyzed VNR samples prepared from commercial pre-vulcanized latex resin, cast in a cylindrical plaster mold.

Figure 2.10(a) shows the adiabatic temperature change as a function of temperature for VNR, directly measured at compression and decompression processes, and different pressure changes, reported by Usuda *et al* [8]. As expected, ΔT enhances with pressure in all the analyzed temperature range. Similar to what was reported for SBR, ΔT increases linearly with temperature and the rate $\frac{d\Delta T}{dT}$ slightly increases when the pressure change increases. The linear behavior observed for VNR is an indication that there is no significant phase transition in the reported pressure and temperature ranges.

In 2018, Bom *et al* [9] extended the temperature and the pressure ranges to study the barocaloric effect in VNR. The adiabatic temperature change (ΔT_S) data for these extended ranges are shown in figure 2.10(b). ΔT_S reaches ~25 K at 314 K for Δp = 390 MPa. We can observe the linear behavior is maintained throughout the

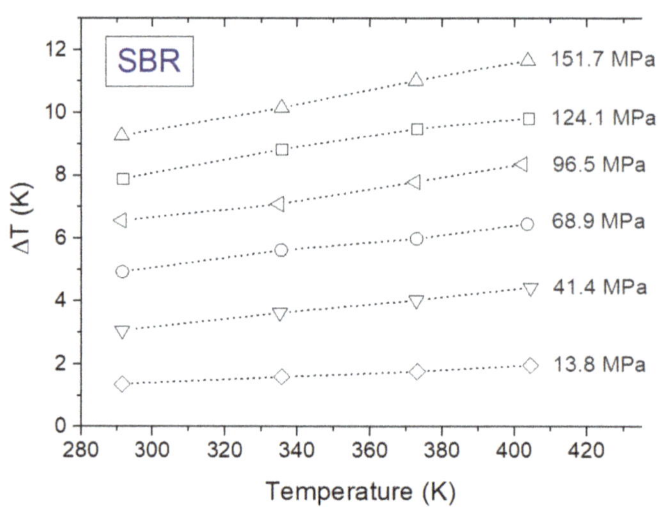

Figure 2.9. Adiabatic temperature change (directly measured) as a function of temperature for non-vulcanized styrene-butadiene rubber (reprinted from [6], copyright (1986), with permission from Elsevier).

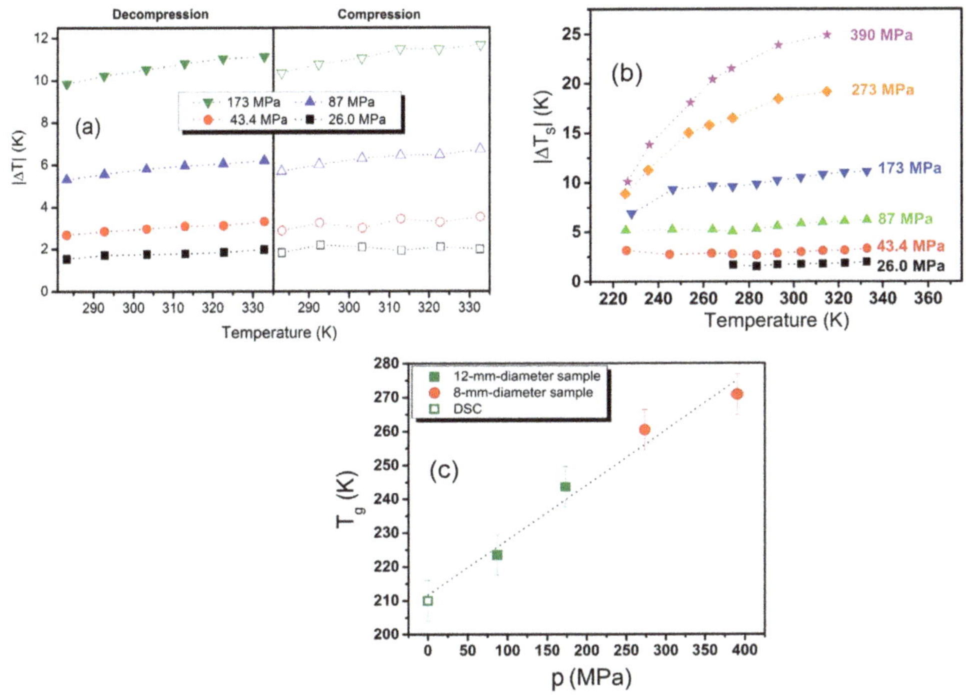

Figure 2.10. Some results for VNR. (a) Adiabatic temperature change directly measured on compression and decompression processes [8]. (b) Adiabatic temperature change directly measured on decompression [9]. (c) Glass transition temperature as a function of pressure. In (c), the original legend was corrected (symbols for 12-mm and 8-mm-diameter sample were swapped). Figure 2.10(a) was reproduced from [8], copyright (2017), with permission from Elsevier; figures 2.10(b) and (c) were reproduced with permission from [9], copyright (2018) American Chemical Society.

temperature range up to a pressure change of 87 MPa. For Δp = 173 MPa, $\Delta T_S(T)$ is not linear at lower temperatures, while for Δp = 390 MPa, $\Delta T_S(T)$ is far from linearity throughout the temperature range. The authors suggested that this non-linear behavior is due to the shift of the glass transition to higher temperatures as the pressure variation increases. The glass transition temperature (T_g) for VNR is around 210 K at atmospheric pressure. Considering the deleterious influence of T_g on the barocaloric effect, Bom *et al* used strain versus temperature data to obtain T_g at higher pressures. T_g as a function of applied pressure is shown in figure 2.10(c).

The barocaloric effect in polydimethylsiloxane (PDMS) rubber was studied by Carvalho *et al* [11]. The authors analyzed rubbery cylindrical PDMS samples prepared from commercial components. The glass transition temperature for this commercial PDMS (prepared according to the manufacturer) is ~148 K [41], at atmospheric pressure, which is below the temperature range analyzed by the authors.

The adiabatic temperature change (ΔT_S) as a function of temperature for PDMS, obtained at decompression processes for different pressure changes, is shown in figure 2.11(a). ΔT_S data was obtained directly through the same method used for VNR. For pressure changes up to 173 MPa, we observe an almost constant ΔT_S

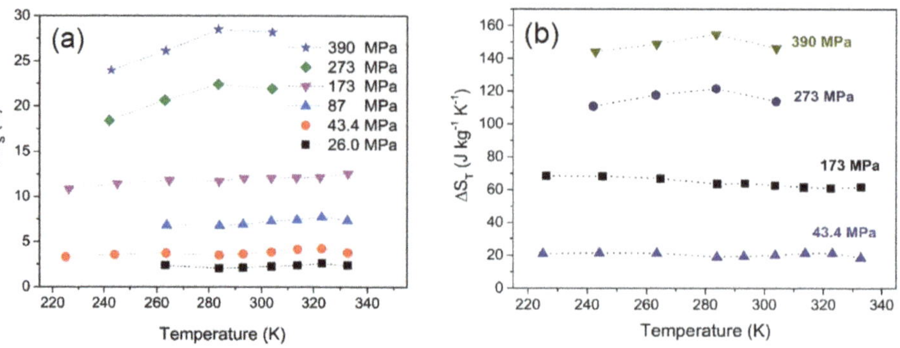

Figure 2.11. (a) Adiabatic temperature change (directly measured) as a function of temperature for polydimethylsiloxane (PDMS) rubber. (b) Isothermal entropy change indirectly obtained from ΔT_S data [11]. Figure 2.11(a) and (b) were reproduced from [11], copyright (2018), with permission from Elsevier.

throughout the temperature range (223–333 K). For Δp of 273 and 390 MPa, we can see a trend of ΔT_S reduction as temperature decreases. Possibly, this reduction is due to the influence of T_g, which is shifted to higher temperatures as the pressure increases. PDMS presents colossal ΔT_S values ($\geqslant$25 K) for Δp of 390 MPa, achieving 28.5 K at ~284 K.

Figure 2.11(b) shows the isothermal entropy change (ΔS_T) as a function of temperature for PDMS obtained for selected pressure changes. ΔS_T values were obtained from ΔT_S data using the following approximation: $\Delta S_T \cong \left(\frac{c_p}{T}\right)\Delta T_S$. This approximation is appropriate throughout the temperature range, since there is no phase transition (especially up to 173 MPa) and considering c_p does not change significantly with pressure. We can notice that ΔS_T is almost constant throughout the temperature range, even at higher pressures, reaching colossal values ($\geqslant$100 J kg^{-1} K^{-1}) for $\Delta p \geqslant 273$ MPa. ΔS_T achieves 154.7 J kg^{-1} K^{-1} for $\Delta p = 390$ MPa at ~284 K.

The barocaloric effect in nitrile butadiene rubber (NBR) was studied by Usuda *et al* [10]. The authors analyzed a commercial vulcanized NBR sample reinforced with carbon black fillers. Typically, the glass transition temperature for pure NBR is around 243 K at atmospheric pressure [42]. Nevertheless, T_g changes depending on the composition, amounts of fillers and applied pressure.

Figure 2.12(a) presents the volumetric strain as a function of temperature for NBR at selected pressures. Considering the heating and cooling curves at 52 MPa, we see that there are significant changes in the derivatives, which occur around 248 K. These derivative variations in strain versus temperature curves are due to the glass transition. Besides, there is also significant thermal hysteresis (for instance, 8.7 K at the strain of −0.024, just above T_g for that pressure of 52 MPa). As pressure increases, T_g and hysteresis increase. For 234 MPa, there is a huge thermal hysteresis of 28.7 K at the strain of −0.057, just above T_g for that pressure.

The shift of T_g towards higher temperatures, as pressure increases, influences the barocaloric effect, and this can easily be observed in ΔT_S data (figure 2.12(b)). Above 43 MPa, ΔT_S significantly reduces at lower temperatures, and onset for this reduction shifts towards higher temperatures as well as it occurs with T_g. The largest

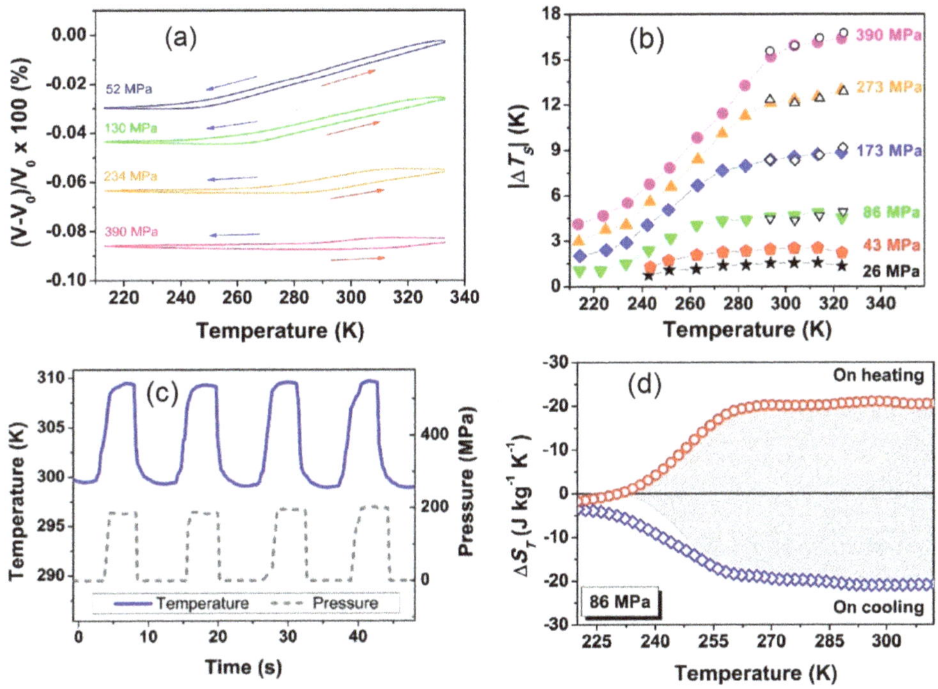

Figure 2.12. Some results for NBR. (a) Strain as a function of temperature measured on heating and on cooling processes. (b) Adiabatic temperature change (directly measured) as a function of temperature. (c) Temperature and pressure as a function of time. (d) Isothermal entropy change obtained from strain versus temperature data [10]. All figures were reproduced with permission from [10], copyright (2019) American Chemical Society.

ΔT_S values directly measured were ~16.8 K and 16.4 K, for compression and decompression processes, respectively, for pressure change of 390 MPa at ~324 K.

The large thermal hysteresis seen specially at higher pressures changes seems not to be a drawback. When we observed compression and decompression data in figure 2.12(b) and the compression–decompression cycles shown in figure 2.12(c), we notice there is reversibility with large values of ΔT_S. The isothermal entropy change also presents reversibility in wide temperature ranges, as exemplified in figure 2.12(d) for $\Delta p = 86$ MPa.

Imamura *et al* [13] investigated the barocaloric effect in acetoxy silicone rubber (ASR). The authors analyzed cylindrical ASR samples prepared from commercial silicone sealant resin. X-ray diffraction shows that ASR is amorphous at room temperature (and atmospheric pressure) and semi-crystalline at 200 K, e.g., part of the material undergoes structural phase transition. Besides, it was reported significant thermal hysteresis and pressure hysteresis. At 173 MPa, for instance, thermal hysteresis is about 31.2 K; at 249 K, pressure hysteresis is about 104.1 MPa.

The adiabatic temperature change for ASR is almost constant up to $\Delta p = 87$ MPa, throughout the temperature range of 223–333 K (figure 2.13(a)). For $\Delta p \geqslant 173$ MPa, we observe a peak in ΔT_S, which becomes wider and shifts to higher

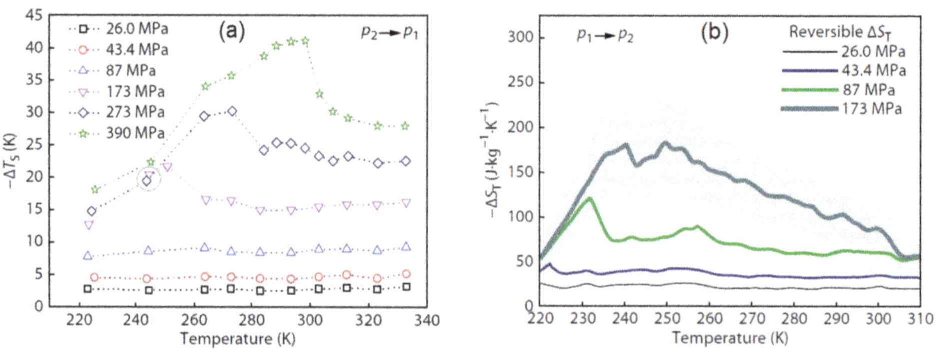

Figure 2.13. (a) Adiabatic temperature change (directly measured) as a function of temperature for acetoxy silicone rubber (ASR). (b) Isothermal entropy change indirectly obtained from strain versus temperature data [13]. Figures 2.13(a) and (b) were reproduced from reference [13], copyright (2020), with permission from Springer.

temperatures as pressure change increases. This peak is due to the structural transition present in ASR. For $\Delta p = 273$ MPa and $\Delta p = 390$ MPa, ΔT_S achieves 30.2 K and 41.1 K, respectively, which are colossal ΔT_S values ($\geqslant$25 K). It is worth noticing that ASR present giant and colossal ΔT_S values even within temperature ranges above the regions of peaks, where ASR is totally amorphous. At ~244 K, ΔT_S is slightly higher for $\Delta p = 173$ MPa (20.4 K) than for $\Delta p = 273$ MPa (19.5 K).

Figure 2.13(b) shows the reversible isothermal entropy change as a function of temperature for ASR. Similar to what occurs with ΔT_S, we see the ΔS_T peak also shifts to higher temperatures as pressure change increases. ΔS_T presents colossal values ($\geqslant$100 J kg^{-1} K^{-1}) already for $\Delta p = 87$ MPa, reaching 182 J kg^{-1} K^{-1} for $\Delta p = 173$ MPa.

Bocca *et al* [12] studied the barocaloric effect in polyurethane (PU). The authors analyzed cylindrical PU samples prepared from commercial components. The glass transition temperature is around 213 K.

ΔT_S and ΔS_T for PU are almost constant throughout the temperature range (293–333 K) and the pressure range (up to 218 MPa) as shown in figure 2.14. This is valid both applying and removing pressure. This is an indication that the pressure of 218 MPa is not enough to shift T_g above 293 K. ΔT_S data were directly obtained while ΔS_T were obtained through the approximation [37]. In this case, the approximation is appropriate throughout the temperature range, since there is no phase transition and considering c_p does not change significantly with pressure. The largest reported ΔT_S is 14.9 K on compression and −14.0 K on decompression; the largest ΔS_T is −95.6 J kg^{-1} K^{-1} on compression and 89.8 J kg^{-1} K^{-1} on decompression.

Recently, Zhang *et al* [43] reported significant barocaloric effects that result from pressure-induced phase transitions between liquid and solid states in elastic copolymers consisting of ethylene and vinyl acetate (EVA). When rapidly compressed (decompressed) the copolymer EVA-28 near the phase transition, the authors directly observed remarkable ΔT_S of 29.6 K (26.9 K) for $\Delta p = 400$ MPa, and concluded copolymers exhibit giant barocaloric effects on both the solid and liquid phases, resulting in a broad refrigeration temperature range of 110 K.

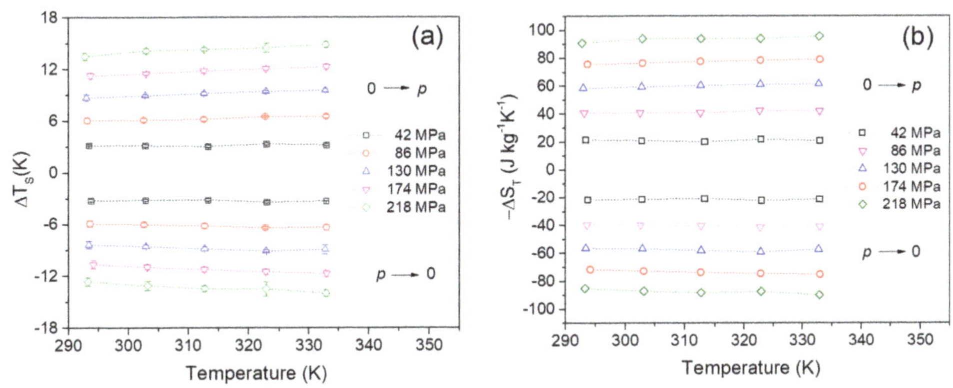

Figure 2.14. (a) Adiabatic temperature change as a function of temperature for polyurethane directly measured under different pressure changes, in compression ($0 \rightarrow p$) and decompression ($p \rightarrow 0$) processes. In the original chart (figure 2.1(a) in reference [12]), the decompression process was mistakenly labeled as '$0 \rightarrow p$'. (b) Isothermal entropy change as a function of temperature for polyurethane, obtained from direct ΔT_S data and using approximation [37]. Figure 2.14(a) and (b) were reproduced from reference [12], CC BY 4.0.

Weerasekera *et al* [44] characterized and evaluated ten commercially available block copolymer thermoplastic elastomers with different mechanical, thermal, and barocaloric properties. The authors showed the isothermal compressibility ranged from 0.1 to 0.4 GPa^{-1}, while the normalized refrigeration capacity ranged from 13.2 to 41.9 kJ K^{-1} GPa^{-1} for a temperature span of 65 K and an applied pressure of 100 MPa. For an applied pressure of 434.1 MPa at 293 K, it was observed a small ΔT_S of 2.2 K for ZYTEL, and a giant ΔT_S of 28 K for ELVAX.

2.7 Barocaloric effects in polymeric composites and other polymeric materials

Barocaloric effects in polymeric composites are very recent topics. In general, polymers are insulating materials with low thermal conductivity and low density, which can compromise their efficiency for thermal energy storage applications. To overcome these limitations, reinforcements such as flakes, particulates, or fibers can be embedded in the polymer, resulting in a composite material; this composite material combines two or more components with very distinct physical and chemical properties, resulting in a material that exhibits a combination of the properties from its constituent phases.

Categorically, the nitrile butadiene rubber (NBR) [10] previously described in section 1.6 was the first composite reported with barocaloric effects. We choose not to address it here because NBR was measured from a commercial vulcanized NBR reinforced with carbon black fillers provided by Elastim Co, which is typically called as an elastomeric rubber. In addition, the true influence of the carbon black fillers on the barocaloric effects and other properties was not evaluated.

In 2020, Bom *et al* [14] investigated the barocaloric effects in ground waste tire rubber (WTR) and WTR/VNR samples as an upcycling process. While the paper refers to the WTR/VNR materials as polymer blends, they are more properly designated as composite materials since the ground WTR acts as particulates dispersed in a matrix of VNR. The WTR/VNR composites offer an innovative solution to the environmental issue of post-consumer tires by creating novel materials for solid-state cooling devices. In general, as shown in figure 2.15, the WTR/VNR compounds presented a slight reduction in the barocaloric effects compared to pure VNR, and the pure WTR also displayed interesting barocaloric effects. Moreover, by observing thermal relaxation in ΔT_S measurements, it was possible to infer that both pure WTR and WTR-based samples also exhibited faster thermal exchange rates than pure VNR.

In 2022, Imamura *et al* [15] reported a systematic study on barocaloric effects in polymeric composites by varying the amount of thermally conductive fillers, in order to investigate its influence on the thermal and mechanical properties. Their work characterized natural graphite/polydimethylsiloxane rubbers composites (NG/PDMS),

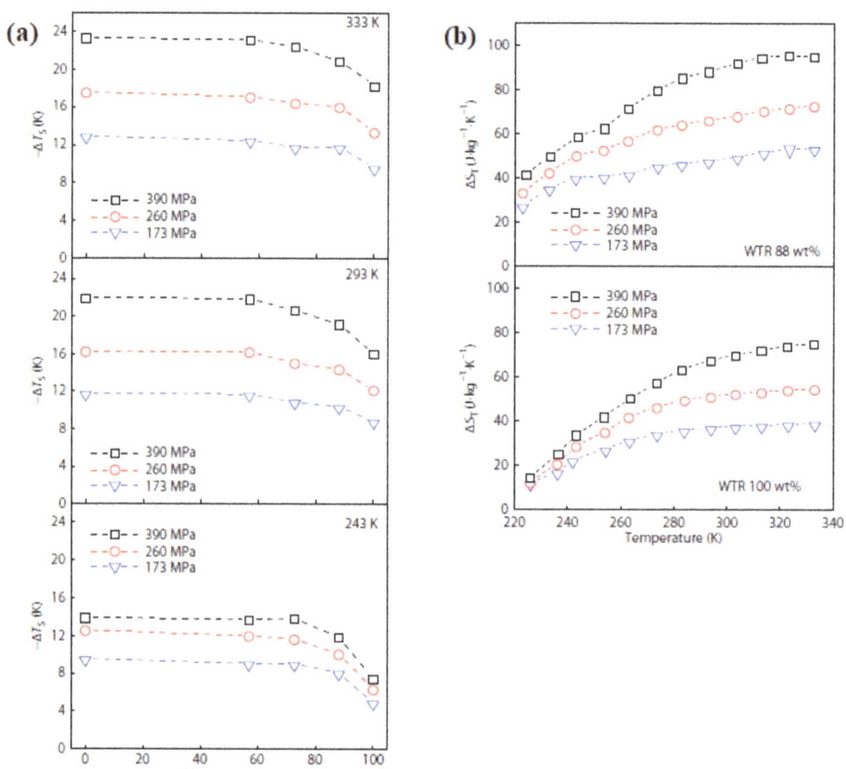

Figure 2.15. (a) Direct measurements of ΔT_S in decompression as a function of WTR weight percentages within the samples; the experiments were performed at the initial temperatures of 333 K, 293 K, and 243 K, for pressure changes of 173, 260, and 390 MPa. (b) ΔS_T as a function of temperature calculated using equation (2.37) and ΔT_S data; the curves correspond to the WTR 88 wt% and 100 wt% samples, within the 223–333 K temperature range for pressure changes of 173, 260, and 390 MPa [14]. Figure 2.15(a) and (b) were reproduced from [14], copyright (2020), with permission from Springer.

in 10, 20, 30 and 40 wt% of NG. In summary, NG/PDMS rubber composites combine giant barocaloric effects over a wide temperature range and at low and moderate pressure changes, in addition to significantly improving the thermal conductivity and thermal diffusivity (up to ~500%), as shown in tables 2.1 and 2.2.

Recently, Yu *et al* [45] investigated the potentially huge barocaloric effect that could be achieved by utilizing the high entropy of amorphous polyethylene glycol (PEG) in the liquid phase, which is transformed into a solid by incorporating 5 wt% polyethylene terephthalate (PET). The authors mentioned that, although amorphous solid PEG loses its –OH end mobility, its molecular chains still retain a high

Table 2.1. Properties of PDMS and NG/PDMS composites.

Sample	Hardness (shore A)	ρ (kg m^{-3})	c_p (J kg^{-1} K^{-1})	$CTE \times 10^4$ (K^{-1})	$\alpha \times 10^7$ (m^2 s^{-1})	κ (W m^{-1} K^{-1})
PDMS	50(1)	1022(4)	1363(7)	8.3(1)	1.18(2)	0.164(3)
10 wt% NG/PDMS	58(2)	1073(2)	1312(9)	7.6(2)	1.85(5)	0.260(7)
20 wt% NG/PDMS	62(2)	1139(2)	1266(14)	7.3(2)	2.57(6)	0.37(1)
30 wt% NG/PDMS	67(2)	1158(2)	1221(28)	6.8(3)	4.9(2)	0.69(3)
40 wt% NG/PDMS	69(2)	1212(5)	1091(29)	6.1(2)	7.3(5)	0.97(7)

Note: ρ is density at ~295 K and ambient pressure; c_p is specific heat at 220–330 K and ambient pressure; CTE is coefficient of thermal expansion at 223–333 K and ambient pressure; α is thermal diffusivity at 298 K and ambient pressure; κ is thermal conductivity calculated by $\kappa = \alpha\rho c_p$ [15].

Table 2.2. Barocaloric properties of PDMS and NG/PDMS composites for $\Delta p = -390$ MPa.

Barocaloric material	T (K)	ΔT_S (K)	$\left\|\frac{\Delta T_S}{\Delta p}\right\|$ (K GPa^{-1})	$\|\Delta S_T\|$ (J kg^{-1} K^{-1})	$\left\|\frac{\Delta S_T}{\Delta p}\right\|$ (J kg^{-1} K^{-1} GPa^{-1})
PDMS	323	−28.6	73.3	120.6	309.3
10 wt% NG/PDMS	323	−28.0	71.8	113.7	291.5
20 wt% NG/PDMS	323	−24.6	63.1	96.2	246.7
30 wt% NG/PDMS	323	−22.1	56.7	83.3	213.6
40 wt% NG/PDMS	323	−18.7	47.9	63.5	162.8
PDMS	273	−24.4	62.6	121.6	311.7
10 wt% NG/PDMS	273	−25.2	64.6	121.1	310.5
20 wt% NG/PDMS	273	−21.2	54.4	98.8	253.2
30 wt% NG/PDMS	273	−19.2	49.2	86.0	220.5
40 wt% NG/PDMS	273	−16.9	43.3	67.8	173.7
PDMS	223	−13.2	33.8	79.9	204.9
10 wt% NG/PDMS	223	−18.4	47.2	109.2	279.9
20 wt% NG/PDMS	223	−16.4	42.1	92.9	238.3
30 wt% NG/PDMS	223	−15.2	39.0	83.5	214.1
40 wt% NG/PDMS	223	−13.8	35.4	67.4	172.7

Notes: T is the initial temperature. Δp the is pressure change (negative for decompression); ΔT_S is the adiabatic temperature change at T and Δp. ΔS_T is the isothermal entropy change at T and Δp. $|\Delta T_S/\Delta p|$ and $|\Delta S_T/\Delta p|$ are the normalized adiabatic temperature change and the normalized isothermal entropy change, respectively [15].

degree of freedom, leading to the preservation of high entropy. Indirect (reversible) ΔS_T of ~426 J kg^{-1} K^{-1} and indirect (reversible) ΔT_S of ~12.6 were reported around 340 K for Δp of 400 MPa. Nevertheless, both ΔS_T and ΔT_S are null at 300 K and below, which may be a limitation for using this material as a refrigerant in a solid-state cooling device.

2.8 Conclusions and outlooks

Remarkably large barocaloric effects are observed in polymers, reaching a maximum adiabatic temperature change of ~41 K (ASR for $|\Delta p|$ = 390 MPa) and isothermal entropy change of 426 J kg^{-1} K^{-1} (5%PET/PEG for $|\Delta p|$ = 400 MPa). Table 2.3 summarizes key properties of the materials discussed here. In spite of the nonlinear nature of barocaloric response, normalized adiabatic temperature change ($\Delta T_S/\Delta p$) and normalized isothermal entropy change ($\Delta S_T/\Delta p$) are helpful to

Table 2.3. Properties of polymeric materials.

Polymeric materials	T (K)	c_p (J kg^{-1} K^{-1})	$\|\Delta p\|$ (MPa)	$\|\Delta T_S\|$ (K)	$\|\Delta S_T\|$ (J kg^{-1} K^{-1})	$\|\Delta T_S/\Delta p\|$ (K GPa^{-1})	$\|\Delta S_T/\Delta p\|$ (J kg^{-1} K^{-1} GPa^{-1})
PMMA	298	1370^a	166	4.6^b	21.1*	27.7	127
PTFE	302	1372^c	166	9.8^d	44.5*	59.0	268
HPDE	294	1665^e	124	5.9^f	33.4*	47.6	269
LPDE	297	2079^g	124	5.5^f	38.5*	44.4	310
PVDF	306	1500^h	200	7.9*	38.7^i	39.5	194
SBR	291	1869^j	152	9.3^k	59.7*	61.2	393
VNR	293	1841^l	173	10.2^m	64.1	59.0	371
			390	23.9^m	150.2	61.3	385
PDMS	284	1535^n	173	11.7^p	63.2*	67.6	365
			390	28.5^p	154.0*	73.1	395
NBR	~303	—	173	8.3^q	36.4†	48.0	210
			390	15.9^q	58.8†	40.8	151
ASR	293	—	173	15.0^r	98.5†	86.7	569
			390	41.0^r	—	105.1	—
	~250	—	173	21.4^r	182.5†	123.7	1055
PU	294	~1972§	174	11.3^s	75.8*	64.9	436
88%WTR/VNR	293	1350^t	173	10.4^u	48.0*	60.1	277
			390	19.2^u	88.5*	49.2	227
10%NG/PDMS	293	~1313§	150	13.6^v	61.1*	90.7	407
			390	28.9^v	129.1*	74.1	331
EVA-28	336	—	400	29.6^w	—	74.0	—
ELVAX	~294	1792^x	434	28.0	170.7	64.5	393
5%PET/PEG	~340	883^y	400	12.6§,‡	31.5	426§,‡	1065

Notes: $^a c_p$ = 137 J mol^{-1} K^{-1} [46], converted using molar mass of 0.1 kg mol^{-1} [47]. ^{b}Measured in compression [3]. ^{c}At 301 K [4]. ^{d}Measured in compression [4]. ^{e}At 290 K [48]. ^{f}Measured in compression [5]. ^{g}At 300 K [48]. ^{h}From [38]. ^{i}From strain data [39]. ^{j}At 290 K [49]. ^{k}Measured in compression [6]. ^{l}At 290 K [50]. ^{m}Measured in decompression [9]. ^{n}At 285 K [51]. ^{p}Measured in decompression [11]. ^{q}Measured in compression [10]. ^{r}Measured in decompression [13]. ^{s}Measured in compression [12]. ^{t}Unpublished data used to calculate ΔS_T in reference [14]. ^{u}Measured in decompression [14]. ^{v}Measured in compression [15]. ^{w}Measured in compression [43]. ^{x}At 308 K [44]. ^{y}At 340 K [45]. *Calculated using the approximation (37). †Reversible ΔS_T obtained from the overlap between ΔS_T curves, calculated using equation (2.36), on heating and cooling processes. §Obtained through data interpolation or extrapolation. ‡Indirectly obtained from differential scanning calorimetry.

compare different materials characterized with different pressure changes. We can highlight normalized temperature change of 123.7 K GPa^{-1} for ASR, and normalized entropy changes of 1065 and 1055 J kg^{-1} K^{-1} GPa^{-1} for 5%PET/PEG and ASR, respectively. Such values should encourage future research in both fundamental and applied fields.

Most polymers present large barocaloric effects over a wide temperature range, especially around room temperature (see table 2.3), which favors refrigeration applications. In general, giant barocaloric effects in polymers do not rely on phase transitions, their large effects arise from molecular level interactions; this mechanism is now being elucidated via molecular dynamics simulations, whose results predict experimental data well. When a phase transition is present, an extra boost to the temperature and entropy changes is observed.

Despite promising materials properties, refrigeration based on barocaloric effect is still an incipient field of research. However, there are numerical studies simulating the performance of barocaloric devices based on polymers as working materials [52, 53]. Aprea *et al* used finite element analysis [53] to simulate a barocaloric refrigerator working with Acetoxy Silicone Rubber (ASR) as refrigerant material. Their simulated device is based on an active regenerative design. The authors also explored the use of nanofluids to enhance the heat transfer rates. The nanofluids were a 50%/50% water ethylene glycol solution with the addition of various volume fractions of Cu nanoparticles. In their study, the authors compared the coefficient of performance (COP) of the simulated device with a domestic vapor compression device [52] using the refrigerant fluid HFC134a [54], both working in the temperature range of 255–290 K. The simulated barocaloric refrigerator outperformed the HFC134a device under all tested conditions, with a maximum improve in COP of 50%.

To conclude, several polymers present favorable properties as caloric materials, making them good candidates for barocaloric devices. While materials design is advancing to improve polymers properties, applied research is still in early development. Despite the challenges still to be overcome, barocaloric refrigeration in future technology seems plausible.

References

[1] Gough J 1805 *Mem. Lit. Philos. Soc. Manchester.* **1** 288–95

[2] Joule J P 1859 *Phil. Trans. R. Soc. Lond.* **149** 413–73

[3] Rodriguez E L and Filisko F E 1982 *J. Appl. Phys.* **53** 6536–40

[4] Rodriguez E L 1986 *J. Appl. Polym. Sci.* **32** 4049–61

[5] Rodriguez E L and Filisko F E 1987 *J. Mater. Sci.* **22** 1934–40

[6] Rodriguez E L and Filisko F E 1986 *Polymer* **27** 1943–7

[7] Bom N M, Usuda E O, Guimarães G M, Coelho A A and Carvalho A M G 2017 *Rev. Sci. Instrum.* **88** 046103

[8] Usuda E O, Bom N M and Carvalho A M G 2017 *Eur. Polym.* J **92** 287–93

[9] Bom N M, Imamura W, Usuda E O, Paixão L S and Carvalho A M G 2018 *ACS Macro Lett.* **7** 31–6

[10] Usuda E O, Imamura W, Bom N M, Paixão L S and Carvalho A M G 2019 *ACS Appl. Polym. Mater.* **1** 1991–7

[11] Carvalho A M G, Imamura W, Usuda E O and Bom N M 2018 *Eur. Polym. J.* **99** 212–21
[12] Bocca J R, Favaro S L, Alves C S, Carvalho A M G, Barbosa J R, Santos A, Colman F C, Conceição W A S, Caglioni C and Radovanovic E 2021 *Polym. Test.* **100** 107251
[13] Imamura W, Usuda E O, Paixão L S, Bom N M, Gomes A M and Carvalho A M G 2020 *Chin. J. Polym. Sci.* **38** 999–1005
[14] Bom N M, Usuda E O, Gigliotti M S, Aguiar D J M, Imamura W, Paixão L S and Carvalho A M G 2020 *Chin. J. Polym. Sci.* **38** 769–75
[15] Imamura W, Usuda E O, Lopes E S N and Carvalho A M G 2022 *J. Mater. Sci.* **57** 311–23
[16] Berzelius J J 1833 *Jahresber. Fortsch. Phys. Wiss.* 11 64
[17] Jensen W B 2008 *J. Chem. Educ.* **85** 624
[18] Polymer. Merriam-Webster.com Dictionary, Merriam-Webster, https://merriam-webster.com/dictionary/polymer (accessed 23 November 2021)
[19] Staudinger H 1920 *Ber. Dtsch. Chem. Ges.* **53** 1073–85
[20] Jones R G, Kahovec J, Stepto R, Wilks E S, Hess M, Kitayama T and Metanomski W V 2009 *Compendium of Polymer Terminology and Nomenclature—IUPAC Recommendations 2008* (Cambridge: RSC Publishing)
[21] Malik M I, Mays J and Shah M R 2021 *Molecular Characterization of Polymers: A Fundamental Guide* 1st edn (Amsterdam: Elsevier)
[22] Wiley-VCH 2016 *Ullmann's Polymers and Plastics: Products and Processes* (Weinheim: Wiley) 4 Volume Set
[23] Gedde U W and Hedenqvist M S 2019 *Fundamental Polymer Science* 2nd edn (Cham: Springer)
[24] Saldívar-Guerra S and Vivaldo-Lima E 2013 *Handbook of Polymer Synthesis, Characterization, and Processing* (New York: Wiley)
[25] Landau L D and Lifshitz E M 1970 *Theory of Elasticity: Volume 7 of Course of Theoretical Physics* (Oxford: Pergamon)
[26] Barrat J-L, Baschnagel J and Jyulin A 2010 *Soft Matter* **6** 3430–46
[27] Abbott L J, Hart K E and Colina C M 2013 *Theor. Chem. Acc.* **132** 1334
[28] Miliante C M, Christmann A M, Usuda E O, Imamura W, Paixão L S, Carvalho A M G and Muniz A R 2020 *Macromolecules* **53** 2606–15
[29] Vanommeslaeghe K, Hetcher E, Acharya C, Kundu S, Zhong S, Shim J, Darian E, Guvench O, Lopes P, Vorobyov I and MacKerell Jr A D 2009 *J. Comput. Chem.* **31** 671–90
[30] Lei M, Reid C G and Zolle P 1988 *Polymer* **29** 1784–8
[31] Calleja G, Jourdan A, Ameduri B and Habas J-P 2013 *Eur. Polym.* J **49** 2214–22
[32] Rae P J and Dattelbaum D M 2004 *Polymer* **45** 7615–25
[33] Fakirov S and Krasteva B 2000 *J. Macromol. Sci.* B **39** 297–301
[34] DSC measurement of polyethylene—the correlation of polyethylene density and melting. Application Brief TA N0.26 1986. Hitachi High-Tech Science Corporation. https://hitachi-hightech.com/file/global/pdf/products/science/appli/ana/thermal/application_TA_026e.pdf. (accessed 9 March 2022)
[35] Ainbinder S B, Tsirule K I and Dzenis A A 1976 *Polym. Mech.* **12** 367–70
[36] Jordan J L, Rowland R L, Greenhall J, Moss E K, Huber R C, Willis E C, Hrubiak R, Kenney-Benson C, Bartram B and Sturtevant B T 2021 *Polymer* **212** 123164
[37] Boldrin D 2021 *Appl. Phys. Lett.* **118** 170502
[38] Patel S, Chauhan A, Vaish R and Thomas P 2016 *Appl. Phys. Lett.* **108** 072903

[39] Roland C M, Garrett J T, Casalini R, Roland D F, Santangelo P G and Qadri S B 2004 *Chem. Mater.* **16** 857–61
[40] Sisanth K S, Thomas M G, Abraham J and Thomas S 2017 General introduction to rubber compounding *Progress in Rubber Nanocomposites* (Woodhead Publishing Series in Composites Science and Engineering) ed S Thomas and H J Maria (Sawston: Woodhead Publishing) ch 1 pp 1–39
[41] Wolf M P, Salieb-Beugelaar G B and Hunziker P 2018 *Prog. Polym. Sci.* **83** 97–134
[42] Essawy H and El-Nashar D 2004 *Polym. Test.* **23** 803–7
[43] Zhang C, Wang D, Qian S, Zhang Z, Liang X, Wu L, Long L, Shi H and Han Z 2022 *Mater. Horiz* **9** 1293
[44] Weerasekera N, Ajjarapu K P K, Sudan K, Sumanasekera G, Kate K and Bhatia B 2022 *Front. Energy Res.* **10** 887006
[45] Yu Z 2022 *NPG Asia Mater.* **14** 96
[46] Gaur U, Lau S, Wunderlich B B and Wunderlich B 1982 *J. Phys. Chem. Ref. Data* **11** 1065–89
[47] NIST Chemistry WebBook, SRD 69 Poly(methyl methacrylate) https://webbook.nist.gov/cgi/formula?ID=C87210320&Mask=2# (accessed 14 May 2022)
[48] Passaglia E and Kevorkian H K 1963 *J. Appl. Polym. Sci.* **7** 119–32
[49] Rands R D, Perguson W J and Prather J L 1944 *J. Res. Natl. Bur. Stand.* **33** RP1595
[50] Warfield R W and Petree M C 1965 *Makromol. Chem.* **84** 1–8
[51] Wang B and Krause S 1987 *Macromolecules* **20** 2201–8
[52] Aprea C, Greco A and Maiorino A 2016 *Appl. Therm. Eng.* **106** 959–67
[53] Aprea C, Greco A, Maiorino A and Masselli C 2020 *Energy* **190** 116404
[54] Franklin J 1993 *Chemosphere* **27** 1565–601

Chapter 3

Barocaloric effect in magnetic and non-magnetic solid materials: theoretical approach and applications

Nilson Antunes de Oliveira

List of symbols

B	Magnetic field
C_p	Specific heat at constant pressure
g	Landé factor
$g_{ii\sigma}^{d}$	Electron local Green function
G	Total Gibbs energy
g^{vib}	Vibrational Gibbs energy
g^{rot}	Rotational Gibbs energy
J	Total angular momentum
$\mathcal{J}_0$	Exchange interaction parameter
$\mathcal{J}_1$	Magnetoelastic interaction parameter
k_B	Boltzmann constant
N_m	Number of magnetic ions per unit formula
N_i	Number of total ions per unit formula
P	Pressure
$\mathfrak{R}$	Gas Constant
S	Total entropy
S_{el}^{spd}	Entropy for spd electrons
S_{el}^{4f}	Entropy of 4f electrons
S_{lat}	Crystalline lattice entropy

doi:10.1088/978-0-7503-4690-0ch3

(*Continued*)

T	Temperature
U^{d}	Coulomb interaction parameter
V	Volume
Z^{vib}	Vibrational partition function
Z^{rot}	Rotational partition function
ΔS_{iso}	Isothermal entropy change
ΔT_{ad}	Adiabatic temperature change
γ	Sommerfeld coefficient
μ_B	Bohr magneton
ω	Phonon frequency
$\rho(\omega)$	Density of phonons frequency
$\rho(\varepsilon)$	Electronic density of states
Θ_D	Debye temperature

3.1 Introduction

The study of the caloric effects and their application in solid state refrigerators started a very long time ago [1–3]. The magnetocaloric effect [4–12], which is the heating/cooling of a magnetic material upon magnetic field variation was the first and most studied effect. However, despite the great advance reached in the study of the magnetocaloric effect, there are still challenges to be overcome until solid state refrigerators [13–19] can be adopted on a commercial scale. For example, the large entropy change in the most studied magnetic materials is associated with magnetic field change as large as 5 T and phase transition on the verge of the first order. These difficulties motivate the search for novel compounds and alternative processes to produce caloric effects in solid state materials.

Besides the magnetocaloric effect, there are several other studies on the cooling/heating of solid materials through the variation of pressure [20–52] (barocaloric effect), uniaxial stress [53–59] (elastocaloric effect) and electric field [60–65] (electrocaloric effect). The focus of this text is on the barocaloric effect. For readers interested in more details about the other caloric effects we recommend the papers mentioned in the reference list and the works cited therein.

The first quantitative works on the barocaloric effect in solid state materials were done by Furrer [20–25] and collaborators in a series of rare earth based compounds with a crystalline electric field. In those materials, they showed that the barocaloric effect was due to the splitting of the energy levels induced by the applied pressure. Later on, de Oliveira [26, 27] theoretically predicted a sizeable barocaloric effect in the series RCo_2 (R = Er, Dy, Ho), which exhibits mixed magnetism and first order phase transition. In these compounds, the large values of the barocaloric quantities are due to the change of the critical temperature upon applied pressure variation. In the following years, Mañosa and co-workers [28–30] measured the barocaloric effect in the compounds of the series NiMnIn, LaFeSiCo and $R_5Si_2Ge_2$. Moreover, the barocaloric effect was experimentally discussed in ferroelectric materials [66–69] and in superionic conductors [70–72].

In 2016, Karl Sandeman [73] proposed that metal complexes could exhibit a large barocaloric effect due to spin crossover transition that is controlled by applied pressure. In the following year, P J von Ranke [74] theoretically described the barocaloric effect in spin crossover material. After these pioneering works, many other papers [75–82] have also discussed the barocaloric effect in metal complexes.

In 2019 Li *et al* [83] and Lloveras and collaborators [84] have almost simultaneously discovered very large values of the barocaloric effect at ambient temperature in plastic crystals. Further studies [85–88] on the barocaloric effect in plastic crystals were published more recently. Other materials including natural rubber [89–91], polymers [92, 93] and n-alkane compounds [94, 95] have also been investigated.

This chapter aims to theoretically discuss the aspects of the barocaloric effect in solid state materials. Firstly, we present the thermodynamics of the effect and its application in solid state refrigerators. After that, we present theoretical models to describe the barocaloric effect in different materials.

3.2 Caloric effects and refrigerators

The caloric effects are usually characterized by the isothermal entropy change (ΔS_{iso}) and the adiabatic temperature (ΔT_{ad}) change upon variation of the external stimuli, such as magnetic and electric field, pressure, or stress. The largest values of the caloric quantities ΔS_{iso} and ΔT_{ad} usually take place around the transition temperature between two different phases. Unlike the magnetocaloric effect, which is supposed to occur only in magnetic materials, the barocaloric effect can exist in both magnetic and non-magnetic materials. This feature makes the barocaloric effect very interesting to be used in solid state refrigeration technology.

The basic working principle of a refrigeration machine is to remove heat from a volume to be refrigerated (cold sink) and release it to an external medium (hot source). In order to make this task, a refrigerant material should perform a thermodynamical cycle upon variation of the driven field. For example, in the current vapor-cycle refrigerator, the refrigerant material, usually hydrofluorocarbons (HFCs), is compressed and decompressed in pipes inside the refrigerator. During these processes of compression and decompression, the refrigerant material undergoes a transition from the liquid phase to the vapor one with a huge temperature change. The great disadvantage of this technology is the fact that the refrigerant material is harmful to the environment and contributes to the enhancement of the greenhouse effect and the depletion of the ozone shell.

In order to illustrate the thermodynamical process involved in the conventional vapor-cycle refrigerator, we plot in figure 3.1 a volume versus pressure diagram. In this thermodynamical cycle, composed of two adiabatic and two isobaric processes upon pressure variation, the fluid extracts heat from the cold sink and releases it to the hot source. Hence, the working principle of the conventional vapor-cycle refrigerators is based on the barocaloric effect in the fluid refrigerant. Now, we ask the following question. Is it possible to build a refrigerator based on the

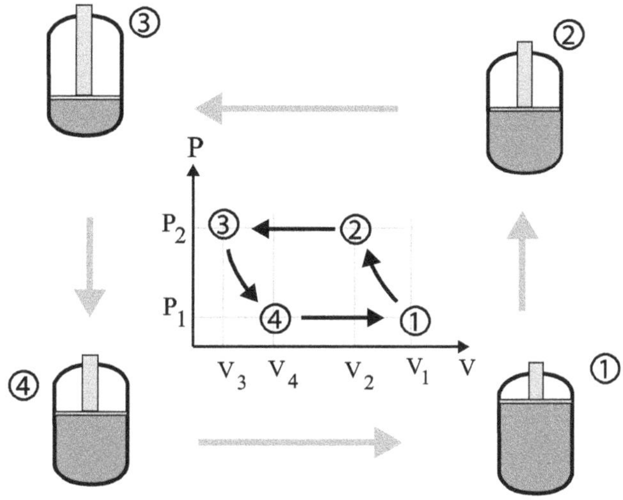

Figure 3.1. Pressure versus volume diagram showing the thermodynamical cycle of a conventional refrigerator based on the compression/decompression of the refrigerant fluid.

barocaloric effect using solid compounds as the refrigerant material? The direct answer to this question is yes.

The thermodynamical cycle for solid state refrigerators based on the barocaloric effect of non-magnetic materials, can be very similar to the one shown in figure 3.1, with a solid material replacing the fluid one. However, it should be mentioned that to get a high performance it is necessary to induce in the solid refrigerant material a kind of phase transition with a large volume change produced by temperature and controlled by pressure.

We can also construct a refrigerator using solid magnetic materials whose magnetizations change with pressure, even if their volumes do not change much. In this case, the working principle is based on a thermodynamical cycle, similar to the one previously discussed, replacing the volume by the magnetization as is shown in figure 3.2. Thus, solid magnetic compounds, in which the transition temperature and/or the nature of the magnetic phase transition can be modified by applied pressure, are suitable to be used as refrigerant material in refrigerator machines having the barocaloric effect as their physical working principle.

3.3 Thermodynamics of the barocaloric effect

In order to present the thermodynamic of the barocaloric effect in solid state material, let us first consider non-magnetic materials where the barocaloric effect is directly associated with the thermal expansion of the crystalline lattice. In figure 3.3, we plot the entropy for a non-magnetic material, as a function of temperature for two values of applied pressure P_1 and P_2. In this figure, the barocaloric quantities $\Delta S_{iso}(T, \Delta P)$ and $\Delta T_{ad}(T, \Delta P)$, where $\Delta P = P_2 - P_1$, are shown by the vertical and horizontal arrows, respectively.

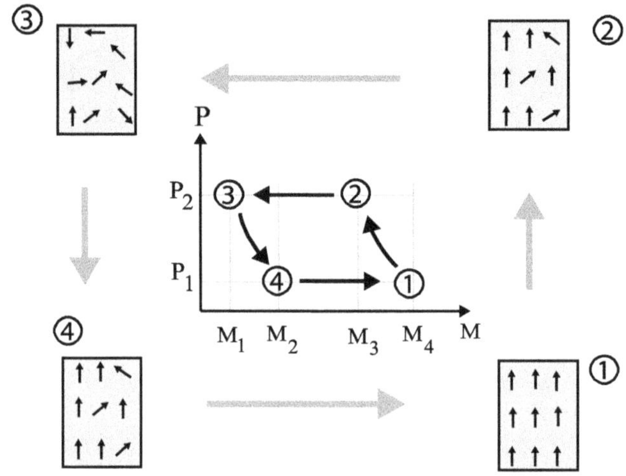

Figure 3.2. Pressure versus magnetization diagram showing the thermodynamical cycle for a refrigerator based on the magnetization/demagnetization of the solid state refrigerant material.

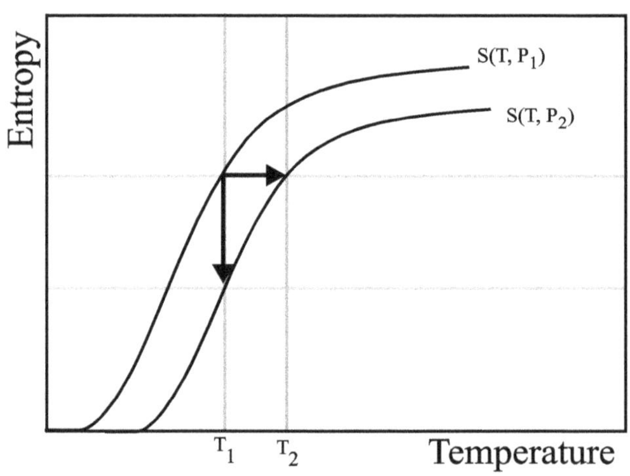

Figure 3.3. Entropy versus temperature diagram for a non-magnetic material for two values of applied pressure. The vertical and horizontal arrows show the caloric quantities $\Delta S_{\text{iso}}(T, \Delta P)$ and $\Delta T_{\text{ad}}(T, \Delta P)$ respectively.

In the case of magnetic materials, it is also necessary to include a magnetic field dependence of the entropy. In figure 3.4, we plot the entropy versus temperature diagram for two different values of magnetic field and pressure. In this figure, the barocaloric quantities are obtained by taking the difference between two entropy curves with the same magnetic field and different values of the external pressure.

According to figure 3.4, the entropy change upon pressure variation in an isothermal process at constant magnetic field is:

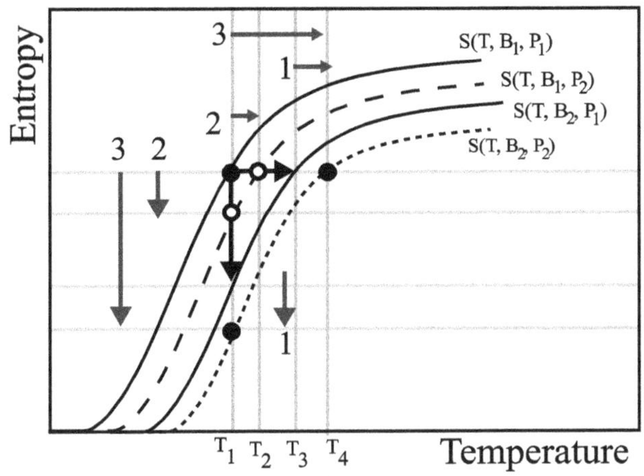

Figure 3.4. Entropy versus temperature diagram for magnetic system for two values of magnetic field and applied pressure. The vertical and horizontal arrows show the caloric quantities $\Delta S_{\text{iso}}(T, \Delta P)$ and $\Delta T_{\text{ad}}(T, \Delta P)$. See text for details.

$$\Delta S_{\text{iso}}(T, B, \Delta P) = S(T, B, P_2) - S(T, B, P_1) \tag{3.1}$$

The temperature change in an adiabatic process upon pressure variation at constant magnetic field is then

$$\Delta T_{\text{ad}}(T, B, \Delta P) = T_2 - T_1 \tag{3.2}$$

upon the adiabatic condition $S(T, P_2) = S(T, P_1)$. On the other hand, in the case where both magnetic field and pressure change simultaneously (multicaloric effect), the isothermal entropy change is $\Delta S_{\text{iso}}(T, \Delta B, \Delta P) = S(T, B_2, P_2) - S(T, B_1, P_1)$ and the temperature change is $\Delta T_{\text{ad}}(T, \Delta B, \Delta P) = T_2 - T_1$ upon the adiabatic condition $S(T, B_2, P_2) = S(T, B_1, P_1)$. Alternatively, considering the entropy as a function of temperature, magnetic field and pressure, we can show that the barocaloric quantities can also be determined by the following relations:

$$\Delta S_{\text{iso}}(T, B, \Delta P) = \int_{P_1}^{P_2} \frac{\partial S(T, B, P)}{\partial P} dP \tag{3.3}$$

$$\Delta T_{\text{ad}}(T, B, \Delta P) = -\int_{P_1}^{P_2} \frac{T}{C_p(T, B, \Delta P)} \left[\frac{\partial S(T, B, P)}{\partial P} \right]_{T,B} dP \tag{3.4}$$

where $C_p(T, B, \Delta P) = T[\partial S(T, B, P)/\partial T]_{P,B}$ is the heat capacity at constant pressure. Using the Maxwell relation $[\partial S(T, B, P)/\partial P]_{T,B} = -[\partial V(T, B, P)/\partial T]_{P,B}$, the above mentioned barocaloric quantities can also be written in terms of the volume variation.

Therefore, in order to calculate the barocaloric quantities, we need to know the specific heat capacity (or entropy) as a function of temperature for different values of the applied pressure.

3.4 Barocaloric effect in magnetic materials

In this section, we present the theoretical description of the barocaloric effect in magnetic compounds. In order to have a clear understanding of the physical mechanisms involved in the barocaloric effect in magnetic materials we separate the discussion in two parts. In the first one, we consider magnetic materials with localized magnetic moments and in the second part, we deal with compounds with itinerant electron magnetism.

3.4.1 Localized magnetic moments

This section is devoted to describing the barocaloric effect in compounds with local magnetic moments. Rare earth metals and their alloys, whose magnetism is mainly associated with 4f electrons that have a localized character, are examples of such compounds. It should be mentioned that although the 'spd' conduction electrons do not directly contribute to the magnetization of this class of compounds, they play a fundamental role in the interaction between the local 4f magnetic moments.

The behavior of the magnetization curve and the values of the critical temperature of the magnetic phase transition depend on the crystalline structure, the type of magnetic interactions, the internal crystalline electric field. For example, the critical temperature decreases in RCo_2 as the applied pressure increases, whereas it increases in $R_5Si_2Ge_2$ (R = Gd, Tb) with increasing pressure [30, 96].

In order to describe the magnetic, thermodynamic and the caloric effects in rare based compounds, we use the following Hamiltonian $\mathcal{H} = \mathcal{H}_{\text{lat}} + \mathcal{H}_{\text{el}}^{\text{spd}} + \mathcal{H}_{\text{el}}^{4\text{f}}$. Here $\mathcal{H}_{\text{lat}} = \sum_q \varepsilon_{\text{vib}} a_q^+ a_q$ describes the vibrations of the ions into crystalline lattice where $\varepsilon_{\text{vib}} = (n + 1/2)\hbar\omega$; $\mathcal{H}_{\text{el}}^{\text{spd}}$ describes the conduction electrons, which are supposed to be non-magnetic and $\mathcal{H}_{\text{el}}^{4\text{f}}$ given by:

$$\mathcal{H}_{\text{el}}^{4\text{f}} = -\sum_{i,l} \mathcal{J}_0 J_i \cdot J_l - \sum_{i,l} \mathcal{J}_1 (J_i \cdot J_l)^2 - \sum_i g\mu_B B \cdot J_i + \sum_i \mathcal{H}_i^{\text{CF}}, \tag{3.5}$$

describes the 4f electrons that are responsible for the magnetic order of the system. In this magnetic Hamiltonian, the first term gives the interaction between two localized magnetic moments and the second term accounts for the magnetoelastic interaction, where $\mathcal{J}_0$ and $\mathcal{J}_1$ are model parameters that control the nature of the magnetic phase transition [97]. In the present model, the phase transition becomes of first order if $\mathcal{J}_1/\mathcal{J}_0$ is larger than a given critical value and it is of second order otherwise. The third term is the Zeeman interaction between the local 4f magnetic moments and the applied magnetic field. Finally, $\mathcal{H}_i^{\text{CF}}$ describes the crystalline electric field that depends on the symmetry of the crystalline structure [98–100].

In this text, we consider that the effect of the external pressure is indirectly incorporated in the renormalization of the model parameters $\mathcal{J}_0$ and $\mathcal{J}_1$. The total entropy is $S(T, B, P) = S_{\text{lat}}(T, B, P) + S_{\text{el}}^{\text{spd}}(T, B, P) + S_{\text{el}}^{4\text{f}}(T, B, P)$, where $S_{\text{lat}}(T, B, P)$ represents the contribution from the lattice vibrations, $S_{\text{el}}^{\text{spd}}(T, B, P)$ is the contribution from the non-magnetic conduction electrons and $S_{el}^{4f}(T, B, P)$ is the contribution from the local magnetic moments. In this section, we consider $S_{el}^{spd}(T, B, P) = \gamma T$ where γ is the Sommerfeld coefficient. The effect of the applied field and pressure on the magnetic part of the entropy can be determined by:

$$S_{el}^{4f}(T, B, P) = N_m \mathfrak{R}\left[\ln \sum_{m=-J}^{m=J} e^{-E_m/k_B T} + \frac{1}{k_B T}\frac{\sum\limits_{m=-J}^{m=J} E_m e^{-E_m/k_B T}}{\sum\limits_{m=-J}^{m=J} e^{-E_m/k_B T}}\right] \tag{3.6}$$

where N_m is the number of magnetic ions; E_m are the energy eigenvalues of the Hamiltonian $\mathcal{H}_{el}^{4f}$; $\mathfrak{R} = 8.31$ J molK^{-1} is the gas constant and k_B is the Boltzmann constant.

In order to consider the effect of pressure on the crystalline lattice, we extend the Debye approximation and indirectly consider that the magnetization, which depends on pressure, renormalizes the Debye temperature. Within this consideration, the crystalline lattice contribution to the total entropy is given by:

$$\begin{aligned} S_{lat}(T, B, P) = N_i\Bigg[& -3\mathfrak{R}\ln\left(1 - e^{-\frac{\tilde{\Theta}_D}{T}}\right) \\ & + 12\mathfrak{R}\left(\frac{T}{\tilde{\Theta}_D}\right)^3 \int_0^{\tilde{\Theta}_D/T} \frac{x^3}{e^x - 1}dx\Bigg] \end{aligned} \tag{3.7}$$

In this expression, $\tilde{\Theta}_D = \Theta_0(1 - \gamma^{ph}M^2)$ where Θ_0 is the bare value of the Debye temperature, M is the magnetization and γ^{ph} is a model parameter that takes into account the magnetoelastic interaction. Once the total entropy is known, the barocaloric quantities can be determined from relations (3.1) and (3.2).

We apply the model to calculate the barocaloric effect in $Gd_5Si_2Ge_2$. Experimental data [30] show that this compound undergoes a first order magnetic phase transition around 265 K at ambient pressure and around 268 K; 271 K and 274 K for applied pressures of 1.0 kbar; 2.0 kbar and 3.0 kbar, respectively. These experimental data of the critical temperature were used to select the values of the model parameters $\mathcal{J}_0$ and $\mathcal{J}_1$. In figure 3.5, we plot the total entropy as a function of temperature for ambient pressure ($P = 0.0$) and for $P = 1.0$ kbar and $P = 2.0$ kbar. The isothermal entropy changes calculated upon pressure variation from 1.0 kbar to 0 kbar and from 2.0 kbar to 0 kbar are shown in figure 3.6. In this figure, symbols represent experimental data [30]. The theoretically calculated adiabatic temperature changes for the same pressure variations are shown in figure 3.7. It should be mentioned that in $Gd_5Si_2Ge_2$ the values of the barocaloric quantities around the temperature of the phase transition are very similar to the magnetocaloric counterparts upon magnetic field variation of 1 T.

In principle, the model presented in this section can also be applied to calculate the barocaloric effect in other rare earth based compounds, where the major contribution to the magnetism comes from the 4f electrons.

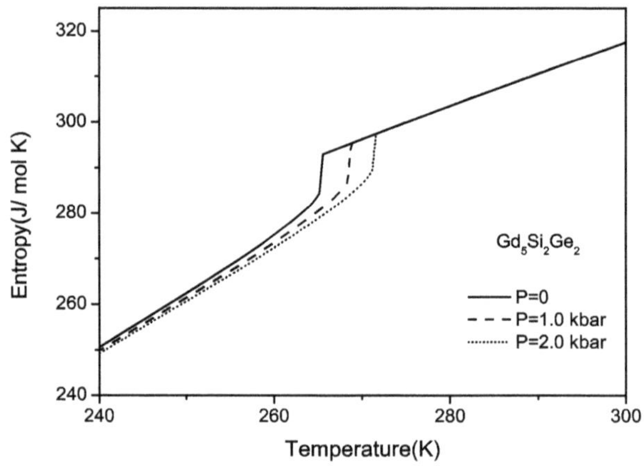

Figure 3.5. Entropy as a function of temperature in $Gd_5Si_2Ge_2$ for ambient pressure (solid line), $P = 1.0$ kbar (dashed line) and $P = 2.0$ kbar (dotted line).

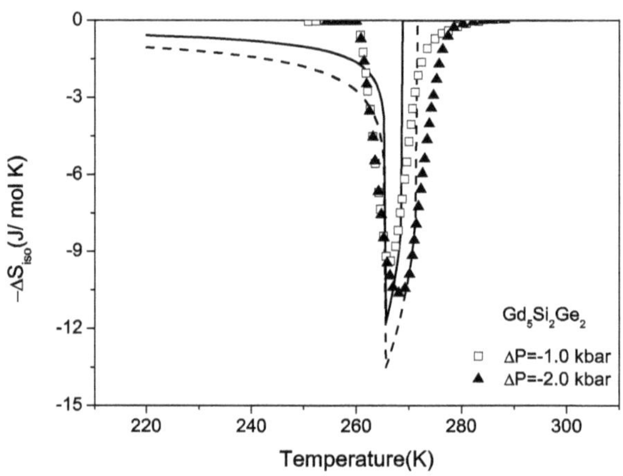

Figure 3.6. Isothermal entropy change in $Gd_5Si_2Ge_2$ for pressure variation $\Delta P = -1.0$ kbar (solid line), and $\Delta P = -2.0$ kbar (dotted line). Symbols are experimental data [30].

3.4.2 Itinerant electrons systems

The theoretical description of the caloric effects in compounds whose electrons responsible for the magnetic order have an itinerant character [101–105] is somewhat different from the one presented in the previous section. In these compounds, the barocaloric effect can be generated by different physical mechanisms. For instance, it can be due to the magnetization change induced by the applied pressure or it can be produced by a transfer of electrons between different sub-bands [106, 107]. A general discussion of itinerant electron systems should consider many sub-bands and more than one magnetic sublattice. This is a very complicated problem to solve and approximations are usually done in order to simplify the calculations. For the sake of simplicity, we separate the discussion of this problem into two categories, namely: (1) one-band model (2) two-sub-band model. The simplest case of one-band model is

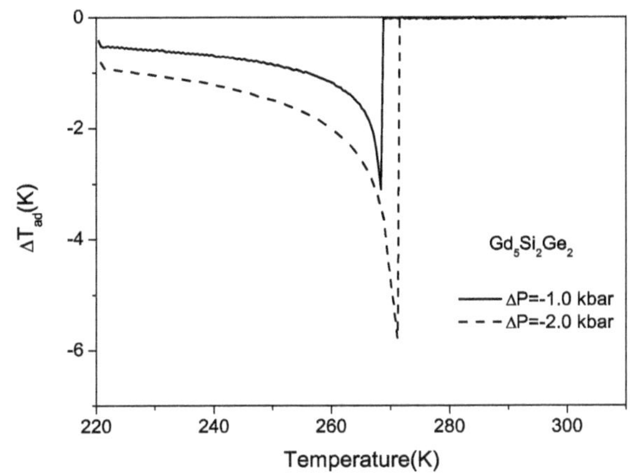

Figure 3.7. Adiabatic temperature change in $Gd_5Si_2Ge_2$ for pressure variation $\Delta P = -1.0$ kbar (solid line), and $\Delta P = -2.0$ kbar (dotted line).

presented in the next section and the case of two-sub-band model is discussed in the sequence.

3.4.2.1 One-band model

In this section, we discuss the barocaloric effect in a system of itinerant electrons considering only one d-band. In this description the contribution from the s and p electrons to the magnetism is neglected. Within this consideration we use the following Hamiltonian to describe the system $\mathcal{H} = \mathcal{H}_{\text{lat}} + \mathcal{H}_{\text{el}}^{\text{sp}} + \mathcal{H}_{\text{el}}^{\text{d}}$, where $\mathcal{H}_{\text{lat}} = \sum_q \varepsilon_{\text{vib}} a_q^+ a_q$, with $\varepsilon_{\text{vib}} = (n + 1/2)\hbar\omega$, describes the vibrations of the crystalline lattice. The term $\mathcal{H}_{\text{el}}^{\text{sp}}$ describes the non-magnetic conduction electrons and the term $\mathcal{H}_{\text{el}}^{\text{d}}$ given by:

$$\mathcal{H}_{\text{el}}^{\text{d}} = \sum_i \varepsilon_0^{\text{d}} d_i^+ d_i + \sum_{ij} T_{ij}^{\text{dd}} d_i^+ d_j + \sum_i U^{\text{d}} n_{i\uparrow} n_{i\downarrow} - \sum_i g\mu_{\text{B}} S_i^{\text{d}} B \tag{3.8}$$

is the Hubbard Hamiltonian [108] that describes the d-itinerant electrons which are responsible for the appearance of the magnetism in the system. In this Hamiltonian, the first term is a reference energy, the second term accounts for the electron hopping, where $T_{ij}^{\text{dd}} = \sum_k \lambda_{\text{p}}^{\text{d}} \varepsilon_k^{\text{d}} e^{ik(R_i - R_j)}$ with ε_k^{d} being the electron dispersion relation and $\lambda_{\text{p}}^{\text{d}}$ a model parameter which takes into account the pressure effect on the width of the energy band. The third term represents the local Coulomb interaction between electrons, where U^{d} is a model parameter and the last term is the Zeeman interaction with the applied magnetic field. In this model, the magnetization and the critical temperature, which depend on the width of the energy band and on the density of states at the Fermi energy level $\rho^{\text{d}}(\varepsilon_{\text{F}})$, can be controlled by the applied pressure. For example, when an external pressure is applied the lattice parameters are reduced and the hopping energy T_{ij}^{dd} increases. As a result, the electronic density of states $\rho^{\text{d}}(\varepsilon_{\text{F}})$ is reduced and the system tends to be demagnetized.

In the mean field approximation, the Hamiltonian $\mathcal{H}_{\text{el}}^{\text{d}}$ turns out to be:

$$\mathcal{H}_{\text{el}}^{\text{d}} = \sum_{i\sigma} \varepsilon_{\sigma}^{\text{d}} d_{i\sigma}^{+} d_{i\sigma} + \sum_{ij} T_{ij}^{\text{dd}} d_i^{+} d_j \tag{3.9}$$

where $\varepsilon_{\sigma}^{\text{d}} = \varepsilon_{0\sigma}^{\text{d}} + U^{\text{d}}\langle n_{i-\sigma}^{\text{d}}\rangle - \sigma\mu_{\text{B}}B$ is a renormalized energy, with $\sigma = \uparrow$ or $\downarrow$ represents the up and down spins. Using the Green function technique, the density of states for the Hamiltonian $\mathcal{H}_{\text{el}}^{\text{d}}$ is:

$$\rho_{\sigma}^{\text{d}}(\varepsilon) = -\frac{1}{\pi} Img_{ii\sigma}^{\text{d}}(\varepsilon) \tag{3.10}$$

where $g_{ii\sigma}^{\text{d}}(\varepsilon)$ is the local Green function given by:

$$g_{ii\sigma}^{\text{d}}(\varepsilon) = \int_{-\infty}^{\infty} \frac{\rho_0(\varepsilon')}{\varepsilon - \lambda_{\text{p}}^{\text{d}}\varepsilon' - \left(\varepsilon_{0\sigma}^{\text{d}} + U^{\text{d}}\langle n_{i-\sigma}^{\text{d}}\rangle - \sigma\mu_{\text{B}}B\right)} d\varepsilon' \tag{3.11}$$

where $\varepsilon = \varepsilon + i0$. The magnetization as a function of temperature, magnetic field and pressure is calculated by $M(T, B, P) = \langle n_{i\uparrow}^{\text{d}}\rangle - \langle n_{i\downarrow}^{\text{d}}\rangle$ where $\langle n_{i\sigma}^{\text{d}}\rangle$ is the electron occupation number per spin direction which is determined by:

$$\langle n_{i\sigma}^{\text{d}}\rangle = \int_{-\infty}^{\infty} \rho_{\sigma}^{\text{d}}(\varepsilon) f(\varepsilon) d\varepsilon \tag{3.12}$$

where $f(\varepsilon) = 1.0/[1.0 + \text{e}^{(\varepsilon-\mu)/k_{\text{B}}T}]$ is the Fermi distribution function. The equations (3.10)–(3.12) should be self-consistently solved.

The total entropy is $S(T, B, P) = S_{\text{lat}}(T, B, P) + S_{\text{el}}^{\text{sp}}(T) + S_{\text{el}}^{\text{d}}(T, B, P)$ where $S_{\text{el}}^{\text{sp}}(T) = \gamma T$ is the entropy of the non-magnetic conduction electrons and $S_{\text{lat}}(T, B, P)$ which is given by relation (3.7) represents the contribution from the crystalline lattice. The contribution from the magnetic d-electrons is given by [8]:

$$S_{\text{el}}^{\text{d}}(T, B, P) = \Re \sum_{\sigma} \int_{-\infty}^{\infty} \left[\ln\left[1 + \text{e}^{-(\varepsilon_{\sigma}-\mu)/k_{\text{B}}T}\right] + \frac{(\varepsilon_{\sigma} - \mu) f(\varepsilon)}{k_{\text{B}}T} \right] \rho_{\sigma}(\varepsilon) d\varepsilon \tag{3.13}$$

As a simple illustration of this calculation, we determine the barocaloric effect in $La(Fe_{0.89}Si_{0.11})_{13}$ [109]. Experimental data [110] show that this compound exhibits at ambient pressure a first order phase transition around 200 K. Moreover, the critical temperature of the phase transition moves to around 120 K upon applied pressure of 6.5 kbar. The model parameters U^{d} and $\lambda_{\text{p}}^{\text{d}}$ were properly chosen to reproduce the experimental data of the saturation magnetization at 0 K and the critical temperature. After solving the self-consistent set of equations (3.10)–(3.12), we determine the magnetization, entropy, and barocaloric quantities. In figure 3.8, we plot the magnetization as a function of temperature for some values of applied pressure. In figure 3.9, we plot the entropy change for ambient pressure and for applied pressure of 6.5 kbar. From these entropy curves, we determine the barocaloric quantities according to relations (3.1) and (3.2). The calculated isothermal entropy changes upon pressure variation from $P_1 = 6.5$ kbar to ambient pressure (P_0) are shown in

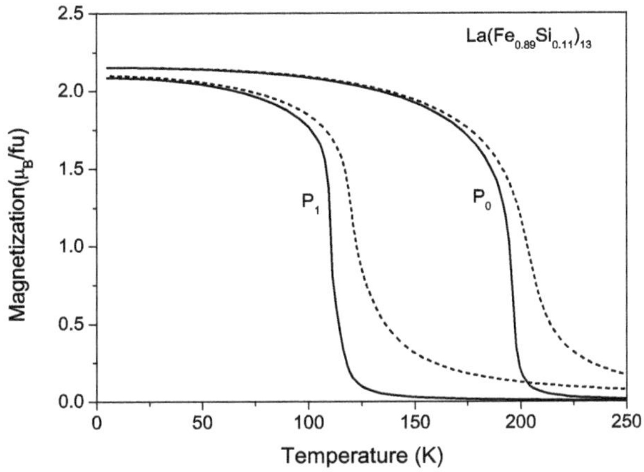

Figure 3.8. Magnetization in $La(Fe_{0.89}Si_{0.11})_{13}$ for ambient pressure (P_0) and for pressure $P_1 = 6.5$ kbar. Solid and dashed lines represent the calculations for $B = 0$ and $B = 5$ T, respectively.

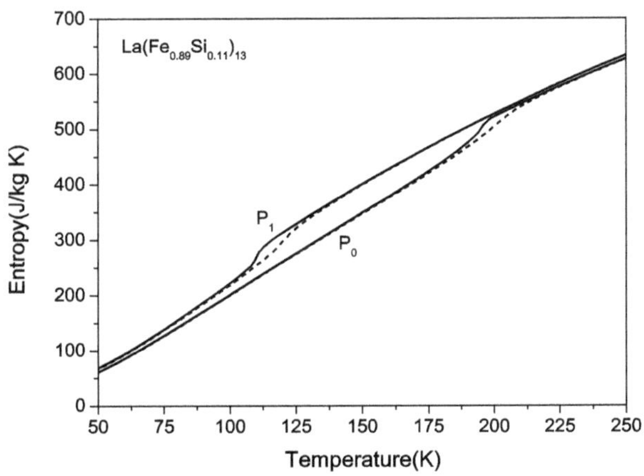

Figure 3.9. Entropy in $La(Fe_{0.89}Si_{0.11})_{13}$ for ambient pressure (P_0) and for pressure $P_1 = 6.5$ kbar. Solid and dashed lines represent the calculations for $B = 0$ and $B = 5$ T, respectively.

figure 3.10 by the solid line for $B = 0$ and by the dashed line for $B = 5$ T. The corresponding curves for the adiabatic temperature changes are plotted in figure 3.11. For the sake of comparison, we also plot in these figures the magnetocaloric quantities upon magnetic field variation of 5 T and fixed pressure (dotted lines). The large values of the barocaloric quantities calculated for $La(Fe_{0.89}Si_{0.11})_{13}$, is because the applied pressure yields a big shift in the temperature of the phase transition as is shown in figure 3.8.

3.4.2.2 Two-sub-bands model

In order to describe the effect of electrons transfer on the barocaloric properties of magnetic materials, it is necessary to consider more than one band. For the sake of

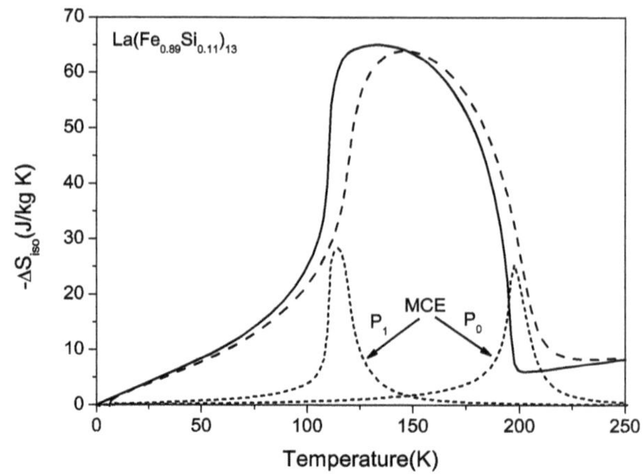

Figure 3.10. Isothermal entropy change in $La(Fe_{0.89}Si_{0.11})_{13}$ upon pressure variation $\Delta P = -6.5$ kbar and fixed magnetic field $B = 0$ (solid line) and $B = 5$ T (dashed line). The dotted lines represents the magnetocaloric quantities upon $\Delta B = 5$ T and fixed pressure ($P_0 = 0$) and $P_1 = 6.5$ kbar.

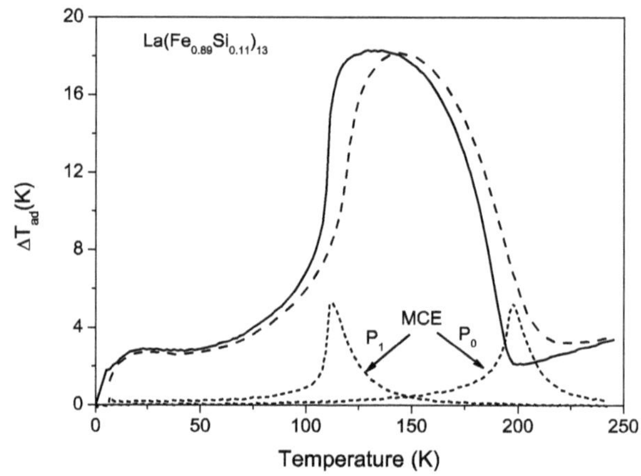

Figure 3.11. Adiabatic temperature change in $La(Fe_{0.89}Si_{0.11})_{13}$ upon pressure variation $\Delta P = -6.5$ kbar and fixed magnetic field $B = 0$ (solid line) and $B = 5$ T (dashed line). The dotted lines represent the magnetocaloric quantities upon $\Delta B = 5$ T and fixed pressure ($P_0 = 0$) and $P_1 = 6.5$ kbar.

simplicity, we consider a model containing just two sub-bands, since the general description involving more than two sub-bands is very complicated.

Different physical scenarios can be discussed in the context of the two-sub-band model. For example, we can use such a formulation to describe a system where the degeneracy of the d-bands is partially removed forming the set t_{2g} and e_g separated by an energy gap. A system where a conduction band of sp electrons is hybridized with a fivefold degenerated d-band can also be discussed in the formalism of a two-sub-band model. Another possible physical scenario for the two-sub-band model can be a narrower band of 4f or 5f electrons coupled with a wider band of spd-conduction electrons.

In order to present a general description of a two-sub-band model which can characterize different physical scenarios, we use the Greek letters α and δ as labels to represent the sub-bands. The complete Hamiltonian for this physical situation including the crystalline lattice is $\mathcal{H} = \mathcal{H}_{\text{lat}} + \mathcal{H}_{\text{el}}^{\alpha} + \mathcal{H}_{\text{el}}^{\delta} + \mathcal{H}_{\text{el}}^{\alpha\delta}$, where $\mathcal{H}_{\text{lat}} = \sum_q (n + 1/2)\hbar\omega a_q^+ a_q$, describes the crystalline lattice; $\mathcal{H}_{\text{el}}^{\alpha}$ describes one band of itinerant electrons, $\mathcal{H}_{\text{el}}^{\delta}$ describes the other band and $\mathcal{H}_{\text{el}}^{\alpha\delta}$ accounts for the hybridization between them. The electronic part of this Hamiltonian $\mathcal{H}_{\text{el}} = \mathcal{H}_{\text{el}}^{\alpha} + \mathcal{H}_{\text{el}}^{\delta} + \mathcal{H}_{\text{el}}^{\alpha\delta}$ is written in the mean field approximation as:

$$\begin{aligned}\mathcal{H}_{\text{el}} = &\sum_{i\sigma} \varepsilon_{0\sigma}^{\alpha} \alpha_i^+ \alpha_i + \sum_{ij} T_{ij}^{\alpha\alpha} \alpha_i^+ \alpha_j + \sum_{ij} T_{ij}^{\alpha\delta} \alpha_i^+ \delta_j \\ &+ \sum_{i\sigma} \varepsilon_{0\sigma}^{\delta} \delta_i^+ \delta_i + \sum_{ij} T_{ij}^{\delta\delta} \delta_i^+ \delta_j + \sum_{ij} T_{ij}^{\delta\alpha} \delta_i^+ \alpha_j\end{aligned} \tag{3.14}$$

In this Hamiltonian, $\varepsilon_{0\sigma}^{\alpha} = \varepsilon_0^{\alpha} + U^{\alpha}\langle n_{-\sigma}^{\alpha}\rangle - \sigma\mu_{\text{B}}B$, where B is the applied magnetic field, ε_0^{α} gives the center of the first energy band and $\langle n_{-\sigma}^{\alpha}\rangle$ represents its occupation. Similarly, $\varepsilon_{0\sigma}^{\delta} = \varepsilon_0^{\delta} + U^{\delta}\langle n_{-\sigma}^{\delta}\rangle - \sigma\mu_{\text{B}}B$, where $\langle n_{-\sigma}^{\delta}\rangle$ gives number of electrons at the second energy band, U^{α} and U^{δ} are model parameters associated with the Coulomb interaction. The terms $T_{ij}^{\delta\delta} = \sum_k \varepsilon_k^{\delta} e^{ik(R_i - R_j)}$ and $T_{ij}^{\alpha\alpha} = \sum_k \varepsilon_k^{\alpha} e^{ik(R_i - R_j)}$ account for the electron hopping inside the same sub-band while the terms $T_{ij}^{\delta\alpha} = T_{ij}^{\alpha\delta} = \sum_k \Gamma_k e^{ik(R_i - R_j)}$ account for the charge transfer between sub-bands.

In order to simplify the calculations, we use the homothetic band approximation [111] to deal with the electron dispersion relations. In this approach, we set the dispersion relation ε_k^{α} as the standard reference and take the dispersion relation ε_k^{δ} and the hybridization term Γ_k proportionally. Thus, we consider $\varepsilon_k^{\alpha} = \lambda_{\alpha}\varepsilon_k$, where the parameter λ_{α} simulates the effect of an external pressure. Thus, $\varepsilon_k^{\delta} = \lambda_{\delta}\varepsilon_k^{\alpha} = \lambda_{\alpha}\lambda_{\delta}\varepsilon_k$ and $\Gamma_k = \lambda_{\delta\alpha}\varepsilon_k$. Using this approach and changing the sum in k into an integral in energy, the local Green functions $g_{ii\sigma}^{\alpha\alpha}$ and $g_{ii\sigma}^{\delta\delta}$ are:

$$\begin{aligned}g_{ii\sigma}^{\alpha\alpha}(z) = \frac{1}{(\lambda_{\alpha}\lambda_{\delta} - \lambda_{\delta\alpha}^2)} &\left[\frac{(z - \varepsilon_{0\sigma}^{\delta} - \lambda_{\alpha}\lambda_{\delta}E_-^{\sigma})F(E_-^{\sigma})}{(E_+^{\sigma} - E_-^{\sigma})} \right. \\ &\left. - \frac{(z - \varepsilon_{0\sigma}^{\delta} - \lambda_{\alpha}\lambda_{\delta}E_+^{\sigma})F(E_+^{\sigma})}{(E_+^{\sigma} - E_-^{\sigma})} \right]\end{aligned} \tag{3.15}$$

$$\begin{aligned}g_{ii\sigma}^{\delta\delta}(z) = \frac{1}{(\lambda_{\alpha}\lambda_{\delta} - \lambda_{\delta\alpha}^2)} &\left[\frac{(z - \varepsilon_{0\sigma}^{\alpha} - E_-^{\sigma})F(E_-^{\sigma})}{(E_+^{\sigma} - E_-^{\sigma})} \right. \\ &\left. - \frac{(z - \varepsilon_{0\sigma}^{\alpha} - E_+^{\sigma})F(E_+^{\sigma})}{(E_+^{\sigma} - E_-^{\sigma})} \right]\end{aligned} \tag{3.16}$$

In these expressions $F(E_{\pm}^{\sigma}) = \int d\varepsilon \rho(\varepsilon)/(E_{\pm}^{\sigma} - \varepsilon)$ where $\rho(\varepsilon)$ is a model density of states and

$$E_{\pm}^{\sigma} = \frac{1}{2(\lambda_{\alpha}^{2}\lambda_{\delta} - \lambda_{\delta\alpha}^{2})}\{[(z - \varepsilon_{0\sigma}^{\alpha})\lambda_{\alpha}\lambda_{\delta} + \lambda_{\alpha}(z - \varepsilon_{0\sigma}^{\delta})] \pm \left\{[(z - \varepsilon_{0\sigma}^{\alpha})\lambda_{\alpha}\lambda_{\delta} - \lambda_{\alpha}(z - \varepsilon_{0\sigma}^{\delta})]^{2} + 4(z - \varepsilon_{0\sigma}^{\alpha})(z - \varepsilon_{0\sigma}^{\delta})\lambda_{\delta\alpha}^{2}\right\}^{1/2}\right\} \tag{3.17}$$

The electron occupation number is determined by $\langle n^{c}\rangle = \int_{-\infty}^{\mu}[\rho_{\uparrow}^{c}(\varepsilon) + \rho_{\downarrow}^{c}(\varepsilon)]f(\varepsilon)d\varepsilon$, where $\rho_{\sigma}^{c}(\varepsilon) = -\frac{1}{\pi}Img_{ii\sigma}^{cc}(z)$ ($c = \alpha$ or δ), with $z = \varepsilon + i0$ and $f(\varepsilon) = 1.0/[1.0 + e^{(\varepsilon-\mu)/k_{B}T}]$ is the Fermi function. It should be mentioned that the chemical potential (μ) should be self-consistently determined upon the condition in which the total number of electrons $\langle n^{\alpha}\rangle + \langle n^{\delta}\rangle$ is kept constant. Therefore, any reduction of $\langle n^{\alpha}\rangle$ implies in an increase of $\langle n^{\delta}\rangle$ and vice versa, characterizing the charge transfer between sub-bands. The total entropy is $S(T, B, P) = S_{lat}(T, B, P) + S_{el}(T, B, P)$ where $S_{lat}(T, B, P)$ represents the contribution from the crystalline lattice [see relation (3.7)]. The electronic entropy is calculated by [8]:

$$S_{el}(T, B, P) = \Re\left\{\sum_{\sigma}\int_{-\infty}^{\infty}\left[\ln[1 + e^{-(\varepsilon_{\sigma}^{\alpha}-\mu)/k_{B}T}] + \frac{(\varepsilon_{\sigma}^{\alpha} - \mu)f(\varepsilon)}{k_{B}T}\right]\rho_{\sigma}^{\alpha}(\varepsilon)d\varepsilon + \sum_{\sigma}\int_{-\infty}^{\infty}\left[\ln[1 + e^{-(\varepsilon_{\sigma}^{\delta}-\mu)/k_{B}T}] + \frac{(\varepsilon_{\sigma}^{\delta} - \mu)f(\varepsilon)}{k_{B}T}\right]\rho_{\sigma}^{\delta}(\varepsilon)d\varepsilon\right\} \tag{3.18}$$

The barocaloric quantities ΔS_{iso} and ΔT_{ad} are calculated by relations (3.1) and (3.2).

3.4.2.3 Application to $YbInCu_4$

We use the model described earlier to calculate the barocaloric quantities in $YbInCu_4$. Experimental data [112] of magnetic susceptibility show that this compound exhibits Fermi liquid behavior at low temperatures and local magnetic moment behavior at higher temperatures. This is a clear indication that the electronic configuration of Yb ions changes from [Xe]$4f^{14}5d^{0}6s^{2}$ to [Xe]$4f^{13}5d^{1}6s^{2}$ with increasing temperature. This implies that one electron from the narrower 4f-electron band moves to the wide conduction electron band. As a consequence, the valence of the Yb ions changes modifying the physics of the compound. In figure 3.12 is schematically shown this physical mechanism of charge transfer between bands in $YbInCu_4$. At ambient pressure, the valence transition takes place at around 42 K. Moreover, experimental data also show that the temperature of the valence transition (T_V) is shifted to lower temperatures with applied pressure. Therefore, it is expected that this compound exhibits a sizeable barocaloric effect due to the charge transfer mechanism. In the compound $YbInCu_4$, a narrower 4f-electron band is coupled with a wider spd-conduction electron band. Thus, in our two-sub-band model, the labels α and δ represent the 4f and spd electrons, respectively. To get numerical results, we choose a set of model parameters and calculate the entropy change at ambient pressure P_0 and for pressures P_1 and P_2 that move the valence

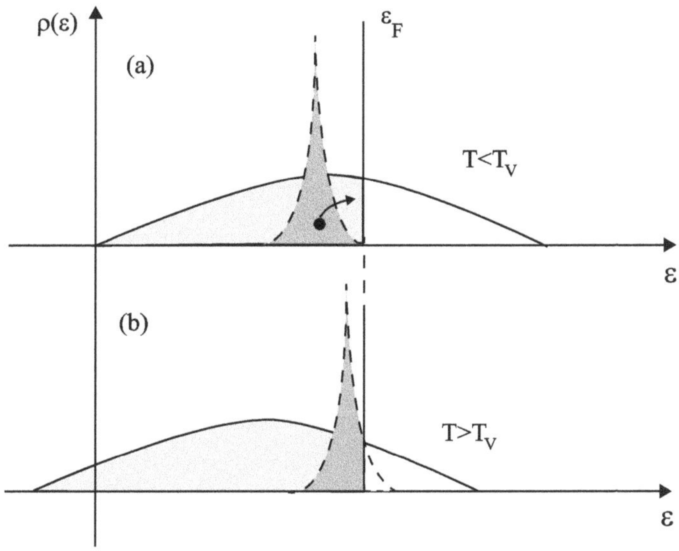

Figure 3.12. Wide and narrow densities of states for $T < T_V$ (a) and for $T > T_V$ (b). The vertical lines are the Fermi energies and the small black circle represents the electron that moves from one band to the other.

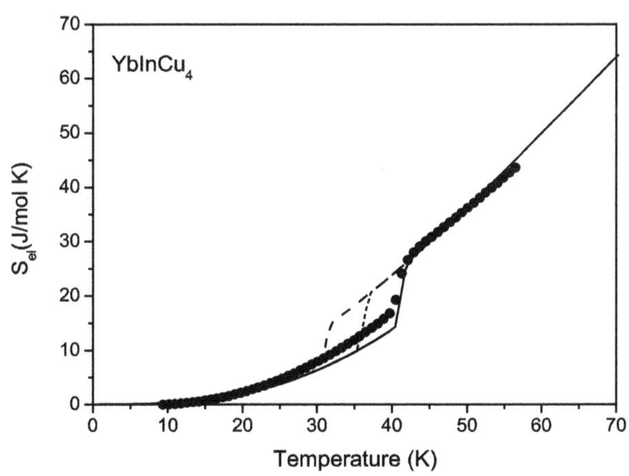

Figure 3.13. Entropy change as a function of temperature in $YbInCu_4$. Solid, dotted and dashed lines are the calculations for ambient pressure and for applied pressure P_1 and P_2 respectively. Symbols are experimental data inferred from heat capacity measurements [113].

transition temperature to 35 and 30 K, respectively. In figure 3.13, we plot the theoretically calculated entropy change as a function of temperature. In this figure, symbols represent the entropy data inferred from specific heat measurements at ambient pressure [113]. From the entropy curves shown in figure 3.13, we calculate the isothermal entropy change and the adiabatic temperature change upon pressure variation. The results of these calculations are shown in figures 3.14 and 3.15, respectively.

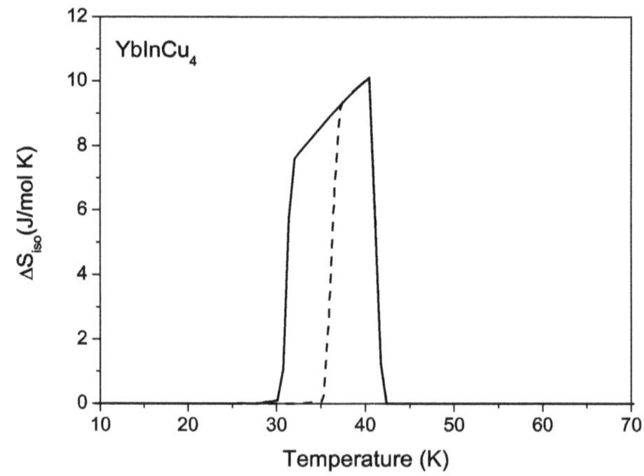

Figure 3.14. Isothermal entropy change in *YbInCu*$_4$ upon pressure variation from P_0 to P_1 (dotted line) from P_0 to P_2 (solid line).

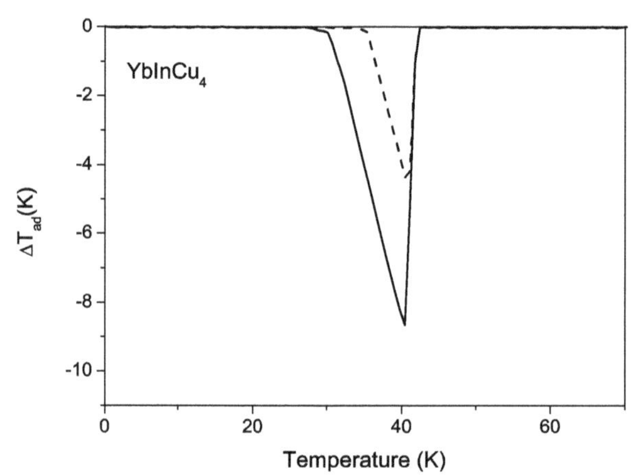

Figure 3.15. Adiabatic temperature change in $YbInCu_4$ upon pressure variation from P_0 to P_1 (dotted line) from P_0 to P_2 (solid line).

Before ending this section we should mention that Ce based compounds are also good candidates to exhibit large values of the barocaloric effect induced by this mechanism of charge transfer. This is because one electron of Ce atom, which has electronic configuration $[Xe]4f^15d^06s^2$, is likely to move to the conduction band as the temperature is increased. It is also worth mentioning that this charge transfer mechanism can be very important in several transition metals based compounds with non-equivalent lattice sites. In such compounds, the charge transfer between ions can lead the system to undergo a structural phase transition and consequently exhibit large values of the barocaloric quantities. In fact, recently Y Kosugi *et al* [114] have measured such an effect in $NdCu_3Fe_4O_{12}$ compound.

3.4.3 System of mixed magnetism

The magnetism in compounds formed by rare earth atoms and *n*d-transition elements ($n = 3, 4, 5$) is associated with both local magnetic moments and itinerant electrons. Compounds such as RT_2, RT_5 (R = rare earth ions and T = Fe, Co, Ni, Mn) can be cast into this category. Therefore, the complete Hamiltonian to describe the caloric effects in this kind of compounds, can be $\mathcal{H} = \mathcal{H}_{\text{el}}^{4\text{f}} + \mathcal{H}_{\text{el}}^{\text{d}} + \mathcal{H}_{\text{el}}^{\text{sp}} + \mathcal{H}_{\text{lat}}$, where $\mathcal{H}_{\text{lat}}$ describes the crystalline lattice; $\mathcal{H}_{\text{el}}^{4\text{f}}$ given by

$$\mathcal{H}_{\text{el}}^{4\text{f}} = -\sum_{i,l}\mathcal{J}_0 J_i^f J_l^f - \sum_{i,l}\mathcal{J}_1(J_i^f J_l^f)^2 - \sum_{i} g\mu_{\text{B}}\left(B + J_{df}\langle S_i^{\text{d}}\rangle\right)J_i^f + \mathcal{H}_i^{\text{CF}}, \tag{3.19}$$

describes the local magnetic moments of the rare earth ions which are coupled with the itinerant magnetization $\langle S_i^{\text{d}}\rangle$ where J_{df} is a coupling parameter. The term $\mathcal{H}_{\text{el}}^{\text{d}}$ given by:

$$\mathcal{H}_{\text{el}}^{\text{d}} = \sum_{i}\varepsilon_0^{\text{d}} d_i^+ d_i + \sum_{ij} T_{ij}^{\text{dd}} d_i^+ d_j + \sum_{i} U^{\text{d}} n_{i\uparrow} n_{i\downarrow} - \sum_{i} g\mu_{\text{B}} S_i^{\text{d}} B + J_{df}\langle J^f\rangle S_i^{\text{d}} \tag{3.20}$$

describes the itinerant electrons under the influence of an effective field $\langle J^f\rangle$ from the rare earth sublattice. Finally, the term $\mathcal{H}_{\text{el}}^{\text{sp}}$ describes the non-magnetic conduction electrons of the compounds.

The total entropy is $S(T, B, P) = S_{\text{el}}^{4\text{f}}(T, B, P) + S_{\text{el}}^{\text{d}}(T, B, P) + S_{\text{lat}}^{4\text{f}}(T, B, P)$ where $S_{\text{el}}^{4\text{f}}(T, B, P)$ is the contribution from the local magnetic moments; $S_{\text{el}}^{\text{d}}(T, B, P)$ is the contribution from the magnetic itinerant electrons and $S_{\text{lat}}^{4\text{f}}(T, B, P)$ as is given in relation (3.7) is the contribution from the crystalline lattice. Following the procedure used before in sections 3.4.1 and 3.4.2.1, the itinerant electrons entropy is

$$S_{\text{el}}^{\text{d}}(T, B, P) = \Re\sum_{\sigma}\int_{-\infty}^{\infty}\left[\ln\left[1 + e^{-(\varepsilon_\sigma^{\text{d}}-\mu)/k_{\text{B}}T}\right] + \frac{(\varepsilon_\sigma^{\text{d}} - \mu)f(\varepsilon)}{k_{\text{B}}T}\right]\rho_\sigma^{\text{d}}(\varepsilon)d\varepsilon \tag{3.21}$$

where now the renormalized energy $\varepsilon_{0\sigma}^{\text{d}} = \varepsilon_0^{\text{d}} + U^{\text{d}}\langle n_{-\sigma}^{\text{d}}\rangle - \sigma\mu_{\text{B}}B + J_{df}\langle J^f\rangle$ contains the effect of 4f electrons. Similarly, the contribution from the localized magnetic moments to the entropy is given by:

$$S_{\text{el}}^{4\text{f}}(T, B, P) = N_m\Re\left[\ln\sum_{m=-J}^{m=J} e^{-E_m/k_{\text{B}}T} + \frac{1}{k_{\text{B}}T}\frac{\sum_{m=-J}^{m=J} E_m e^{-E_m/k_{\text{B}}T}}{\sum_{m=-J}^{m=J} e^{-E_m/k_{\text{B}}T}}\right] \tag{3.22}$$

where E_m are the energy eigenvalues of the Hamiltonian $\mathcal{H}_{\text{el}}^{4\text{f}}$, that now brings the effect of the itinerant magnetization.

We calculate the barocaloric effect in the compound $ErCo_2$ [115], which undergoes a first order magnetic phase transition around the critical temperature 33 K. Moreover, the critical temperature is systematically shifted towards lower temperatures with increasing applied pressure. We select a proper set of model parameters

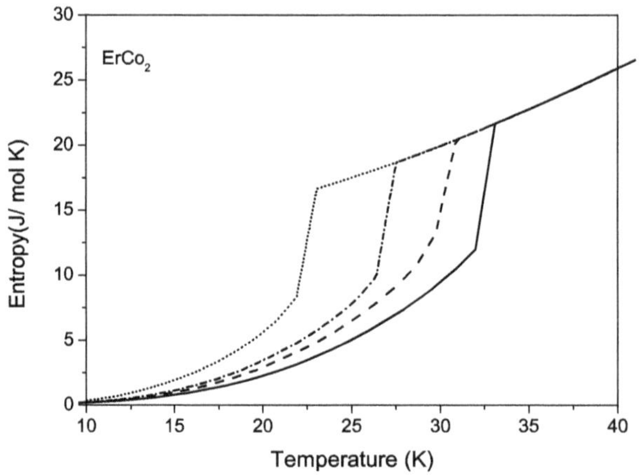

Figure 3.16. Entropy in $ErCo_2$ for ambient pressure P_0 (solid line) and for P_1 (dashed line), P_2 (dashed dotted) and P_3 (dotted line).

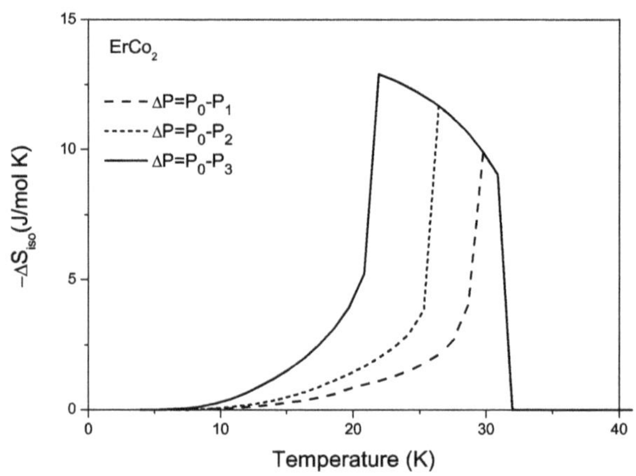

Figure 3.17. Isothermal entropy change in $ErCo_2$ upon pressure variation from P_1 to P_0 (dashed line), from P_2 to P_0 (dotted line) and from P_3 to P_0 (solid line).

to reproduce the critical temperature as a function of pressure and self-consistently calculated the magnetization and the total entropy. In figure 3.16, we plot the entropy as a function of temperature for ambient pressure P_0 and for pressures P_1, P_2 and P_3 that shift the critical temperature of the phase transition from 33 K to 30 K, 26 K and 22 K, respectively.

From the entropy versus temperature diagram shown in figure 3.16, we determine the barocaloric quantities ΔS_{iso} and ΔT_{ad}. In figure 3.17, we plot the isothermal entropy changes as a function of temperature upon pressure variation from P_0 to P_1 and from P_0 to P_2. The corresponding curves for the adiabatic temperature change are shown in figure 3.18.

It should be mentioned that the value of the theoretically calculated barocaloric quantities in $ErCo_2$ are comparable to the magnetocaloric counterparts obtained

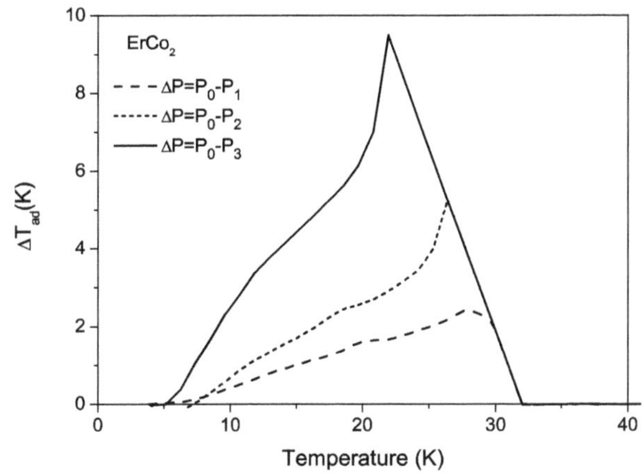

Figure 3.18. Adiabatic temperature change in $ErCo_2$ upon pressure variation from P_1 to P_0(dashed line) from P_2 to P_0 (dotted line) and from P_3 to P_0 (solid line).

upon magnetic field variation of 5 T. It is also worth mentioning that other compounds of this RCo_2 series as well as compounds of the series RFe_2, are also good candidates to exhibit large values of the barocaloric quantities.

3.4.4 Coordination compounds

Some materials formed by 3d transition ions surrounded by molecules, called ligands, exhibit interesting electronic/spin configurations that can be controlled by different driven fields. These materials are usually called metal complexes, or coordination compounds or spin crossover materials [116–121].

In metal complexes, the electric field generated by the ligands splits the 3d-states of the transition element in the set t_{2g} containing the orbitals d_{xy}, d_{xz} and d_{yz} and the set e_g with the orbitals d_{z^2} and $d_{x^2-y^2}$. The energy of the orbitals t_{2g} and e_g as well as the magnitude of the energy gap between them depend on the symmetry of the compound. For example, in the octahedral symmetry, the energy of the set t_{2g} is smaller than the energy of the set e_g, while in compounds with tetrahedral symmetry, the energy of the orbital t_{2g} and e_g is scaled in an opposite way.

The electronic occupation of the 3d-orbitals depends on the relation between the energy gap and the electron pairing energy. For example, let us consider a coordination compound with 5d-electrons represented by d^5 in the octahedral symmetry where the set t_{2g} has the lowest energy. If the energy gap is smaller than the electron pairing energy, the electron distribution follows Hund's rule and the five electrons are arranged as in figure 3.19(a). In this configuration where all electrons are unpaired, the total spin is 5/2, characterizing a high-spin state. The metal complex $[Fe(H_2O)_6]^{3+}$ which is formed by the following chemical reaction $Fe^{3+} + 6H_2O = [Fe(H_2O)_6]^{3+}$ can be cast into this category. On the other hand, if the energy gap is much larger than the Coulomb repulsion between electrons, the occupation of the orbitals no longer follows Hund's rule, so that the five electrons are arranged as in figure 3.19(b). In this case, the total spin is 1/2 and the system is in

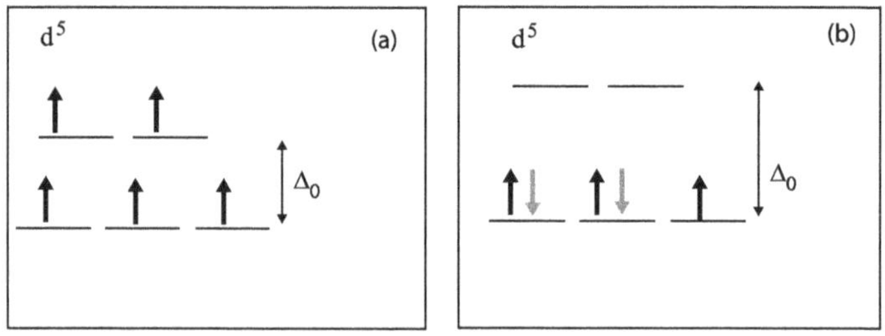

Figure 3.19. Electron distribution in the d^5 state. (a) High-spin (b) low-spin configuration.

the low-spin state. The coordination compound $[Fe(CN)_6]^{3-}$ is an example of this low-spin configuration.

If energy is provided to the system in the low-spin configuration, electrons can be promoted from the t_{2g} set to one of the e_g excited states. As a result, the configuration of spins is modified and the system passes from the low-spin to the high-spin state. This physical process is called spin crossover transition and can be induced by different external stimuli such as heat, light, pressure, electric and magnetic field. The fact that the temperature of the spin crossover transition can be tuned by external pressure, makes the metal complexes very good candidates to exhibit large barocaloric effect, as has been proposed by Karl Sandeman [73] and theoretically confirmed by P J von Ranke [74].

3.4.4.1 Model Hamiltonian

The theoretical description of the caloric properties in spin crossover materials has been intensively done by P J von Ranke and collaborators [74–79]. However, in this present text, we present an alternative description using a model similar to the one used in reference [119] based on the Bragg–Willians approximation for chemical mixtures. In this context, the Gibbs free energy for this system is

$$G(T, P) = (E_0 - k_B T \ln Z_0)n_{ls} - J_0 n_{ls}^2 + k_B T[n_{ls} \ln n_{ls} + (1 - n_{ls}) \ln (1 - n_{ls})] \quad (3.23)$$

where n_{ls} is the number of low-spin states; $Z_0 = \ln \sum_i e^{-(E_l - E_0)/k_B T}$, with E_0 representing the energy of the lowest t_{2g} orbitals and E_1 the energy of e_g orbitals; J_0 is a coupling energy parameter. The temperature dependence of the number n_{ls} can be obtained by minimizing the free energy with respect to n_{ls}. Thus, taking the derivative and after a few algebraic manipulations, we get

$$n_{ls}(T) = \frac{Z_0}{\left(Z_0 + e^{\frac{[E_0 - 2J_0 n_{ls}]}{N k_B T}}\right)} \quad (3.24)$$

Notice that this expression for n_{ls} should be self-consistently solved. The entropy associated with the electronic configuration, calculated by $S_{conf} = -\partial G(T, P)/\partial T$ is given by

$$S_{\text{conf}}(T,\, P) = n_{\text{ls}}\Re\left[\ln Z_0 + \frac{1}{k_{\text{B}}T}\frac{\sum_i (E_1 - E_0)\mathrm{e}^{-(E_1-E_0)/k_{\text{B}}T}}{Z_0}\right] + \Re[n_{\text{ls}}\ln n_{\text{ls}} + (1-n_{\text{ls}})\ln(1-n_{\text{ls}})] \tag{3.25}$$

The total entropy is $S = S_{\text{conf}} + S_{\text{lat}}$ where S_{lat} is the lattice entropy that we consider in the Debye approximation, which has already been discussed before [see relation (3.7)].

In order to illustrate the model, we select a set of parameters to perform a systematic calculation. We take for ambient pressure $E_0 = 0.04$ eV, $E_1 = 0.08$ eV, $J_0 = 0.025$ eV. With these model parameters, the system undergoes at ambient pressure P_0, a first order transition from the low-spin to high-spin state around 52 K. When an external pressure is applied the energy separation between the energy levels as well as the interaction with the neighborhood change. Here, we consider an applied pressure P_1 that renormalizes the parameter to $J_0 = 0.023$ eV and pushes the temperature of the phase transition to 22 K. Within this set of model parameters, we calculate the fraction of the low-spin configuration through the self-consistent solution of equation (3.24) for ambient pressure and for the external pressure P_1. The obtained results are shown in figure 3.20. Using this temperature dependence of n_{ls}, and a fixed Debye temperature of 150 K, we calculate the total entropy via relations (3.7) and (3.25). The results for ambient pressure P_0 and applied pressure P_1 are shown in figure 3.21 by the solid and dashed lines, respectively.

From the entropy versus temperature diagram shown in the previous figure, we estimate the barocaloric quantities. The isothermal entropy change upon pressure variation from P_1 to P_0 is shown in figure 3.22 and the corresponding adiabatic temperature change is shown in figure 3.23.

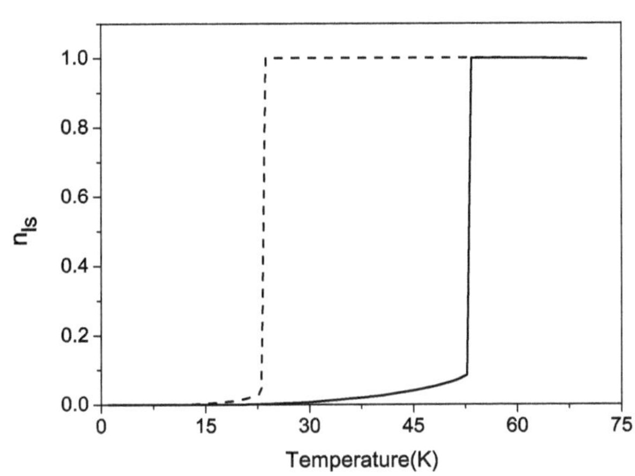

Figure 3.20. n_{ls} as a function of temperature for ambient pressure (solid line) and for an external pressure P_1 (dashed line).

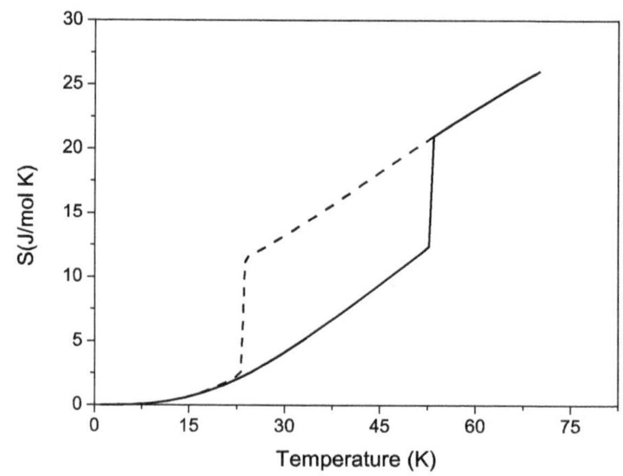

Figure 3.21. Total entropy as a function of temperature for ambient pressure (solid line) and for an external pressure P_1 (dashed line).

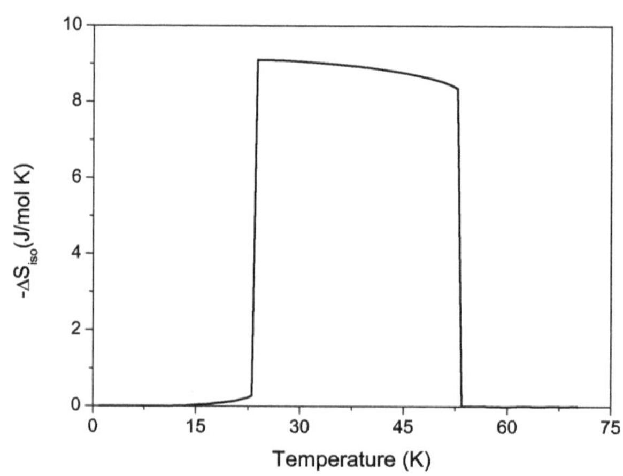

Figure 3.22. Isothermal entropy change as a function of temperature for a pressure variation from P_1 to P_0.

3.5 Barocaloric effect in non-magnetic materials

In this section, we discuss the barocaloric effect in solid state materials without magnetic order, where the main contribution to the entropy change comes from the vibrational and rotational motion of the ions into the crystalline lattice. In the next subsection, we discuss the lattice vibrations and the thermal expansion effects. The rotational degree of freedom will be discussed later.

3.5.1 Lattice vibrations: thermal expansion effects

In order to consider the thermal expansion [122–133] effects on the barocaloric properties of non-magnetic materials it is necessary to go beyond the harmonic approximation to get the volume variation as a function of temperature and pressure.

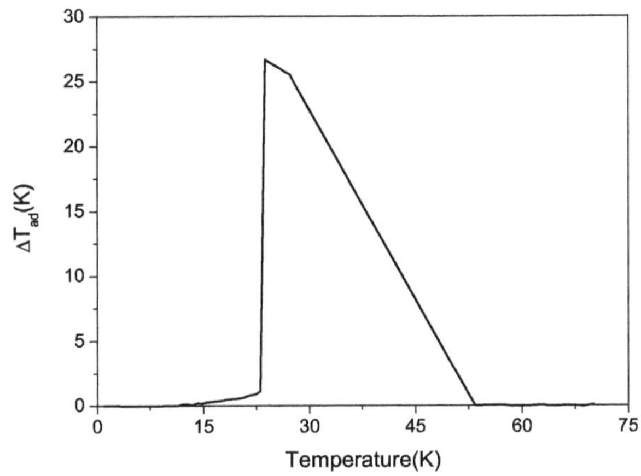

Figure 3.23. Adiabatic temperature change as a function of temperature for a pressure variation from P_1 to P_0.

3.5.1.1 Model Hamiltonian

The lattice vibrations and thermal expansion of the crystalline lattice can be described by using, for example, the Morse potential [134] which takes into account the interactions between ions in the crystalline lattice or by introducing anharmonic terms to the harmonic oscillator energy potential. In this text, for simplicity reasons, we take an approximation where the anharmonic effects are included via the renormalized potential $V = m\tilde{\omega}^2x^2/2$ where $\tilde{\omega}$ represents the temperature and pressure dependent frequency of the lattice vibrations.

In many materials, the volume of the sample usually increases monotonically with increasing temperature. However, in some materials, the increase in temperature can lead to a phase transformation with a substantial increase in the volume around the transition temperature. Moreover, there are also materials where the phase transformation takes place with a sizeable reduction of the volume. Thus, in order to consider all these possible physical scenarios of thermal expansions, we use the following model Hamiltonian including two phases:

$$\mathcal{H} = \sum_q \varepsilon_1^{\text{vib}} a_{q_1}^+ a_{q_1} + \sum_q \varepsilon_2^{\text{vib}} c_{q_2}^+ c_{q_2} \tag{3.26}$$

The first term of this Hamiltonian describes the crystalline phase (phase 1) at low temperatures, where $\varepsilon_1^{\text{vib}} = (n + 1/2)\hbar\tilde{\omega}_1$ represents the vibration energies of the ions with the renormalized frequency $\tilde{\omega}_1$. Similarly, the second term accounts for the vibrational motion of the ions in the crystalline phase (phase 2), at higher temperatures, with frequency $\tilde{\omega}_2$ and energy $\varepsilon_2^{\text{vib}} = (n + 1/2)\hbar\tilde{\omega}_2$. In a first simple approach, we can assume that $\varepsilon_1^{\text{vib}} = \varepsilon_0^{\text{vib}} + \Delta_1$ and $\varepsilon_2^{\text{vib}} = \varepsilon_0^{\text{vib}} + \Delta_2$ where $\varepsilon_0^{\text{vib}} = (n + 1/2)\hbar\omega_0$ is the temperature independent vibrational energy, and Δ_1 and Δ_2 are temperature and pressure dependent parameters which take into account the thermal expansion.

3.5.1.2 Partition function, free energy and entropy

The partition function for the Hamiltonian given in (3.26) is

$$Z^{\text{vib}} = \text{tr}\,e^{-\frac{1}{k_BT}\left[\sum_q \varepsilon_1^{\text{vib}} n_{1q} + \sum_q \varepsilon_2^{\text{vib}} n_{2q}\right]} \tag{3.27}$$

The Gibbs free energy for this system with a mixture of phases is given by:

$$\begin{aligned} G(T, P) = {} & n_1(T, P)\left[g_1^{\text{vib}}(T, P)\right] + n_2(T, P)\left[g_2^{\text{vib}}(T, P)\right] + \Re T\{n_1(T, P)\ln n_1(T, P) \\ & + [1 - n_1(T, P)]\ln[1 - n_1(T, P)]\} \end{aligned} \tag{3.28}$$

where $n_1(T, P)$ and $n_2(T, P)$ represent the fraction of phases 1 and 2 and should obey the following conditions. At 0 K and surrounding temperatures $n_1(T, P) \longrightarrow 1$ and $n_2(T, P) \longrightarrow 0$ while at temperatures above the phase transition the opposite condition, should be fulfilled. Moreover, around the phase transition temperature $n_1(T, P) = n_2(T, P)$ assuring the coexistence of the two phases in equilibrium. In the previous relation, $g_i^{\text{vib}}(T, P)$ $(i = 1, 2)$ are the partial Gibbs free energy associated with lattice vibrations in the i-phase. According to statistical mechanics, $g_i^{\text{vib}}(T, P)$ is given by [8]:

$$g_i^{\text{vib}}(T, P) = \int_0^\infty \frac{\hbar\omega}{2}\rho_i(\omega)d\omega + k_BT\int_0^\infty \ln\left(1 - e^{-\frac{1}{k_BT}\hbar\omega}\right)\rho_i(\omega)d\omega, \tag{3.29}$$

where $\rho_i(\omega)$ is the density of phonons frequency which depends on temperature and pressure. Using the Green's function technique it can be written in the form:

$$\rho_i(\omega) = -\frac{1}{\pi}\,\text{Im}\int_{-\infty}^{\infty} \frac{\rho_0(\omega')}{\omega' - \omega(T, P)}d\omega' \tag{3.30}$$

where $\omega' = \omega' + i0$ and $\rho_0(\omega')$ is a temperature independent model of the density of frequency. The temperature and pressure dependence of the phonon frequency ω makes the solution of this integral equation very difficult. In order to simplify the calculations and get an insight into the physical mechanisms involved in the entropy change, we introduce some approximations. Here, for a fixed temperature, we make an expansion of the frequency around the volume V_0 that is associated with a given pressure P_0, in the form

$$\omega(T, P) = \lambda_0(P)\omega_0[1 + \lambda_1(P)\Delta V + \lambda_2(P)\Delta V^2 + \lambda_3(P)\Delta V^3 + \cdots] \tag{3.31}$$

where ω_0 is the frequency at constant pressure and λ_i are model parameters that depend on pressure. Inserting this expansion of $\omega(T, P)$ into (3.30), the density of phonons frequency can be written as

$$\rho_i(\omega) = -\frac{1}{\pi}\,\text{Im}\int_{-\infty}^{\infty} \frac{\rho_0(\omega')}{\omega' - \lambda_0(P)(1 + \lambda_1\Delta V + \lambda_2(\Delta V)^2 + \lambda_3(\Delta V)^3 + \cdots)\omega_0}d\omega' \tag{3.32}$$

The entropy due to the vibrational motion of the ions is calculated via the thermodynamic relation $S_i^{\text{vib}} = -(\partial g_i^{\text{vib}}/\partial T)$. Thus, the total entropy as a function of temperature and pressure is given by:

$$S(T, P) = n_1(T, P)S_1^{\text{vib}}(T, p) + n_2(T, P)S_2^{\text{vib}}(T, p) + \Re\{n_1(T, P)\ln n_1(T, P) + [1 - n_1(T, P)]\ln[1 - n_1(T, P)]\} \tag{3.33}$$

where the vibrational entropy of the i-crystalline phase ($i = 1, 2$) is:

$$S_i^{\text{vib}}(T, P) = \Re\left[-\int_0^{\infty} \ln[1 - e^{-\hbar\omega/k_BT}]\rho_i(\omega)d\omega + \frac{1}{k_BT}\int_0^{\infty} \frac{\hbar\omega\rho_i(\omega)}{e^{\hbar\omega/k_BT} - 1}d\omega\right] \tag{3.34}$$

Once the entropy is calculated, the isothermal entropy change and the adiabatic temperature change upon pressure variation are calculated by relations (3.1) and (3.2).

3.5.1.3 Order parameter and phase transition

To calculate the total entropy given in (3.33), we need to know the temperature and pressure dependence of the volume variation and the fractions of ions $n_i(T, P)$ of the i-phase. At this point it is convenient, to define an order parameter and write the Gibbs free energy in terms of it. Here, we define $\Delta V_i(T, P) = \pm[V_i(T_C, P) - V_i(T, P)]$ as the order parameter, where $V_i(T_C, P)$ and $V_i(T, P)$ represent the volume of the i-phase at the critical temperature and a given temperature, respectively. Moreover, the plus (minus) sign represents a positive (negative) thermal expansion.

For an isothermal thermodynamic process, the total differential of this free energy is $dG_i^{\text{vib}}(T, \Delta V) = \Delta V(T, P)dP$. Since the volume variation modifies the internal pressure of the system, we can expand the pressure as: $P = P_0 - \gamma_1\Delta V - \gamma_2\Delta V^2 - \gamma_3\Delta V^3 - \gamma_4\Delta V^4 - \gamma_5\Delta V^5$, where P_0 is a reference pressure and γ_i, are model parameters. Thus, plugging the derivative $dP = (-\gamma_1 - 2\gamma_2\Delta V - 3\gamma_3\Delta V^2 - 4\gamma_4\Delta V^3 - 5\gamma_5\Delta V^4)\,d(\Delta V)$ into the total differential dG_i^{vib}, integrating and neglecting the odd terms to get a symmetric function

$$G_i^{\text{vib}}(T, \Delta V) = G_0(T, 0) - \frac{\gamma_1}{2}\Delta V^2 - \frac{3\gamma_3}{4}\Delta V^4 - \frac{5\gamma_5}{6}\Delta V^6 \tag{3.35}$$

This expression provides the Gibbs free energy as a function of temperature and volume variation. The minima of this Gibbs free energy (3.35) give the volume variation of the system for a fixed temperature and pressure. Note that if the volume variation is zero the Gibbs energy reduces to $G_i^{\text{vib}}(T, \Delta V) = G_0(T, 0)$ which is the usual expression in the absence of the thermal expansion. The fraction of phase 1, can be taken as proportional to the volume variation. For the sake of simplicity, we take it as $n_1(T, P) = \alpha\Delta V_1(T, P) = \Delta V_1(T, P)/\Delta V_1(0, P)$.

3.5.1.4 Application to a hypothetical compound

In order to illustrate the calculations, we consider a hypothetical physical system with a first order phase transformation with an abrupt reduction of the volume. The model parameters γ_1, γ_2 and γ_3 that appear in the free energy given in relation (3.35), were chosen to give at ambient pressure, a phase transition temperature around 238 K. With this set of model parameters, we calculate the Gibbs free energy as a function of the volume variation for fixed values of temperature and pressure. In figure 3.24, we plot the Gibbs free energy at ambient pressure (P_0) and for $T < T_C$ (dashed line), $T = T_C$ (solid line) and $T > T_C$ (dotted line). The volume variation of the phase 1 at temperatures below T_c is determined from the lowest minimum of the free energy. Moreover, the temperature of the phase transition diminishes when an external pressure is applied. Here we consider applied pressures P_1, P_2 and P_3 that push the critical temperature of the phase transition from 238 K to 226 K, 222 K and 214 K, respectively. The fraction of the phase 1 is taken proportional to this volume variation in the form $n_1(T, P) = \Delta V_1(T, P)/\Delta V_1(0, P)$. In figure 3.25, we plot $n_1(T, P)$ as a function of temperature for some values of pressure. Using the volume variation obtained via the minimum of the free energy and a model of the density of frequency $\rho_0(\omega')$ with semicircular shape, we calculate the temperature and pressure dependent density of phonons frequency $\rho_i(\omega)$ given in relation (3.30) and the total entropy given in (3.33).

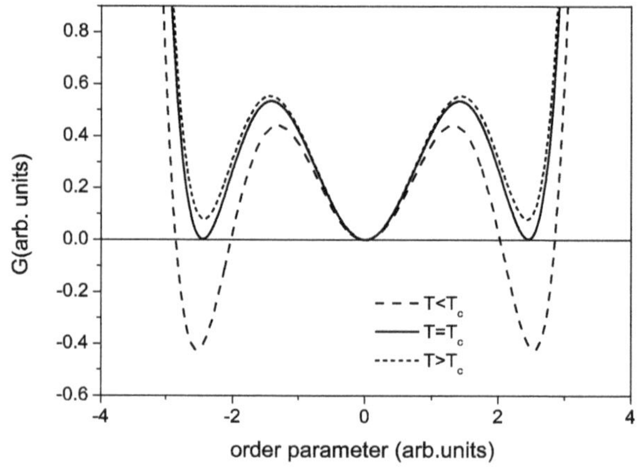

Figure 3.24. Free energy as a function of the order parameter for $T < T_C$ (dashed line), $T = T_C$ (solid line) and $T > T_C$ (dotted line).

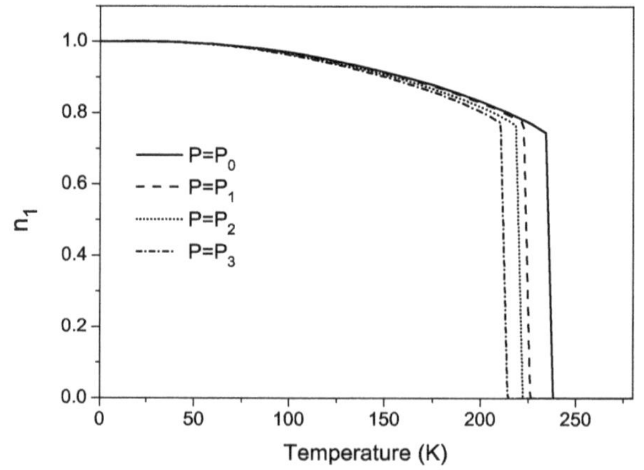

Figure 3.25. n_1 as a function of temperature for some values of pressure.

In figure 3.26, we plot the entropy as a function of temperature for ambient pressure ($P_0 = 0$) and for the pressures P_1, P_2 and P_3. Using these entropy curves, we get the isothermal entropy change (ΔS_{iso}) and the adiabatic temperature change (ΔT_{ad}) upon pressure variations $\Delta P = P_1 - P_0$, $\Delta P = P_2 - P_0$, and $\Delta P = P_3 - P_0$. The results are plotted in figures 3.27 and 3.28, respectively. It should be mentioned that the results presented in this section for the hypothetical compound were obtained with parameters arbitrarily chosen. For real compounds, the model parameters should be taken in order to reproduce the temperature and the nature of the phase transition.

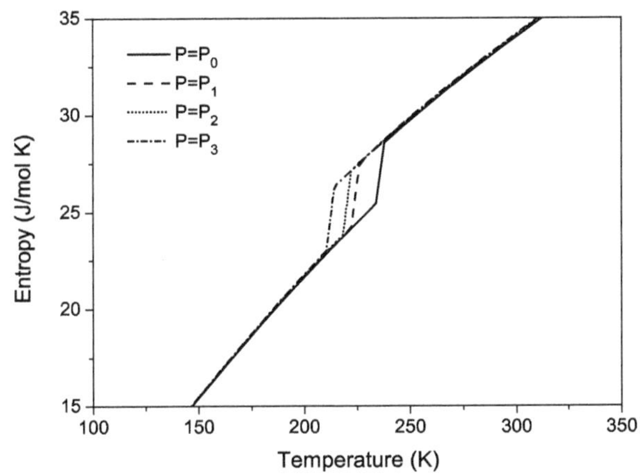

Figure 3.26. Entropy as a function temperature for some values of pressure P_0, P_1, P_2 and P_3. See text for discussion.

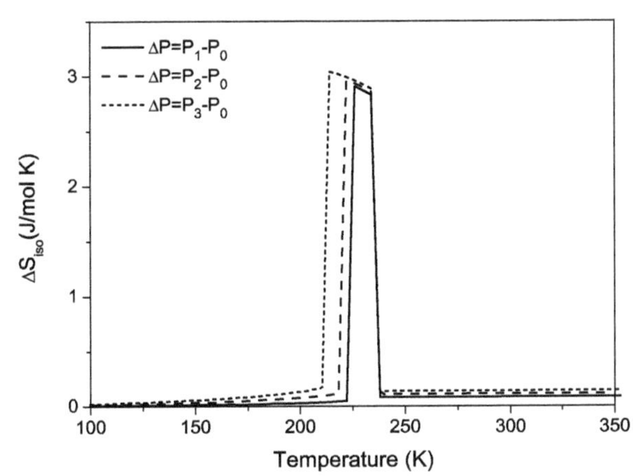

Figure 3.27. Isothermal entropy as a function temperature upon pressure variations $\Delta P = P_1 - P_0$, $\Delta P = P_2 - P_0$ and $\Delta P = P_3 - P_0$.

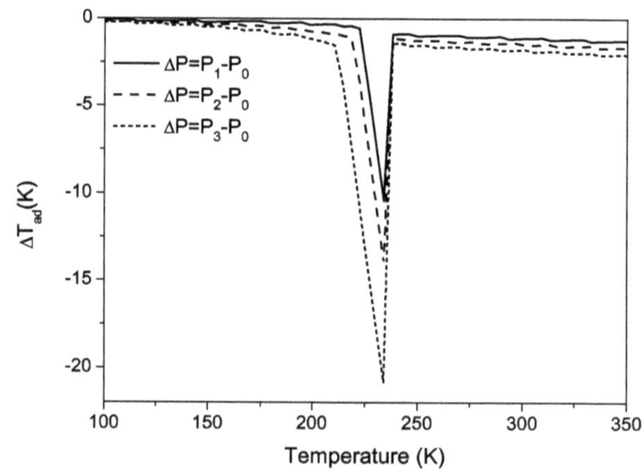

Figure 3.28. Adiabatic temperature change as a function temperature upon pressure variations $\Delta P = P_1 - P_0$, $\Delta P = P_2 - P_0$ and $\Delta P = P_3 - P_0$.

3.5.2 Contribution from rotational motion

In this section, we extend the model of the previous section to include the contribution from the rotational motion. Plastic crystals [135, 136] such as neopentane (NP) whose chemical formula is C_5H_{12}, neopentylalcohol (NPA) represented by $C_5H_{12}O$, pentaglycerine (PG) which is represented by $C_5H_{12}O_3$, pentaerythritol (PE) with the chemical formula $C_5H_{12}O_4$ and the neopentylglycol (NPG) whose chemical representation is $C_5H_{12}O_2$ are examples of compounds where the rotational degree of freedom is also important to generate large values of the entropy change. In these plastic crystals, the molecules at low temperatures exhibit only vibrational motion. However, at a given critical temperature they undergo a phase transformation and the molecules have both vibrational and rotational motions. Moreover, the temperatures of the phase transition in plastic crystals are in the range between 100 K and 500 K. This fact makes these materials interesting candidates to be studied for solid state refrigeration around room temperature.

3.5.2.1 Model Hamiltonian

In order to discuss the effect of vibrational and rotational motion on the barocaloric properties of non-magnetic materials, we consider the following model Hamiltonian

$$\mathcal{H} = \sum_{q_1} \varepsilon_1^{\text{vib}} a_{q_1}^+ a_{q_1} + \sum_{q_2} \varepsilon_2^{\text{vib}} c_{q_2}^+ c_{q_2} + \sum_{q_2} \varepsilon_2^{\text{rot}} c_{q_2}^+ c_{q_2} \tag{3.36}$$

In this Hamiltonian, $\varepsilon_i^{\text{vib}} = \varepsilon_0^{\text{vib}} + \Delta_i$ $(i = 1, 2)$ represents the vibrational energy where $\varepsilon_0^{\text{vib}} = (n + 1/2)\hbar\omega_0$ is independent of temperature, and Δ_i is a temperature and pressure dependent parameter which takes into account the thermal expansion. The term $\varepsilon_2^{\text{rot}}$ is the rotational energy which is present only in the more disordered phase at higher temperatures.

3.5.2.2 *Free energy and entropy*

The Gibbs free energy for this system with a mixture of phases is given by:

$$\begin{aligned} G(T, P) = n_1(T, P)\left[g_1^{\text{vib}}(T, P)\right] + n_2(T, P)\left[g_2^{\text{vib}}(T, P) + g_2^{\text{rot}}(T, P)\right] \\ + \Re T\{n_1(T, P) \ln n_1(T, P) + [1 - n_1(T, P)] \ln [1 - n_1(T, P)]\} \end{aligned} \tag{3.37}$$

where $g_i^{\text{vib}}(T, P)$ given by

$$g_i^{\text{vib}}(T, P) = \int_0^\infty \frac{\hbar\omega}{2}\rho_i(\omega)d\omega + k_B T \int_0^\infty \ln(1 - e^{-\hbar\omega/k_B T})\rho_i(\omega)d\omega,$$

is the vibrational entropy of the i-phase, where ω is the frequency of vibrations and $\rho_i(\omega)$ is the density of phonons frequency. Using the Green's function technique and the expansion of the phonon frequency in terms of the volume variation, the density of phonons frequency can be written in the form of relation (3.32) (see section 3.5.1.2 for details).

The entropy due to the vibrational motion of the molecules, calculated from $S_i^{\text{vib}} = -(\partial g_i^{\text{vib}}/\partial T)$, is given by:

$$S_i^{\text{vib}}(T, P) = \Re\left[-\int_0^\infty \ln[1 - e^{-\hbar\omega/k_B T}]\rho_i(\omega)d\omega + \frac{1}{k_B T}\int_0^\infty \frac{\hbar\omega\rho_i(\omega)}{e^{\hbar\omega/k_B T} - 1}d\omega\right] \tag{3.38}$$

In order to calculate the rotational contribution to the entropy let us consider a symmetric non-linear molecule where the momenta of inertia fulfill the condition $I_x = I_y \neq I_z$. In this case, the rotational energy is given by $E_{l,m} = l(l + 1)hB + h(C - B)m^2$, where l and m are quantum numbers associated with the total angular momentum and its z component; h is the Planck constant; $B = h/8\pi^2 I_x$ and $C = h/8\pi^2 I_z$. The partition function due to the rotational motion of one molecule is

$$Z_2^{\text{rot}} = \sum_{l=0}^{\infty}\sum_{m=-l}^{m=l} g_l e^{-\beta E_{l,m}} \tag{3.39}$$

where $g_l = (2l + 1)$ is the degeneracy of the energy levels. Substituting the value of the energy $E_{l,m} = l(l + 1)hB + h(C - B)m^2$ and after an algebra, the rotational partition function reads:

$$Z_2^{\text{rot}} = \frac{8\pi^2}{\sigma h^3}(2\pi k_B T)^{3/2}\left(I_x I_y I_z\right)^{1/2} \tag{3.40}$$

where σ is a parameter which takes into account the equivalent spatial orientations of the molecules under rotation. It is $\sigma = 1$ for non-symmetric diatomic molecules and $\sigma = 2$ for symmetric diatomic molecules. The rotational free energy, calculated by $g_2^{\text{rot}}(T, P) = -k_B T \ln Z_2^{(rot)}$, is then

$$g_2^{\text{rot}}(T, P) = -k_B T \ln\left[\frac{8\pi^2}{\sigma h^3}(2\pi k_B T)^{3/2}\left(I_x I_y I_z\right)^{1/2}\right] \tag{3.41}$$

The rotational entropy of the molecules, calculated from $S_2^{\rm rot}(T,\,P) = -(\partial g_2^{\rm rot}/\partial t)$, is given by:

$$S_2^{\rm rot}(T,\,P) = \Re\left\{\frac{3}{2} + \ln\left[\frac{8\pi^2}{\sigma h^3}(2\pi k_{\rm B}T)^{3/2}\left(I_xI_yI_z\right)^{1/2}\right]\right\} \tag{3.42}$$

The total entropy, calculated by $S(T,\,P) = -\partial G(T,\,P)/\partial T$, is then

$$\begin{aligned} S(T,\,P) = {} & n_1(T,\,P)S_1^{\rm vib}(T,\,P) + n_2(T,\,P)[S_2^{\rm vib}(T,\,P) + S_2^{\rm rot}(T,\,P)] \\ & + \Re T\{n_1(T,\,P)\ln n_1(T,\,P) + [1 - n_1(T,\,P)]\ln[1 - n_1(T,\,P)]\} \end{aligned} \tag{3.43}$$

The volume variation, as well as the number of mols in the phase 1, is estimated by the minima of the Gibbs free energy given in relation (3.35).

3.5.2.3 Application to neopentylglycol plastic crystal

In this section, we calculate the barocaloric quantities $\Delta S_{\rm iso}$ and $\Delta T_{\rm ad}$ in the plastic crystal neopentylglycol (NPG) with the chemical formula $C_5H_{12}O_2$. In NPG plastic crystal, the monoclinic phase is stable at low temperatures, while at higher temperatures the cubic phase is the most stable one. Besides that, at lower temperatures, the molecules have only vibrational motion around an equilibrium position, whereas at higher temperatures they have both vibrational and rotational motion. This additional contribution associated with the rotational motion of the molecules is responsible for the large entropy change at the phase transition. Moreover, the variation of the phase transition temperature by an applied pressure gives rise to large values of the barocaloric quantities.

In order to obtain numerical results, we choose some parameters. Here we take $\lambda_1 = 0.0132$ and $\lambda_2 = \lambda_3 = 0.001$. The parameter λ_0 was taken as $\lambda_0 = 1$ for ambient pressure; $\lambda_0 = 0.98$ for $P = 0.09$ GPa; $\lambda_0 = 0.95$ for $P = 0.25$ and $\lambda_0 = 0.93$ for 0.35 GPa. The volume variation was taken from the lowest minimum of the Gibbs free energy (3.35) and the inertia momenta were taken as: $I_x = I_y = 1.15 \times 10^{-47}$ kgm^2 and $I_z = 1.25 \times 10^{-47}$ kgm^2. Using this set of model parameters, we calculate the total entropy as a function of temperature. The obtained results for ambient pressure and applied pressure of 0.09 GPa, 0.25 GPa and 0.35 GPa are shown in figure 3.29 by the solid lines. In this figure, symbols represent experimental data [84]. The jump in the entropy at the phase transition temperature comes mainly from the extra term associated with the rotational entropy.

From the entropy versus temperature diagram shown in figure 3.29, we calculate the barocaloric quantities $\Delta S_{\rm iso}$ and $\Delta T_{\rm ad}$ upon pressure variation from 0 to 0.09 GPa, from 0 to 0.25 GPa and from 0 to 0.35 GPa. The obtained results are, respectively, shown in figures 3.30 and 3.31 together with the available experimental data [84].

3.6 Conclusions and outlook

In this text, we used specific models to treat the different physical mechanisms involved in the barocaloric effect in solid state materials. The barocaloric quantities

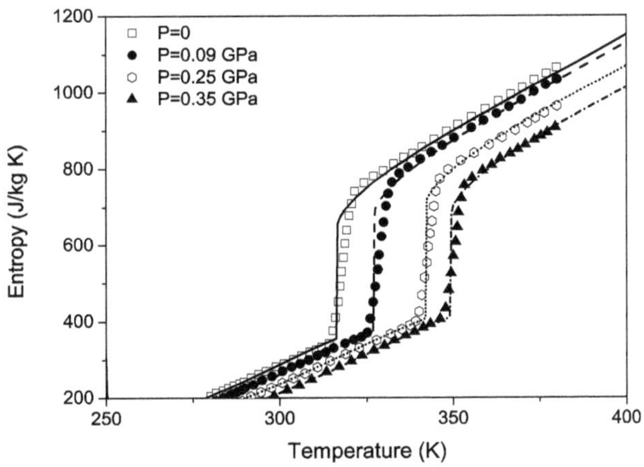

Figure 3.29. Entropy as a function of temperature in NPG plastic crystal. Solid lines are our calculations and symbols are experimental data [84].

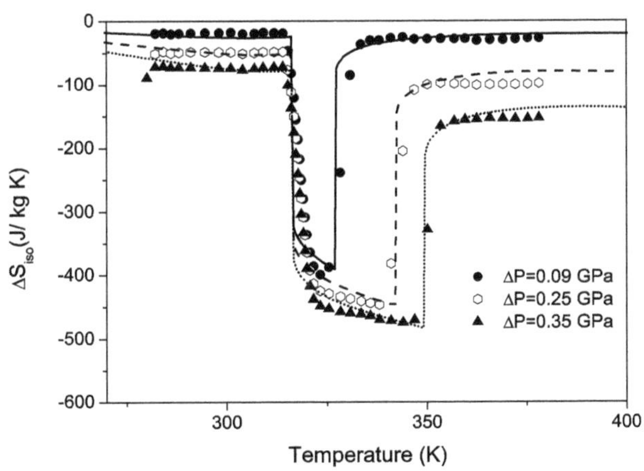

Figure 3.30. Isothermal entropy change in NPG plastic crystal, for some pressure variations. Solid lines are our calculations and symbols are experimental data inferred from the entropy curves shown in figure 3.29.

in materials with localized magnetism, were calculated using a Heisenberg-like Hamiltonian of localized interacting magnetic moments. However, the barocaloric effect in materials with itinerant electron magnetism was described by a Hubbard-like Hamiltonian in the picture of the band theory of solids. On the other hand, to calculate the barocaloric effect in materials without magnetic order, it is necessary to go beyond the harmonic approximation to treat the thermal expansion and the temperature and pressure dependence of the volume variation.

In the theoretical description presented in this work, we used some approaches to simplify the calculations and get the barocaloric quantities. For a more rigorous treatment, it is necessary to perform further analysis. For example, the lattice vibrations, which were treated in this text via the renormalization of the harmonic potential, as well as the evaluation of the temperature and pressure dependence of

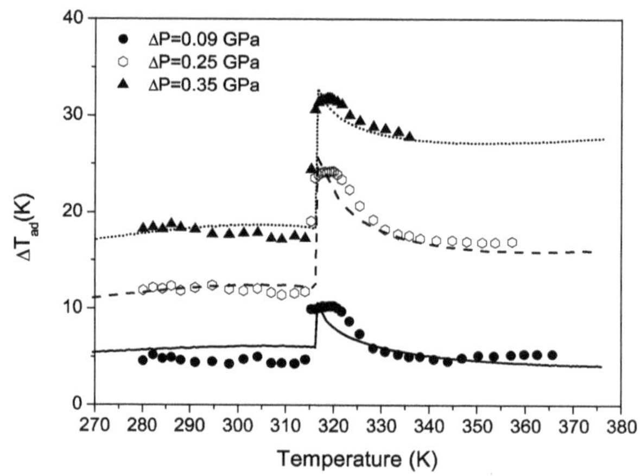

Figure 3.31. Adiabatic temperature change in NPG plastic crystal, for some pressure variations. Solid lines are our calculations and symbols are experimental data inferred from the entropy curves shown in figure 3.29.

the phonons frequency can be described by first principles calculations. Besides that, the inclusion of the optical modes and the effect of gaps in the density of phonons frequency, which are supposed to yield a large barocaloric effect, are also important to be investigated. The hysteresis effect on the barocaloric properties is another point that should be theoretically investigated in more detail. The charge transfer mechanism, the barocaloric effect in materials with a metal–insulator transition and in materials with a solid–liquid phase transition are also important issues that deserve more attention in the near future.

Acknowledgments

We acknowledge the Brazilian agencies, Conselho Nacional de Desenvolvimento Científico e Tecnológico-CNPq and Fundação Carlos Chagas Filho de Amparo à Pesquisa do Estado do Rio de Janeiro-FAPERJ, for financial support.

References

[1] Weiss P and Piccard A 1917 *J. Phys.* **7** 103
[2] Debye P 1926 *Ann. Phys.* **386** 1154
[3] Giauque W F and MacDougall D P 1933 *Phys. Rev.* **43** 768
[4] Tishin A M and Spichkin Y 2003 *The Magnetocaloric Effect and Its Application* (Bristol: IOP Publishing)
[5] Gschneidner K A Jr and Pecharsky V K 2000 *Annu. Rev. Mater. Sci.* **30** 387
[6] Gschneidner K A Jr, Pecharsky V K and Tsokol A O 2005 *Rep. Prog. Phys.* **68** 1479
[7] Brück E 2005 *J. Phys. D: Appl. Phys.* **38** R381
[8] de Oliveira N A and von Ranke P J 2010 *Phys. Rep.* **489** 89
[9] de Oliveira N A, von Ranke P J and Troper A 2014 *Int. J. Refrig.* **37** 237
[10] Smith A, Bahl C R H, Bjørk R, Engelbrecht K, Nielsen K K and Pryds N 2012 *Adv. Energy Mater.* **2** 1288
[11] Lyubina J 2017 *J. Phys. D: Appl. Phys.* **50** 053002

[12] Franco V, Blazquez J S, Ipus J J, Law J Y, Moreno-Ramirez L M and Conde A 2018 *Prog. Mater. Sci.* **93** 112
[13] Brown G V 1976 *J. Appl. Phys.* **47** 3673
[14] Yu B, Liu M, Egolf P W and Kitanovski A 2010 *Int. J. Refrig.* **33** 1029
[15] Tušek J, Zupan S, Šarlah A, Prebil I and Poredoš A 2010 *Int. J. Refrig.* **33** 294
[16] Kitanovski A, Plaznik U, Tušek J and Poredoš A 2014 *Int. J. Refrig.* **37** 28
[17] Arnold D S, Tura A, Ruebsaat-Trott A and Rowe A 2014 *Int. J. Refrig.* **37** 99
[18] Aprea C, Greco A, Maiorino A, Mastrullo R and Tura A 2014 *Int. J. Refrig.* **43** 111
[19] Lozano J A, Capovilla M S, Trevizoli P V, Engelbrecht K, Bahl C R H and Barbosa J R Jr 2016 *Int. J. Refrig.* **68** 187
[20] Müller K A, Fauth F, Fischer S, Koch M, Furrer A and Lacorre P 1998 *Appl. Phys. Lett.* **73** 1056
[21] Strässle Th, Furrer A, Lacorre P and Müller K A 2000 *J. Alloys Comp.* **303** 228
[22] Strässle Th, Furrer A, Hossain Z and Geibel Ch 2003 *Phys. Rev.* B **67** 054407
[23] Strässle Th, Furrer A and Müller K A 2000 *Physica* B **276** 944
[24] Strässle Th, Furrer A, Altorfer F, Mattenberger K, Böhm M and Mutka H 2001 *J. Alloys Compd.* **323-4** 392
[25] Strässle Th, Furrer A, Dönni A and Komatsubara T 2002 *J. Appl. Phys.* **91** 8543
[26] de Oliveira N A 2007 *Appl. Phys. Lett.* **90** 052501
[27] de Oliveira N A 2008 *J. Phys.: Condens. Matter* **20** 175209
[28] Mañosa L, Gonzalez-Alonso D, Planes A, Bonnot E, Barrio M, Tamarit J L, Aksoy S and Acet M 2010 *Nat. Mater.* **9** 478
[29] Mañosa L, Gonzales-Alonso D, Planes A, Barrio M, Tamarit J L, Titov I S, Acet M, Bhattacharyya A and Majumdar S 2011 *Nat. Commun.* **2** 595
[30] Yuce S, Emre B, Stern-Taulats E, Planes A, Mañosa L, Barrio M, Tamariti J L, Mudryk Y, Gschneidner K A Jr and Pecharsky V K 2012 *Appl. Phys. Lett.* **101** 071906
[31] Mañosa L, Planes A and Acet M 2013 *J. Mater. Chem.* A **1** 4925
[32] Moya X, Narayan S K and Mathur N D 2014 *Nat. Mater.* **13** 439
[33] Wu R R *et al* 2015 *Sci. Rep.* **5** 18027
[34] Mañosa L and Planes A 2017 *Adv. Mater.* **29** 1603607
[35] Li J, Dunstan D, Lou X, Planes A, Mañosa L, Barrio M, Tamarit J L and Lloveras P 2020 *J. Mater. Chem.* A **8** 20354
[36] Aznar A, Lloveras P, Barrio M, Negrier P, Planes A, Mañosa L, Mathur D, Moya and Tamarit J L 2020 *J. Mater. Chem.* A **8** 639
[37] Romanini M *et al* 2021 *Adv. Mater.* **33** 2008076
[38] de oliveira N A N A 2011 *J. Appl. Phys.* **109** 053315
[39] de oliveira N A 2013 *J. Appl. Phys.* **113** 033910
[40] Santana R P, de Oliveira N A and von Ranke P J 2011 *J. Phys.: Condens. Matter* **23** 306003
[41] Matsunami D, Fujita A, Takenaka K and Kano M 2015 *Nat. Mater.* **14** 73
[42] Gorev M V, Bogdanov E V and Flerov I N 2017 *J. Phys. D: Appl. Phys.* **50** 384002
[43] Gorev M V, Bogdanov E V and Flerov I N 2017 2017 *Scr. Mater.* **139** 53
[44] Flerov I N, Gorev M V, Kartashev A V, Pogorel'tsev E I and Laptash N M 2019 *J. Mater. Sci.* **54** 14287
[45] Gorev M V, Mikhaleva E A, Flerov I N and Bogdanov Evgeniy V 2019 *J. Alloys Compd.* **806** 1047

[46] He X, Kang Y, Wei S, Zhang Y, Cao Y, Xu K, Li Z, Jing C and Li Z 2018 *J. Alloys Comp.* **741** 821

[47] Li J, Barrio M, Dunstan D J, Dixey R, Lou Xi, Tamarit J L, Phillips A E and Lloveras P 2021 *Adv. Funct. Mater.* **31** 2105154

[48] Sau K, Ikeshoji T, Takagi S, Orimo S I, Errandonea D, Chu D and Cazorla C 2021 *Sci. Rep.* **11** 11915

[49] Liu J, Li Z, Liu H, Yu L, Zhang Y, Cao Y, Xu K and Liu Y 2022 *Metals* **12** 516

[50] Hao J *et al* 2020 *Chem. Mater.* **32** 1807

[51] Cazorla C 2019 *Appl. Phys. Rev.* **6** 041316

[52] Lloveras P and Tamarit J L 2021 *MRS Energy Sustain.* **8** 315

[53] Tušek J, Engelbrecht K, Solsona R M, Mañosa L, Vives E, Mikkelsen L P and Pryds N 2015 *Adv. Energy Mater.* **5** 1500361

[54] Schmidt M, Schüze A and Stefan S 2015 *Int. J. Refrig.* **54** 88

[55] Xiao F, Fukuda T, Tomoyuki K and Jin X 2015 *Acta Mater.* **87** 8

[56] Wu Y, Ertekin E and Sehitoglu H 2017 *Acta Mater.* **135** 158

[57] Cong D *et al* 2019 *Phys. Rev. Lett.* **122** 255703

[58] Zhang J, Xu Y, An S, Sun Y, Li X and Li Y 2020 *Chin. Phys.* B **29** 076202

[59] Chen J, Lei L and Fang G 2021 *Mater. Today Commun.* **28** 102706

[60] Kim D S, Kim B C, Han S H, Kang H W, Kim J S and Cheon C I 2019 *J. Appl. Phys.* **126** 234101

[61] Kimmel A V, Gindele O T, Duffy D M and Cohen R E 2019 *Appl. Phys. Lett.* **115** 023902

[62] Ye C, Wang J B, Li B and Zhong X L 2018 *Sci. Rep.* **8** 293

[63] Lu S G and Zhang Q 2009 *Adv. Mater.* **21** 1983

[64] Ožzbolt M, Kitanovski A, Tušek J and Poredoš A 2014 *Int. J. Refrig.* **40** 174

[65] Valant M 2012 *Prog. Mater. Sci.* **57** 980

[66] Mikhaleva E A, Flerov I N, Gorev M V, Molokeev M S, Cherepakhin A V, Kartashev A V, Mikhashenok N V and Sablina K A 2012 *Phys. Solid State* **54** 1832

[67] Liu Y, Wei J, Janolin P E, Infante I C, Lou X and Dkhil B 2014 *Appl. Phys. Lett.* **104** 162904

[68] Lloveras P *et al* 2015 *Nat. Commun.* **6** 8801

[69] Stern-Taulats E, Lloveras P, Barrio M, Defay E, Egilmez M, Planes A, Tamarit J L, Mañosa L, Mathur N D and Moya X 2016 *Appl. Mater.* **4** 091102

[70] Aznar A *et al* 2017 *Nat. Commun.* **8** 1851

[71] Sagotra A K, Chu D and Cazorla C 2018 *Nat. Commun.* **9** 3337

[72] Min J, Sagotra A K and Cazorla C 2020 *Phys. Rev. Mater.* **4** 015403

[73] Sandeman K G 2016 *Appl. Mater.* **4** 111102

[74] von Ranke P J 2017 *Appl. Phys. Lett.* **110** 181909

[75] von Ranke P J, Alho B P, Ribas R M, Nóbrega E P, Caldas A, de Sousa V S R, Colaço M V, Marques L F, Rocco D L and Ribeiro P O 2018 *Phys. Rev.* B **98** 224408

[76] von Ranke P J, Alho B P and Ribeiro P O 2018 *J. Alloys Compd.* **749** 556

[77] von Ranke P J, Alho B P, Nóbrega E P, Caldas A, de Sousa V S R, Colaço M V, Marques L F, Rocha G M, Rocco D L and Ribeiro P O 2019 *J. Magn. Magn. Mater.* **489** 165421

[78] von Ranke P J 2020 *J. Appl. Phys.* **127** 165104

[79] von Ranke P J, Alho B P and Ribeiro P O 2021 *Phys. Stat. Sol* B **258** 2100108

[80] Vallone S P, Tantillo A N, dos Santos A M, Molaison J J, Kulmaczewski R, Chapoy A, Ahmadi P, Halcrow M A and Sandeman K G 2019 *Adv. Mater.* **31** 1807334

[81] Reis M 2020 *Coord. Chem. Rev.* **417** 213357
[82] Seo J, Braun J D, Dev V M and Mason J A 2022 *J. Am. Chem. Soc.* **144** 6493
[83] Li B *et al* 2019 *Nature* **567** 506
[84] Lloveras P *et al* 2019 *Nat. Commun.* **10** 1803
[85] Li F B, Li M, Xu X, Yang Z C, Xu H, Jia C K, Li K, He J, Li B and Wang H 2021 *Nat. Commun.* **11** 4190
[86] Zhang K, Song R, Qi J, Zhang Z, Zhang Z, Yu C, Li K, Zhang Z and Li B 2022 *Adv. Funct. Mater.* **32** 2112622
[87] Liu Y, Zhou H, Xu Z, Liu D, Li J, Hu F and Ma T 2022 *Mater. Res. Lett.* **10** 675
[88] Gao Y *et al* 2022 *NPG Asia Mater.* **14** 34
[89] Usuda E O, Bom N M and Carvalho A M G 2017 *Eur. Polym. J.* **92** 287
[90] Carvalho A M G, Imamura W, Usuda E O and Bom N M 2018 *Eur. Polym. J.* **99** 212
[91] Miliante C M, Christmann A M, Usuda E O, Imamura W, Paixão L S, Carvalho A M G and Muniz A R 2020 *Macromolecule* **53** 2606
[92] Zhang C, Wang D, Qian S, Zhang Z, Liang X, Wu L, Long L, Shi H and Hand Z 2022 *Mater. Horiz.* **9** 1293
[93] Weerasekera N, Ajjarapu K P K, Sudan K, Sumanasekera G, Kate K and Bhatia B 2022 *Front. Energy Res.* **10** 887006
[94] Lin J, Tong P, Zhang K, Tao K, Lu W, Wang X, Zhang X, Song W and Sun Y 2022 *Nat. Commun.* **13** 596
[95] Miliante C M, Christmann A M, Soares R P, Bocca J R, Alves C S, Carvalho A M G and Muniz A R 2022 *J. Mater. Chem.* A **10** 8344
[96] Morellon L, Arnold Z, Magen C, Ritter C, Prokhnenko O, Skorokhod Y, Algarabel P A, Ibarra M R and Kamarad J 2004 *Phys. Rev. Lett.* **93** 137201
[97] Santana R P, de Oliveira N A and von Ranke P J 2011 *J. Alloys Comp.* **509** 6346
[98] Purwins H G and Lesson A 1990 *Adv. Phys.* **39** 309
[99] Stevens K W H 1952 *Proc. Phys. Soc. Lond., Sect.* A **65** 209
[100] Hutchings M T 1964 *Solid State Phys.* **16** 227
[101] de Oliveira N A 2004 *Eur. Phys. J.* B **40** 259
[102] de Oliveira N A and von Ranke P J 2005 *J. Phys.: Condens. Matter* **17** 3325
[103] de Medeiros L G, de Oliveira N A and Troper A 2008 *J. Appl. Phys.* **103** 113909
[104] de Medeiros L G, de Oliveira N A and Troper A 2010 *J. Magn. Magn. Mater.* **322** 1558
[105] de Medeiros L G, de Oliveira N A and Troper A 2010 *J. Alloys Comp.* **501** 177
[106] de Oliveira N A 2018 *Phys. Stat. Sol.* B **255** 1800218
[107] de Oliveira N A and Troper A 2019 *J. Alloys Comp.* **779** 335
[108] Hubbard J 1963 *Proc. R. Soc. Lond.* A **276** 238
[109] Fujita A, Fujieda S, Hasegawa Y and Fukamichi K 2003 *Phys. Rev.* B **67** 104416
[110] Sun Y, Arnold Z, Kamarad J, Wang G J, Shen B G and Cheng Z H 2006 *Appl. Phys. Lett.* **89** 172513
[111] Giner J, van der Rest J, Brouers F and Gautier F 1976 *J. Phys.* F **6** 1281
[112] Sarrao J L 1999 *Physica* B **259** 128
[113] Sarrao J L, Ramirez A P, Darling T W, Freibert F, Migliori A, Immer C D, Fisk Z and Uwatoko Y 1998 *Phys. Rev.* B **58** 409
[114] Kosugi Y *et al* 2021 *Adv. Funct. Mater.* **31** 2009476
[115] Wada H, Tomekawa S and Shiga M 1999 *Cryogenics* **39** 915

[116] Ridier K, Zhang Y, Piedrahita-Bello M, Quintero C M, Salmon L, Molnár G, Bergaud C and Bousseksou A 2020 *Adv. Mater.* **32** 2000987

[117] Bari R A and Sivardière J 1972 *Phys. Rev.* **5** 4466

[118] Gindulescu A, Rotaru A, Linares J, Dimian M and Nasser J 2011 *J. Phys.: Conf. Ser.* 268 012007

[119] Zimmermann R and König E 1977 *J. Phys. Chem. Solids* **38** 779

[120] Zimmermann R 1983 *J. Phys. Chem. Solids* **44** 151

[121] Soray M and Seki S 1974 *J. Phys. Chem. Solids* **35** 555

[122] Dove M T and Fang H 2016 *Rep. Prog. Phys.* **79** 066503

[123] Barrera G D, Bruno J A O, Barron T H K and Allan N L 2005 *J. Phys.: Condens. Matter* **17** R217

[124] Pryde A K A, Hammonds K D, Dove M T, Heine V, Gale J D and Warrenk M C 1996 *J. Phys.: Condens. Matter* **8** 10973

[125] Wang L, Peng-Fei Y, Wang F, Sun Q, Liang E J and Jia Y 2012 *Mater. Res. Bull.* **47** 1113

[126] Wang J, Xu P, Yuan H, Gao Q, Sun Q and Liang E 2020 *Results Phys.* **19** 103531

[127] Sanson A, Rocca F, Dalba G, Fornasini P, Grisenti R, Dapiaggi M and Artioli G 2006 *Phys. Rev.* B **73** 214305

[128] Rimmer L H N, Dove M T, Winkler B, Wilson D J, Refson K and Goodwin A L 2014 *Phys. Rev.* B **89** 214115

[129] Romao C P 2017 *Phys. Rev.* B **96** 134113

[130] Dove M T, Du J, Wei Z, Keen D A, Tucker M G and Phillips A E 2020 *Phys. Rev.* B **102** 094105

[131] Li B *et al* 2016 *Phys. Rev.* B **93** 224405

[132] Hu F, Shen F, Hao J, Liu Y, Wang J, Sun J and Shen B 2018 *Frontiers in Chem.* **6** 438

[133] Li C W, Tang X, Muñoz J A, Keith J B, Tracy S J, Abernathy D L and Fultz B 2011 *Phys. Rev. Lett.* **107** 195504

[134] Morse P 1929 *Phys. Rev.* **34** 57

[135] Timmermans J 1961 *J. Phys. Chem. Solids* **18** 1

[136] Kaneko Y and Sorai M 2007 *Ph. Transit.* **80** 517

Chapter 4

Barocaloric effect in ferroelastic fluorides and oxyfluorides

Igor N Flerov, Evgeniy V Bogdanov and Mikhail V Gorev

4.1 Introduction

Intensively increasing attention of the scientific and engineering community is attracted by caloric effects (CEs) in solids, which are reversible changes in entropy or temperature under the action of an external field in isothermal and adiabatic conditions, respectively. It is generally recognized that one of the most promising areas for the search and development of new highly efficient methods of cooling is associated with the intensification of studies of CEs in solids undergoing phase transitions of varying physical nature [1–5]. In accordance with the nature of the external field and the conjugated order parameter, electro (ECE)-, magneto (MCE)-, baro (BCE)—and piezo (PCE)—caloric effects are distinguished. For many years, most attention was paid to the first two effects, which led, firstly, to the appearance of remarkable reviews devoted to MCE [5–11] and ECE [12, 13], and, secondly, to the development and implementation of models aimed at the promising use of caloric materials in refrigerators, heat pumps, energy collection and storage devices, medical devices [6, 14, 15].

Over the past decade, interest in BCE and PCE has grown noticeably and continues to grow due to their significant advantage over other CEs [15, 16]. Indeed, both effects are universal, since phase transitions of varying physical nature are very often accompanied by significant changes in linear or bulk deformation, which, in accordance with the Maxwell relations $(\partial S/\partial p)_T = -(\partial V/\partial T)_p$ and, assuming that the compressive stress corresponds to a negative value of σ, $(\partial S/\partial \sigma)_T = (\partial L/\partial T)_\sigma$ are the defining parameters of the baro- and piezocaloric efficiency. Moreover, a comparative analysis showed that materials demonstrating CE of a mechanical nature are the most promising for use as working bodies in refrigerators and heat pumps [16, 17].

doi:10.1088/978-0-7503-4690-0ch4

To implement significant caloric parameters, one of two ways is usually used. The first is related to the search and design of new effective caloric materials [18–24]. The second one is aimed at optimizing the properties of known materials demonstrating noteworthy CEs of varying physical nature [25–32]. In the first case, it is necessary to note the recent discovery of a giant BCE in plastic crystals [18] and perovskites with organic anions [22]. A remarkable illustration of the second approach is the study of PCE in shape memory alloys, which have been very actively carried out over the past few years [21, 27, 33–38]. However, despite the significant caloric parameters of plastic crystals and organic perovskites, the gigantic temperature hysteresis of phase transitions significantly complicates the implementation of an effective reversible cooling cycle [18, 27].

Below we provide a brief overview of extensive and intensive barocaloric parameters and coefficients in relation to the degree of structural disorder and its changes during phase transitions in some fluoro- and oxyfluorometallates containing octahedral or quasi-octahedral anionic groups.

Among substances with significant BCE, complex fluorides and oxyfluorides very often demonstrate high efficient caloric characteristics [25, 34]. These compounds have as usual a strongly disordered initial cubic or orthorhombic phase and, upon cooling, undergo single or successive order–disorder phase transitions close to the tricritical point, and accompanied by a large change in entropy. Another feature of many of these materials is complex T–p phase diagrams with triple points and phase boundaries with different signs of dT/dp, which allows one both conventional and inverse BCE to be implemented in one material in fairly narrow pressure and temperature ranges [25].

The chemical pressure associated with cation–anion substitutions also turned out to be an effective tool for varying the characteristics of phase transitions (temperature, hysteresis, entropy, deformation, sensitivity to external pressures, etc) and, as a consequence barocaloric efficiency parameters.

4.2 Preliminary estimation of barocaloric parameters

Let us consider how some parameters of phase transitions affect the values and features of the temperature and pressure behavior of extensive, $\Delta S_{BCE}(T, p)$, and intensive, $\Delta T_{AD}(T, p)$, BCE in fluorides and oxyfluorides.

Regardless of the order of phase transitions (first or second) and the degree of proximity to the tricritical point, their main characteristics are: temperature, T_0, and its sensitivity to pressure dT_0/dp, the value and temperature dependence of entropy, $\Delta S_0(T)$, coefficient of volumetric thermal expansion of the crystal lattice, β_{Lat}, and others. For transformations of the first kind, the temperature hysteresis, δT_0, is also an important parameter.

The availability of information even on the approximate values of some of the above characteristics makes it possible to estimate the possible values of ΔS_{BCE} and ΔT_{AD} and, thus, to select the most promising materials for detailed studies.

4.2.1 Maximum possible values of BCE at phase transitions

In the case of a single phase transition of the first order, the maximum extensive BCE, ΔS_{BCE}, related only to the transformation (i.e., associated with the appearance of order parameter) is limited by entropy of the phase transition ΔS_0 [39] (figure 4.1).

The maximum value of intensive BCE, $\Delta T_{AD} = -T_0\Delta S_0/C_{Lat}$, is observed at pressure $p > p_{min}$, where C_{Lat} is the heat capacity of the crystal lattice. The minimum pressure p_{min} that produces maximum BCE values is determined by the expression

$$p_{min} = \Delta T_{AD}^{max}/(dT_0/dp) = (T_0\Delta S_0)/(C_{Lat}dT_0/dp). \quad (4.1)$$

The maximum value of the intensive BCE ΔT_{AD}^{max} is also limited by the ΔS_0 value, and depends also on the $(\partial S_{Lat}/\partial T)_p$ derivative, i.e. on the temperature region where phase transition takes place. Figures 4.1(c) and (d) show that maximum values of ΔS_{BCE} and ΔT_{AD}, as a rule, are realized at different pressures.

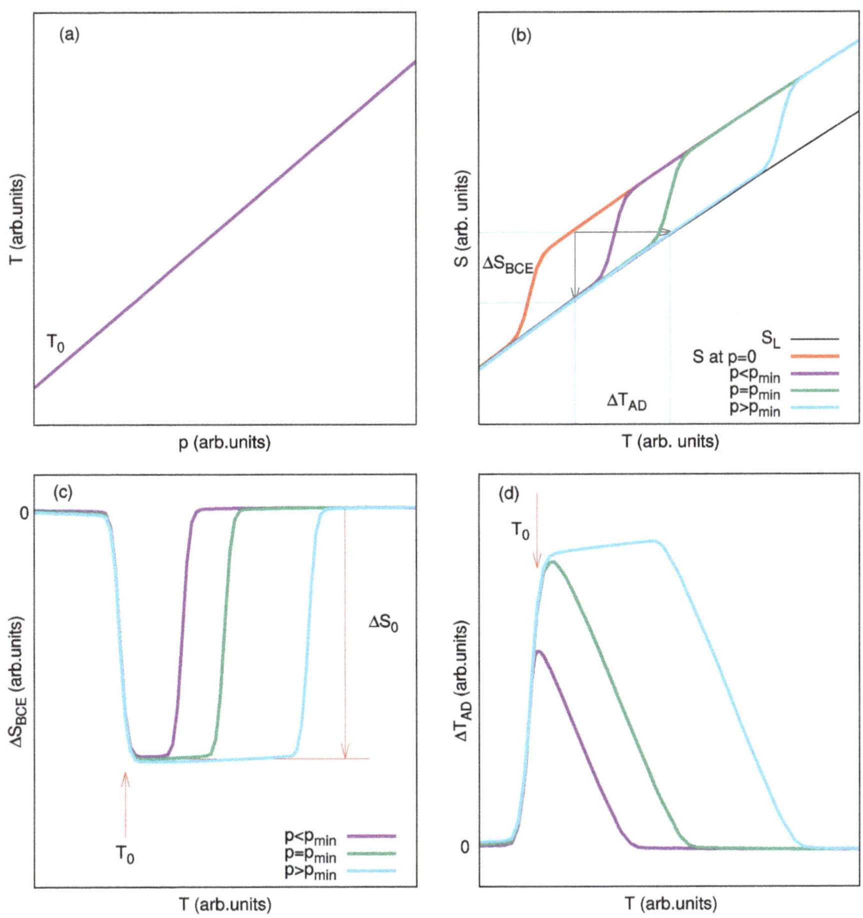

Figure 4.1. Schematic presentation of the *T–p* phase diagram for single phase transition (a) and related temperature dependences of extensive (b) and intensive (c) BCE at different pressures.

4.2.2 BCE and features of phase *T–p* diagrams

BCE studies are carried out on materials undergoing a single phase transition with a large change in entropy. At the same time, materials with successive phase transitions or having triple points on the *T–p* phase diagram are also of interest.

The BCE behavior of some fluorides and oxyfluorides near triple points was analyzed in [25, 40, 41]. The baric coefficients, dT/dp, for adjacent transformations in the same crystal may have the same or different signs. This means that BCE can be conventional and/or inverse in the same material at different temperatures and/or pressures. The *T–p* phase diagrams of some compounds also show triple points initiated by pressure at $p = p_{trp}$, which are associated with splitting or merging of phase boundaries with increase in pressure. So there is the possibility to change the magnitude and sign of BCE in narrow temperature and pressure ranges around the triple points.

One variant of the *T–p* phase diagram is possible in materials undergoing two successive phase transitions: with the close values of dT_1/dp and dT_2/dp characterized by the same (positive or negative) sign. In this case, BCE rises in the temperature range from $T_1(p > p^*)$ to $T_2(p = 0)$, where $p^*=(T_1 - T_2)/(dT_1/dp)$, and ΔS_{BCE}^{max} can reach a value equals to $\Delta S_1 + \Delta S_2$ and remain constant at further increase in pressure. However, the intensive BCE achieves ΔT_{AD}^{max} at higher pressure (figure 4.2).

The presence of one triple point in the *T–p* phase diagram can be associated with two variants of the BCE behavior (figure 4.3). In the first case (figure 4.3(a)), three characteristic pressure regions exist. At $T_1(p = 0) < T < T_{trp}$, increase in pressure $p \to p_{trp}$ is accompanied by an increase of extensive and intensive effects up to maximum values $\Delta S_{BCE1}^{max} = \Delta S_1$ and $\Delta T_{AD1}^{max} = -T\Delta S_1/C_{Lat}$, when $p_{min\,1} < p_{trp}$. At a pressure just above p_{trp}, the temperature region of existence of BCE associated with ΔS_1 narrows and at the same time, BCE associated with ΔS_2 appears (figure 4.3(b) and (c)).

In accordance with $dT_2/dp < 0$, at $p > p^*$ and $T < T_1(p = 0)$, the inverse BCE appears, which very quickly reaches the maximum values ΔS_{BCE} and ΔT_{AD}. Caloric effects at this pressure in the temperature ranges $T < T_1(p = 0)$ and $T > T_1(p = 0)$

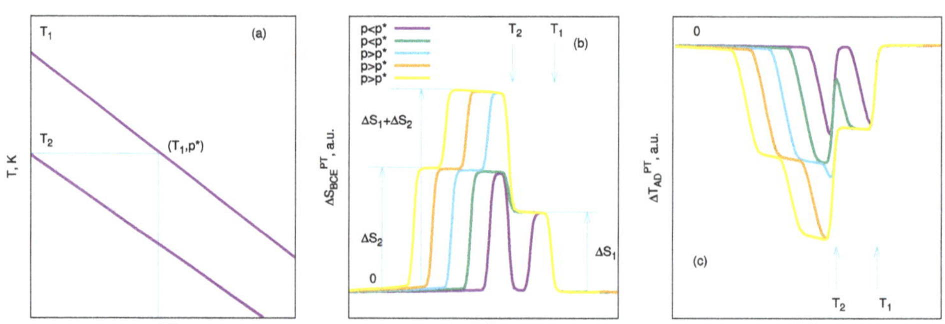

Figure 4.2. Schematic presentation of the *T–p* phase diagram for successive phase transitions (a) and related temperature dependences of extensive (b) and intensive (c) BCE at different pressures.

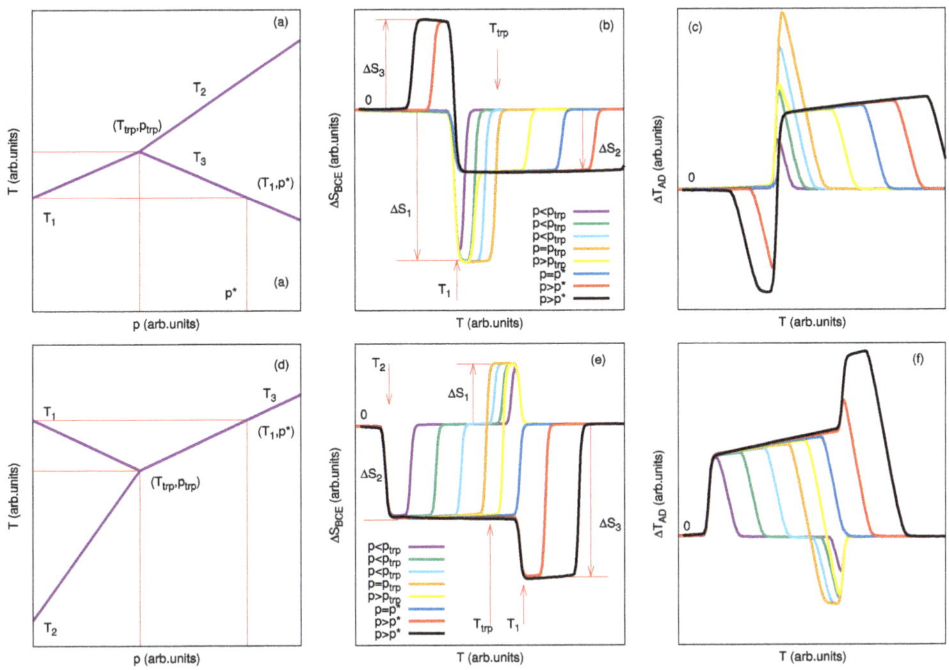

Figure 4.3. Schematic presentation of the T–p phase diagrams with triple points ((a) and (d)) and related temperature dependences of extensive ((b) and (e)) and intensive ((c) and (f)) BCE at different pressures.

depend on the parameters of phase transitions that occur at $p > p_{trp}$ and at higher pressures, $p > p_{trp} + p_{min\,2}$, $p > p_{trp} + p_{min\,3}$, they reach maximum values.

In the situation presented in figure 4.3(e) and (f), extensive and intensive BCE associated with the distinct phase transitions at $p < p_{trp}$ are observed below $T_1(p = 0)$ and above $T_2(p = 0)$. These two temperature regions expand under pressure up to the triple point, and both BCEs can achieve maximum values when p_{trp} is lower than $p_{min\,1}$ and $p_{min\,2}$. At $p > p_{trp}$, the temperature region with conventional BCE continues to expand. The range of existence of inverse BCE narrows and disappears at $p = p^*$. Above $T_1(p = 0)$, further increase in pressure leads to the appearance of BCE with the following values: $\Delta S_{BCE} = \Delta S_1 + \Delta S_2$ and $\Delta T_{AD} = -T\Delta S_{BCE}/C_{Lat}$.

The schematic phase diagrams considered above do not demonstrate all the possible variants of the dependences $T(p)$. However, their analysis is very important and informative because it allows one to see the qualitative behavior of extensive and intensive BCE in conventional and inverse variants at successive structural transformations and to carry out preliminary evaluations of the range of the p and T parameters associated with the most pronounced extensive and intensive BCE.

4.2.3 Thermal hysteresis of phase transition and reversibility of BCE

Thermal hysteresis during first-order phase transitions can prevent the system from recovering to its original state after completion of the thermodynamic cycle. Overcoming this obstacle introduces the minimum pressure threshold required to obtain reversible BC effects, p_{rev}, which leads to excessive input work and, hence, to a subsequent decrease in performance.

Therefore, materials demonstrating transitions with small hysteresis are highly preferable. Using the known values of δT_0 and dT_0/dp, we can determine the minimum pressure to achieve reversible BCEs $p_{rev} = \delta T_0/dT_0/dp$ [18].

In the case of a positive value of dT_0/dp, endothermic transitions occur on heating or pressure reduction, while exothermic transitions occur both on cooling and on pressure increase. This means that the irreversible changes in isothermal entropy that occur when the pressure changes between $p = 0$ and a given pressure $p \neq 0$ are calculated as

$$\Delta S(T, p \to 0) = S_H(T, 0) - S_H(T, p) \tag{4.2}$$

$$\Delta S(T, 0 \to p) = S_C(T, p) - S_C(T, 0) \tag{4.3}$$

$$\Delta T(T_S, p \to 0) = T(S_H, 0) - T_S(S_H, p) \tag{4.4}$$

$$\Delta T(T_S, 0 \to p) = T(S_C, p) - T_S(S_C, 0), \tag{4.5}$$

where S_H, S_C—entropy in heating and cooling processes at constant pressure, T_S—starting temperature of process.

4.3 Experimental determination of barocaloric effect

In contrast to MCE and ECE, direct measurements of the intensive and extensive barocaloric effects are associated with a number of experimental difficulties.

There are various methods for measuring intensive caloric effects of different physical nature. However, the most reliable instrument for this purpose is an adiabatic calorimeter, which makes it possible to control and minimize heat exchange with the environment, providing adiabatic conditions, S = const. From calorimetric studies, in principle, one can also obtain information about the extensive caloric effect by measuring the heat capacity of materials as a function of temperature and the field of the corresponding nature.

Unfortunately, the adiabatic calorimeter is practically unsuitable for BCE studies, since in this case it is necessary to use a massive high-pressure chamber, the heat capacity of which is many times greater than the heat capacity of the sample under study. As a consequence, there is a significant loss in the accuracy of determining both the heat capacity and entropy of the sample, which makes the determination of $\Delta T_{AD}(p)$ and $\Delta S_{BCE}(p)$ practically impossible. Measurements of entropy and its changes under pressure by the DSC method, as a rule, are carried out at sufficiently low pressures and in the absence of the condition S = const.

Alternative indirect methods based on the use of Maxwell's equations are often used to determine extensive MCE and ECE from data on the dependences of phase transition parameter (magnetization, polarization) on the temperature and the strength of the conjugate field (magnetic and electrical). For precise calculations of BCE using this strategy, it is necessary to obtain information about the equation of state (dependence of volume on temperature and the pressure), for example, by high-pressure diffraction methods. Unfortunately, such studies also represent a complex experimental problem and have not yet been carried out for the complex fluorides and oxyfluorides under study. To determine the extensive, ΔS_{BCE}, and intensive, ΔT_{AD}, BCE in these new compounds, we used a previously developed method based on the analysis of the experimental heat capacity at $p = 0$ and the T–p phase diagram [42, 43].

Information about the dependences of the barocaloric entropy, ΔS_{BCE}, on temperature and pressure was obtained by analyzing the total entropy of crystals at $p = 0$ and $p \neq 0$ under isothermal conditions:

$$\Delta S_{BCE}(T, p) = S(T, p) - S(T, 0). \tag{4.6}$$

The baric and temperature behavior of the intensive BCE with pressure change, ΔT_{AD}, was determined using the following relations:

$$S(T + \Delta T_{AD}, p) = S(T, 0) \tag{4.7}$$

or

$$\Delta T_{AD}(T, p) = T(S, p) - T(S, 0). \tag{4.8}$$

The total entropy of solids, $S(T, p)$, can be represented as the sum of the entropy contributions of individual subsystems: the crystal lattice, $S_{Lat}(T, p)$, the anomalous component associated with phase transitions, $\Delta S_0(T, p)$, electrons, nuclear spins, etc. Since the compounds under study are characterized mainly by ionic bonds, we can assume that at rather low pressures ($p < 1$ GPa), usually used in experiments, all contributions to the entropy except $S_{Lat}(T, p)$ and $\Delta S_0(T, p)$ do not change significantly with pressure and are not considered further.

4.3.1 Entropy at $p = 0$, $S(T, 0)$

Total entropy of compounds $S(T, 0)$, as well as anomalous, $\Delta S_0(T, 0)$, and lattice, $S_{Lat}(T, 0)$, contributions were determined by analyzing experimental heat capacity data $C_p(T, 0)$, obtained at $p = 0$ using adiabatic calorimeter. Information on C_{Lat} was obtained by fitting the experimental data $C_p(T, 0)$, taken at $T < T_i$ and $T > T_i$ far from the region of the phase transitions, using combination of Debye's and Einstein's functions: $C_{Lat}(T) = A_1 D(\Theta_D/T) + A_2 E(\Theta_E/T)$. Excess heat capacity was determined as the difference $\Delta C_p(T) = C_p(T) - C_{Lat}(T)$,

$$S(T, 0) = \int [C_p(T, 0)/T]dT = S_{Lat}(T, 0) + \Delta S_0(T, 0) = \int [C_{Lat}(T, 0)/T]dT + \int [\Delta C_p(T, 0)/T]dT. \tag{4.9}$$

4.3.2 Lattice entropy change with pressure

A change in the thermal expansion of a crystal lattice under pressure can lead to a change in the total entropy of solids and, in some cases, play a significant role in the formation of the barocaloric effect. This follows from the Maxwell equation $(\partial S_{Lat}/\partial p)_T = -(\partial V_{Lat}/\partial T)_p$, which shows that the change in isothermal entropy is proportional to the volumetric thermal expansion coefficient β_{Lat},

$$\Delta S_{\mathrm{Lat}}(T,\,p) = -\int \partial V_{\mathrm{Lat}}/\partial T dp \approx -V_{\mathrm{m}}\beta_{\mathrm{Lat}}(T)p. \tag{4.10}$$

Here we assume that the molar volume, V_{m}, and the coefficient $\beta_{\mathrm{Lat}}(T)$ weakly depend on pressure. Thus, if both of these parameters are characterized by large values, the change in entropy under pressure can also be large, even in materials without phase transitions.

4.3.3 Anomalous entropy change with pressure

The effect of hydrostatic pressure on the behavior of anomalous entropies has not been determined experimentally. However, taking into account the results of calorimetric studies by differential thermal analysis (DTA) under pressure for a wide range of compounds [40, 43–45], it can be assumed that the phase transition entropy, ΔS_0, which consists of a temperature-dependent part and, in the case of transformations of the first order, the jump δS_0 at T_0, as well as the degree of proximity of the phase transition to the tricritical point, are practically independent of pressure.

The temperature dependence of the anomalous entropy at various pressures, $\Delta S_0(T,\,p)$, was determined by the shift of the function $\Delta S_0(T,\,0)$ along the temperature scale in accordance with the baric coefficients of individual phase transitions dT_0/dp:

$$\Delta S_0(T,\,p) = \Delta S_0(T + pdT_0/dp,\,0). \tag{4.11}$$

Significant changes in the remaining components of the total entropy associated with electrons, nuclear spins, etc are most likely absent in the studied ranges of pressures and temperatures. Therefore, the temperature dependences of the total entropy at different pressures were determined by summing the entropies of the lattice and the anomalous contributions:

$$S(T,\,p) = S_{\mathrm{Lat}}(T,\,0) + \Delta S_{\mathrm{Lat}}(T,\,p) + \sum \Delta S_i(T,\,p). \tag{4.12}$$

4.3.4 Refrigerant capacity (RC) or relative cooling power (RCP)

Integral parameters are used to assess and compare the suitability of caloric materials for use as solid-state refrigerants. One of them is the refrigerant capacity (RC) or relative cooling power (RCP) [46]. These two quantities take into account the area below the peak associated with the behavior of the function $\Delta S_{\mathrm{BCE}}(T)$ at constant pressure and are defined as $\Delta S_{\mathrm{max}}\times$(FWHM), where FWHM is full width at half maximum. Both RC and RCP refer to the amount of heat, which can be transferred between cold and hot reservoirs in a cycle. However, these quantities do not take into account the input work required to control the cycle and therefore an additional useful parameter is the cooling performance factor CRP (coefficient of refrigeration performance), which can be roughly calculated for BCE as CRP $= (\Delta S \times \Delta T_{\mathrm{rev}})/(p\Delta V_0/2)$, where ΔT_{rev} refers to reversible adiabatic temperature changes and ΔV_0 is the change in volume during the transition [46].

It is not possible to correctly estimate the RC or RCP parameters for all compounds analyzed in this chapter, since many of them exhibit both conventional and reverse caloric effects. There is quite a lot of arbitrariness in determining the width of the peak maximum of the BCE. However, to estimate CRP values, we can use the relation:

$$CRP \approx (\Delta S_{BCE} \times \Delta T_{rev})/(p\Delta V_0/2) \approx 2(\Delta S_{BCE} \times \Delta T_{rev})/(p\Delta S_0 \times dT_0/dp). \quad (4.13)$$

4.4 Barocaloric effect in some fluorides and oxyfluorides

In this section, we will consider two groups of compounds. The first group includes the compounds Rb_2KFeF_6, Rb_2KTiOF_5, and $(NH_4)_2SnF_6$, which undergo one phase transition and are characterized by different signs of the baric coefficients dT/dp and, correspondingly, conventional and reverse BCEs. Compounds of the second group, $(NH_4)_2MoO_2F_4$, $(NH_4)_2WO_2F_4$, $(NH_4)_2NbOF_5$, demonstrate successive phase transitions, for which the influence of features of the complex phase T–p diagrams on the BCE is considered.

Complex fluorides and oxyfluorides are of undoubted interest as barocaloric materials due to a large number of their remarkable properties that manifest themselves at ambient conditions. It is known that some fairly wide crystallographic families of fluorides and oxide-fluorides can significantly change the symmetry of the structure, undergoing phase transitions with temperature and/or pressure changes [47–49].

A detailed study of phase transitions makes it possible to reveal the relationship between the chemical composition, structure and physical properties, which is key knowledge in developing an approach to the targeted creation of materials with desired properties that operate under certain environmental conditions [49].

4.4.1 Structural types, phase transitions and BCE in fluorides and oxyfuorides

One of the important features of compounds containing fluorine and/or mixed fluorine–oxygen ligands combined into six-, seven-, and eight-coordinated anionic polyhedra is related to the possibility of the formation of different crystal lattice symmetries in the initial phase: cubic, orthorhombic, etc.

Fluorides with the general chemical formula $AMF_3/A_2A'MF_6$ often crystallize in cubic form with a perovskite/double perovskite structure (sp. gr. $Pm\bar{3}m$, $Z = 1$ or $Fm\bar{3}m$, $Z = 4$), formed by vertex-linked MF_6 octahedra.

The cubic symmetry of oxyfluorides $A_2A'MO_xF_{6-x}$ is due to the static/dynamic disorder of the fluorine–oxygen ligands in the anion polyhedron, and/or such a relative orientation of the six-coordinated units located next to each other, which leads to the neutralization of the dipole moments of individual pseudo-octahedra. Both types of disorder are responsible for the fact that oxyfluorides, despite polar anions, quite often retain a nonferroelectric structure even in distorted phases resulting from phase transitions.

Materials with a perovskite-like crystal lattice have considerable potential to be disordered due to their great structural flexibility [50–52].

Among the oxyfluorides, there are also compounds $A(A')MO_x F_{6-x}$ ($x = 1, 2, 3$) and $A_2MO_xF_{6-x}$ ($x = 1, 2$) containing isolated octahedra and crystallizing in a rather low local symmetry (trigonal, tetragonal, orthorhombic) [53–55]. Additional disordering of any type of structure of fluorides and oxyfluorides can be achieved by replacing the atomic cation A^+ with the NH_4^+ tetrahedral molecular cation.

The high sensitivity of complex fluorides and oxyfluorides to variations in external (temperature, pressure) and internal (size and shape of ions, cation–anion substitution) parameters makes it possible to trace the physical nature and mechanism of changes in the stability of crystallographic phases, which arise as a result of single or successive phase transitions.

Lattice distortions caused by structural transformations are often considered in terms of two limiting mechanisms, namely displacement and/or order–disorder type mechanisms, which are characterized by a rather large difference in entropy change, ΔS: $\sim 0.1R$ and $\geqslant R \ln 2$, respectively (R is the gas constant). However, among fluorides and oxyfluorides there are a number of compounds in which the mechanism of transformations can be changed using internal chemical pressure.

The structure of fluorides and oxyfluorides with isolated octahedra can be disordered due to both octahedral and/or ammonium tetrahedral sub-subsystem. Significant entropy changes and volume jumps can be observed as a result of one and/or a sequence of phase transformations.

DTA studies under pressure of a number of complex fluorides and oxyfluorides have shown that, on the one hand, the temperatures of phase transitions are shifted to different degrees, and, on the other hand, the entropy of phase transitions practically does not change with increasing pressure [40, 43, 45].

Thermodynamic parameters of the phase transitions in the compounds under study are presented in table 4.1. As can be seen, for a number of materials, large changes in the entropy, significant shifts in the transition temperature under pressure, dT_0/dp, and

Table 4.1. Structural and thermodynamic parameters of the phase transitions in fluorides and oxyfluorides.

Compound	*PT*	T_0, K	δT_0, K	dT/dp, K GPa^{-1}	ΔS_0, J kg K^{-1}	$\Delta V_0/V$, %	Reference
Rb_2KTiOF_5	$Fm\bar{3}m$–$I4/m$	215	6.1	109.6	46.0	1.2	[43, 56, 57]
$(NH_4)_2SnF_6$	$P\bar{3}m1$–$P\bar{1}$	110	0.5	−157	61.0	−1.0	[44]
Rb_2KFeF_6	$Fm\bar{3}m$–$I4/m$	198	4.6	132	41.1	1.8	[45]
$(NH_4)_2NbOF_5$	$Cmc2_1$–$C2$	258	0.6	−45	90.0	−1.1	[26, 57, 58]
	$C2$–Ia	219	1.0	−45	69.2		
$(NH_4)_2MoO_2F_4$	$Cmcm$–$Pnma$	271	0.9	96	75.9	1.8	[59, 60]
	$Pnma$–?	180		17	7.1		
$(NH_4)_2WO_2F_4$	$Cmcm$–$P\bar{1}$	201	1.4	13.4	58.0	0.3	[60]
	$P\bar{1}$–?	161		41.7	4.3		

a small temperature hysteresis, δT_0, are observed. All this makes fluorides and oxyfluorides suitable for exhibiting significant barocaloric efficiency.

4.4.2 Ordering in the octahedral subsystem

Cubic in the initial phase, Rb_2KMF_6 elpasolites with large M cations (Sc, In, Lu, rare earth elements) undergo successive ferroelastic phase transitions $Fm\bar{3}m \leftrightarrow I4/m \leftrightarrow P12_1/n1$ upon cooling [61]. The corresponding structural distortions are considered to be associated with small-angle rotations of the MF6 octahedra. It has been established that an increase in the total entropy change during phase transitions from $0.5R$ (Sc) to $1.1R$ (Ho) with an increase in the ionic radius of M atoms correlates with an increase in the degree of anharmonicity of vibrations of critical F atoms.

A decrease in the size of the M: Fe, Cr, Ga ion leads to the appearance of one pronounced ferroelastic transition of the first order, associated with both the ordering of the octahedra and the displacement of the A^+ cation, and is accompanied by a significant change in the entropy $\Delta S \approx 2R$ [61, 62].

The substitution of the anionic group MF_6^{3-} for $TiOF_5^{3-}$ led to the presence of one phase transition $Fm\bar{3}m \leftrightarrow I4/m$ in the oxyfluoride Rb_2KTiOF_5 [56]. According to structural and IR spectral studies, the cubic phase of Rb_2KTiOF_5 is disordered in the octahedral subsystem, while the tetragonal phase is completely ordered, which corresponds to the entropy change $\Delta S = 17.0$ J mol^{-1}K$^{-1} \approx 2R$.

$A_2A'MO_xF_{6-x}$ compounds with linked octahedra in the structure show a significant susceptibility of transition temperatures from the $Fm\bar{3}m$ phase and a positive sign of the pressure coefficient dT/dp. The loss of stability of the cubic phase is affected by the variation of the anionic group and/or the set of cations. An increase in external pressure leads to an increase in the phase transition temperature at a rate of $dT/dp = 132$ K GPa^{-1} and $dT/dp = 110$ K GPa^{-1} in the Rb_2KFeF_6 and Rb_2KTiOF_5 compounds, respectively (figure 4.4).

4.4.2.1 BCE in Rb_2KFeF_6, Rb_2KTiOF_5

The related compounds Rb_2KFeF_6 and Rb_2KTiOF_5 undergo single phase transition of the first order from the cubic to the tetragonal phase $Fm\bar{3}m \leftrightarrow I4m$ and are characterized by a large change in entropy ΔS_0 and a significant shift in the transition temperature under pressure dT/dp. Figures 4.5 and 4.7 show the $\Delta S_{BCE}(T, p)$ and $\Delta T_{AD}(T, p)$ dependences corresponding to the processes of pressure application and removal, and figures 4.6 and 4.8 demonstrate the maximum BCE values as a function of pressure. Because of the strong first-order phase transition in Rb_2KFeF_6, the barocaloric contributions, ΔS_{BCE}^{PT} and ΔT_{AD}^{PT}, associated with the appearance/disappearance of the order parameter, reach their maximum values already at 0.1 GPa. In Rb_2KTiOF_5, this occurs at a higher pressure, 0.2 GPa, due to the closer proximity of the phase transition to the tricritical point. This feature significantly affects also the difference in the temperature behavior of the extensive and intensive BCEs in both elpasolites. The minimum pressure to achieve reversible BC effects is low, $p_{rev} \approx 0.04$–0.06 GPa.

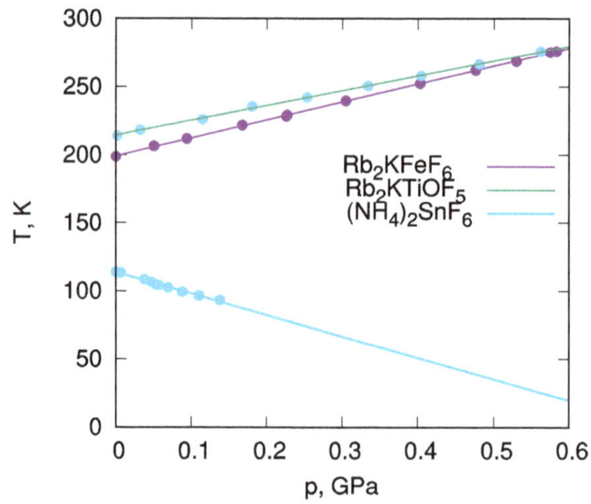

Figure 4.4. Transition temperature as a function of hydrostatic pressure for Rb_2KFeF_6, Rb_2KTiOF_5 and $(NH_4)_2SnF_6$.

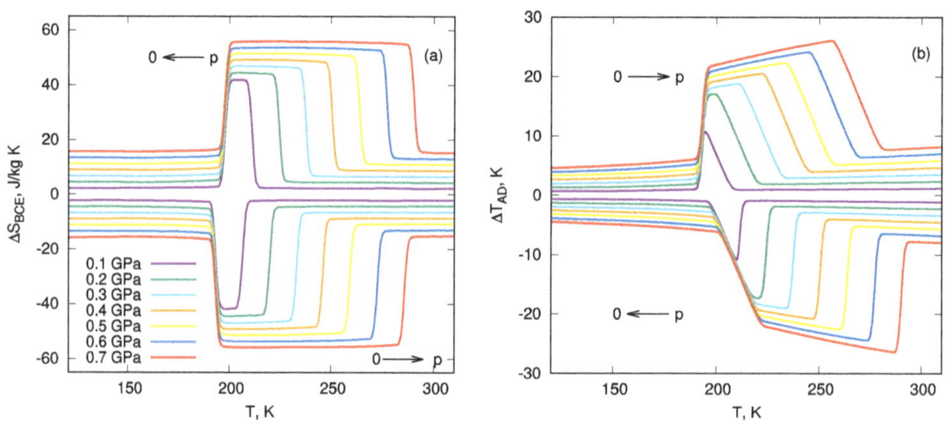

Figure 4.5. Extensive (a) and intensive (b) BCE in heating (pressure removal) and cooling (pressure application) modes in Rb_2KFeF_6.

4.4.3 Ordering of ammonium ions

At room temperature, the $(NH_4)_2SnF_6$ structure (sp. gr. $P\bar{3}m1$, $Z = 1$), which consists of isolated $[SnF_6]^{2-}$ octahedra and $[NH_4]^+$ tetrahedra, is disordered due to three equivalent orientations of two ammonium groups. The ordering of the structure occurs as a result of the phase transition $P\bar{3}m1 \leftrightarrow P\bar{1}(T_0 = 110\ \text{K})$, accompanied by a change in the entropy $\Delta S \approx R \ln 9$, corresponding to the ordering in one position of both NH_4 groups [44]. An increase in pressure leads to a decrease in the phase transition point. The boundary between $P\bar{3}m1$ and $P\bar{1}$ phases can be described by a significantly large negative baric coefficient $dT_0/dp = -157\ \text{K GPa}^{-1}$ (figure 4.4).

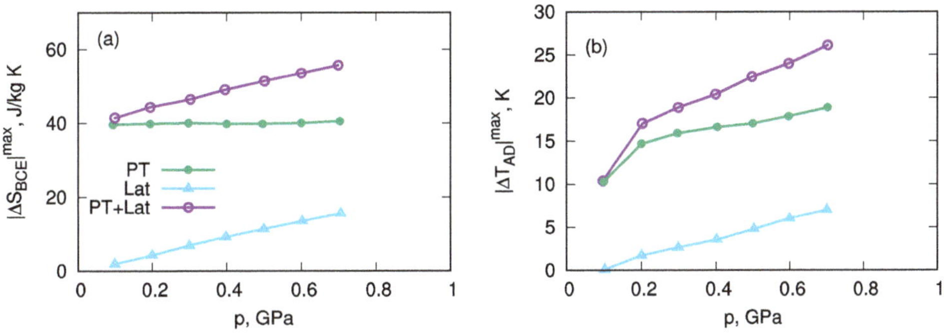

Figure 4.6. Pressure dependences of maximum values of extensive (a) and intensive (b) BCE in Rb_2KFeF_6.

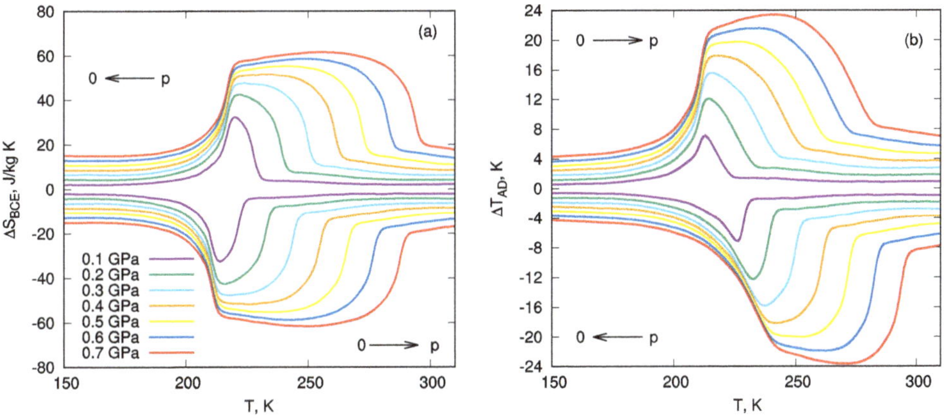

Figure 4.7. Extensive (a) and intensive (b) BCE in heating (pressure removal) and cooling (pressure application) modes in Rb_2KTiOF_5.

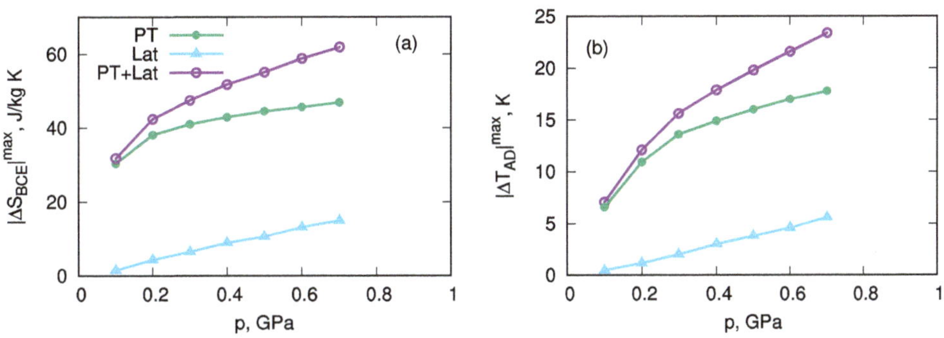

Figure 4.8. Pressure dependences of maximum values of extensive (a) and intensive (b) BCE in Rb_2KTiOF_5.

4.4.3.1 BCE in $(NH_4)_2SnF_6$

In accordance with the negative volume change near the phase transition point (figure 4.4, table 4.1), BCE in $(NH_4)_2SnF_6$ is inverse, i.e., $\Delta T_{AD} < 0$ and $\Delta S_{BCE} > 0$. A very important point is that the maximum BCE values ($\Delta S_{BCE}^{max} \approx 60$ J kg K^{-1} and

$\Delta T_{AD}^{max} \approx 11$ K) can be achieved at low pressures of 0.05–0.1 GPa (figures 4.9 and 4.10). The contributions to the extensive and intensive BCE from changes in the lattice entropy in the pressure range up to 0.2 GPa are insignificant and reach 4 J $kg^{-1}K^{-1}$ and 1 K. Figure 4.9 shows the dependences $\Delta S_{BCE}(T, p)$ and $\Delta T_{AD}(T, p)$ corresponding to the processes of applying and removing pressure for $(NH_4)_2SnF_6$. Since the phase transition in this fluoride is characterized by small value of temperature hysteresis, $\delta T \approx 0.5$ K, the minimum pressure required to achieve reversible effects is very low, $p_{rev} \approx 0.003$ GPa.

4.4.4 Structural ordering in oxyfluorides

The initial crystal structure of oxyfluorides $(NH_4)_2MO_2F_4$ (M = W, Mo) is characterized by orthorhombic symmetry (sp. gr. *Cmcm*, $Z = 4$) and consists of isolated mixed $[MO_2F_4]^{2-}$ octahedra and two nonequivalent ammonium groups [59, 63–65].

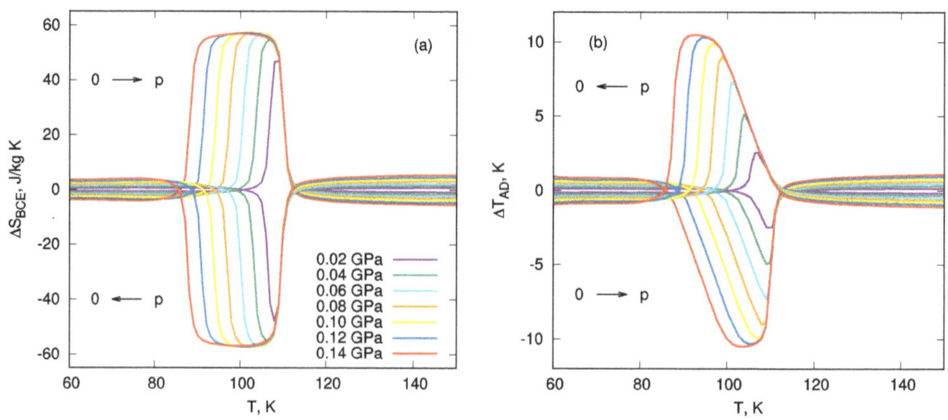

Figure 4.9. Extensive (a) and intensive (b) BCE in heating (pressure removal) and cooling (pressure application) modes in $(NH_4)_2SnF_6$.

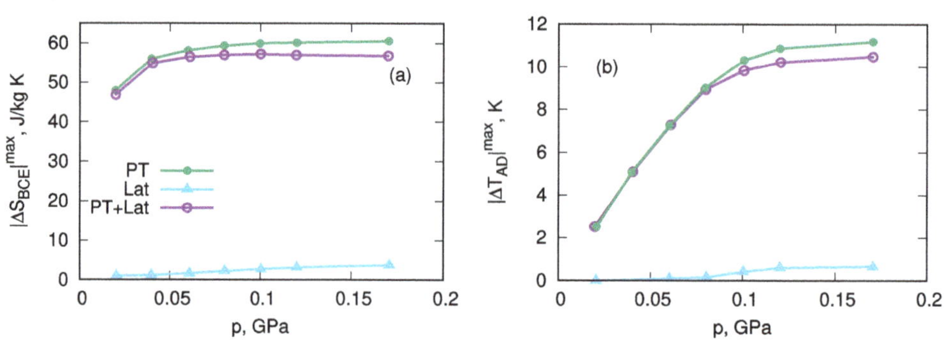

Figure 4.10. Pressure dependences of maximum values of extensive (a) and intensive (b) BCE in $(NH_4)_2SnF_6$.

The chemical pressure resulting from the substitution of the central atom: Mo (0.59 Å) → W (0.60 Å), significantly increases the stability temperature of the initial phase, and also changes the symmetry and origin of the distorted phase (antiferroelectric → ferroelastic) (table 4.1). In both compounds, the structural transformation from the initial phase *Pnma* is a first-order transition, accompanied by similar values of a significant change in entropy: W—ΔS_1=18.9 J mol^{-1}K^{-1}; Mo—ΔS_1=18.2 J mol^{-1}K^{-1}. However, the corresponding volume jumps are different: W—$\Delta V_1/V = 0.3\%$; Mo —$\Delta V_1/V = 1.8\%$. A significantly smaller change in entropy ($\Delta S_2 \approx 1.7$ J mol^{-1}K^{-1}) is observed in both oxyfluorides during the phase transition to the low-symmetry $P\bar{1}$ phase, which is associated with a slight additional displacement of structural elements.

The original orthorhombic structure is much more resistant to pressure changes in the $(NH_4)_2WO_2F_4$ (dT_1/dp = 13 K GPa^{-1}) than in $(NH_4)_2MoO_2F_4$ (dT_1/dp = 93 K GPa^{-1}) (figure 4.11). Intermediate phases in tungstate and molibdate have different symmetries, $P\bar{1}$ and *Pnma*, and are characterized by different baric coefficients dT_2/dp = 42 K GPa^{-1} and dT_2/dp = 17.0 K GPa^{-1}, respectively.

Oxyfluoride $(NH_4)_2NbOF_5$ is another outstanding example of a large structural disorder and undergoes two successive phase transitions $Cmc2_1$ (T_1 = 258 K) → $C2$ (T_2 = 219 K) →Ia [58]. In the initial phase, both octahedral and two nonequivalent tetrahedral structural units occupying 6 and 4 positions, respectively, are disordered. At T_1, the ordering is associated with a simultaneous halving of the number of orientations of each polyhedron, and at T_2, there is a complete structural ordering [66].

Both entropy values calculated within the above structural model, $\Delta S_1 = R \ln 3 + 2R \ln 2 = R \ln 12$=20.7 J (mol K)$^{-1}$ and $\Delta S_2 = R \ln 2 + 2R \ln 2 = R \ln 8 = 17.3$ J (mol K)$^{-1}$, agree well with the experimentally found entropies, ΔS_1 = 21.6 J (mol K)$^{-1}$ and ΔS_2 = 16.6 J (mol K)$^{-1}$.

A unique situation is typical for $(NH_4)_2NbOF_5$: an increase in pressure leads to a decrease in the temperatures of both transitions at the same rate $dT_1/dp = -45.4$ K GPa^{-1} and $dT_2/dp = -45.2$ K GPa^{-1} (figure 4.12) [26]. Thus, the intermediate phase $C2$,

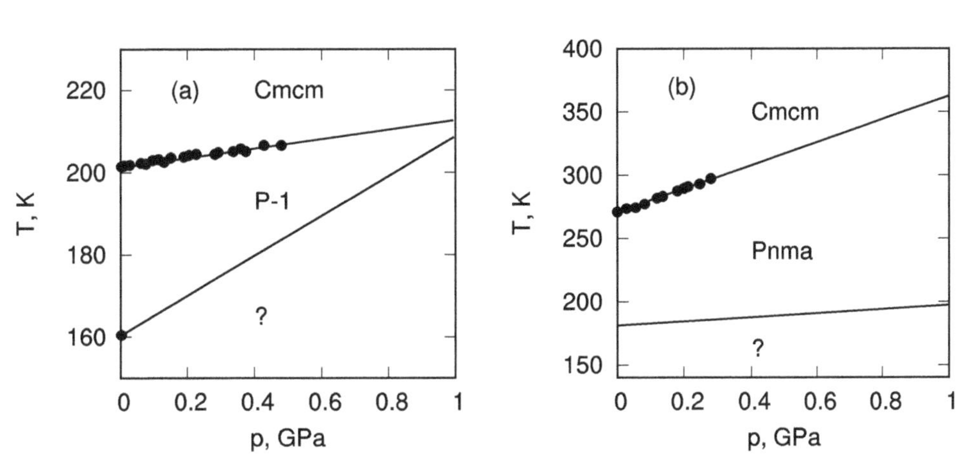

Figure 4.11. Transition temperature as a function of hydrostatic pressure for $(NH_4)_2MoO_2F_4$ (a) and $(NH_4)_2WO_2F_4$ (b).

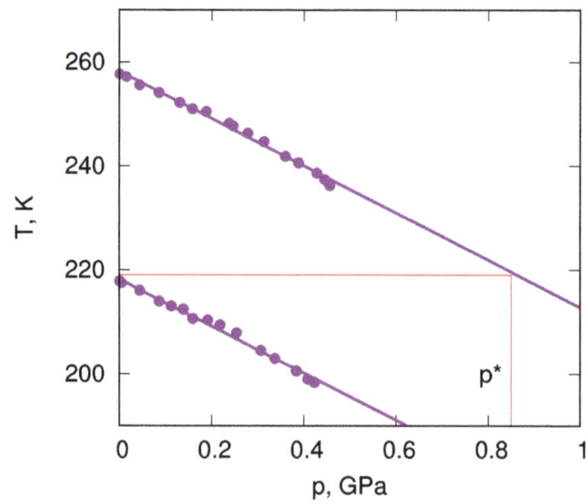

Figure 4.12. Transition temperature (peak in the calorimetric curves) as a function of hydrostatic pressure for $(NH_4)_2NbOF_5$.

for which the temperature range of existence under pressure remains unchanged, turns out to be less sensitive to external influences. In accordance with the Clausius–Clapeyron relation, the negative sign of the baric coefficients indicates a decrease in the volume of $(NH_4)_2NbOF_5$ during structural phase transitions upon heating.

4.4.4.1 BCE in $(NH_4)_2MoO_2F_4$, $(NH_4)_2WO_2F_4$

The magnitudes of the extensive and intensive BCEs in $(NH_4)_2MoO_2F_4$, $(NH_4)_2WO_2F_4$ are presented in figures 4.13 and 4.15, respectively. Despite the close values of the entropy of phase transitions at T_1, significant difference in the baric coefficients dT_1/dp leads to both a slower increase in ΔS_{BCE} and ΔT_{AD} with increasing pressure in the tungstate, and a higher value of p_{min} (figures 4.14 and 4.16). The contribution to BCE from the change in the lattice entropy with increasing pressure is significant in both oxyfluorides and, at $p \sim 1$ GPa, is comparable to the effect associated with the phase transition.

4.4.4.2 BCE in $(NH_4)_2NbOF_5$

Preliminary estimates of the BCE parameters for individual phase transitions $Cmc2_1 \leftrightarrow C2$ and $C2 \leftrightarrow Ia$ showed that, at pressure $p > p_{min} \approx 0.5$ GPa, the corresponding maximum possible values of the extensive and intensive effects ($\Delta S_{BCE}^{T1} = 90$ J (kg K)$^{-1}$, $\Delta S_{BCE}^{T2} = 69$ J (kg K)$^{-1}$ and $\Delta T_{AD}^{T1} = 22$ K, $\Delta T_{AD}^{T2} = 16$ K) can be achieved. For $p > p^* \approx (T_1 - T_2)/(dT_2/dp) \approx 0.9$ GPa, an increase in ΔS_{BCE} and ΔT_{AD} should be observed due to the summation of the effects associated with individual phase transitions (figure 4.17(a and b)). figure 4.17 shows the temperature dependences of the extensive and intensive effects before (a, b) and after (c, d) taking into account the contribution to BCE associated with the change in the entropy of the crystal lattice under pressure. As can be seen from figure 4.17(a), the extensive BCE in the region of T_2 reaches maximum possible value with increasing pressure and begins

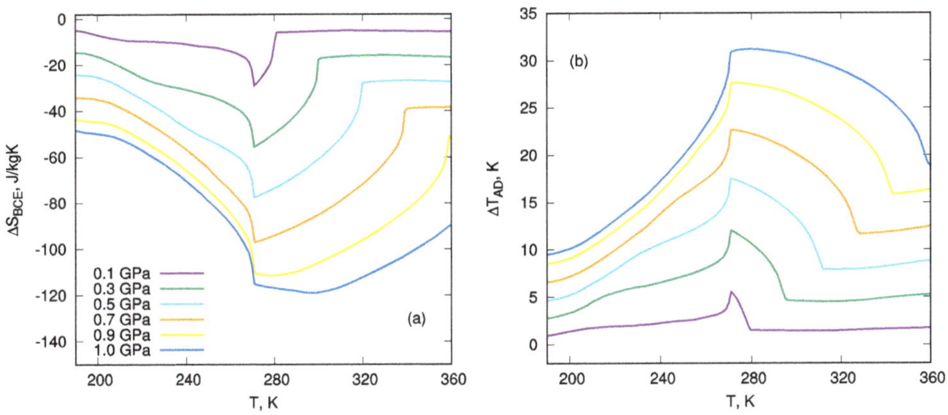

Figure 4.13. Extensive (a) and intensive (b) BCE in heating mode in $(NH_4)_2MoO_2F_4$.

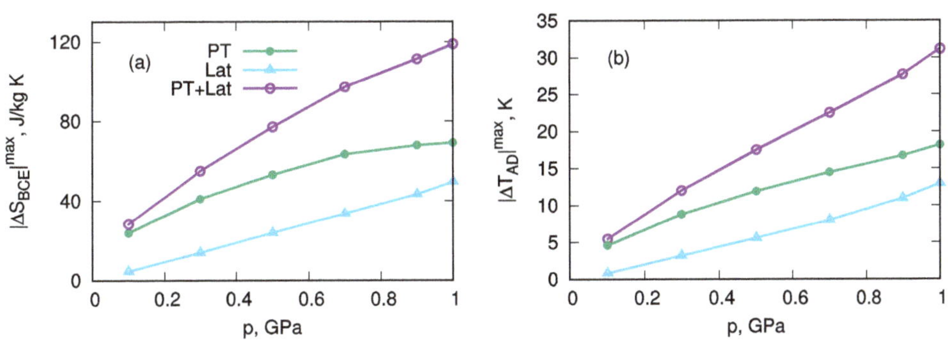

Figure 4.14. Pressure dependences of maximum values of extensive (a) and intensive (b) BCE in $(NH_4)_2MoO_2F_4$.

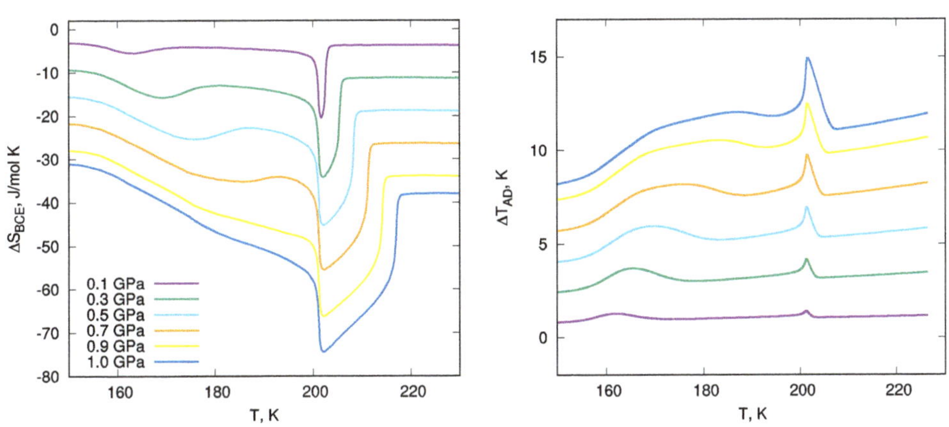

Figure 4.15. Extensive (a) and intensive (b) BCE in heating mode in $(NH_4)_2WO_2F_4$.

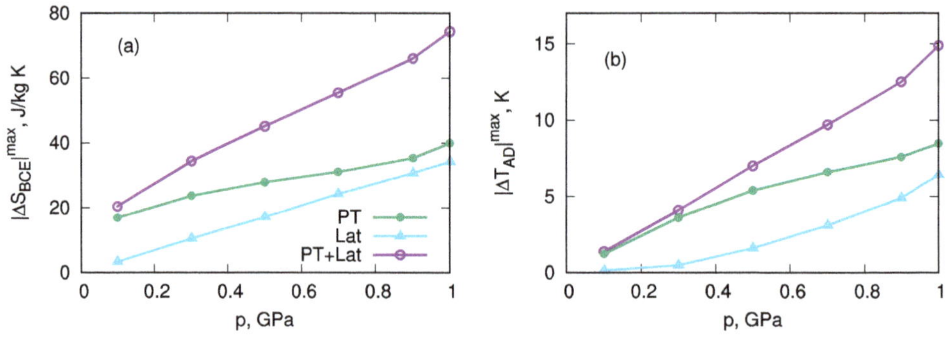

Figure 4.16. Pressure dependences of maximum values of extensive (a) and intensive (b) BCE in $(NH_4)_2WO_2F_4$.

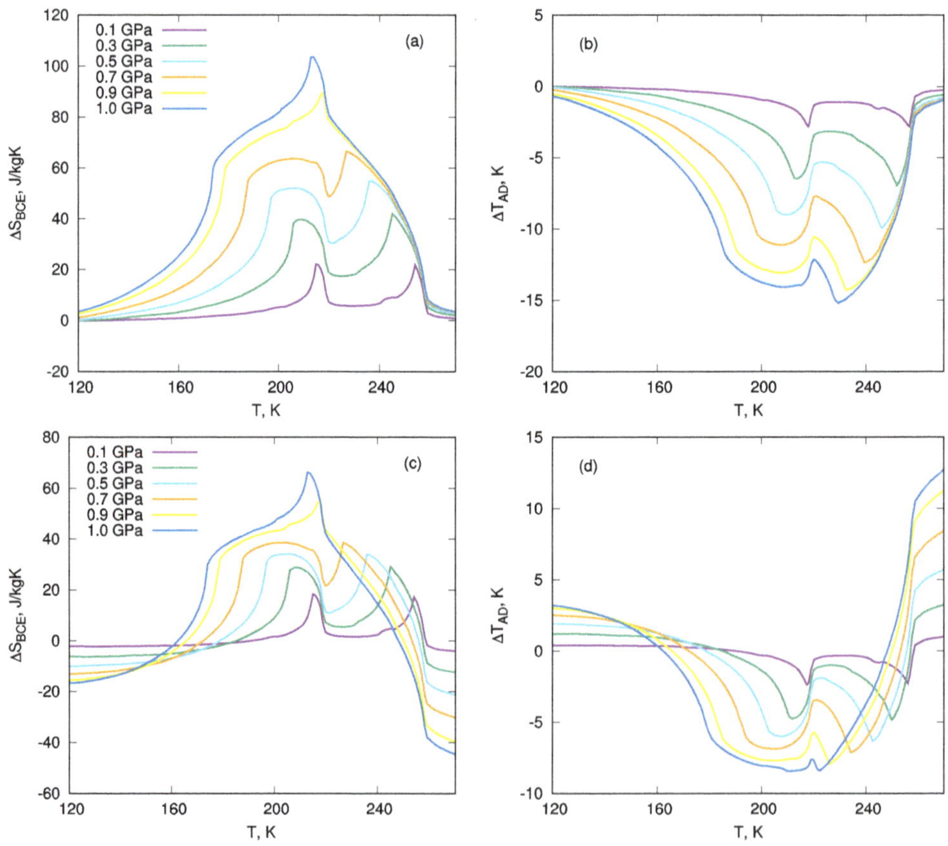

Figure 4.17. Temperature dependencies of barocaloric entropy and temperature changes in $(NH_4)_2NbOF_5$ at different hydrostatic pressures before (a, b) and after (c, d) correction for the additional lattice entropy changes with pressure.

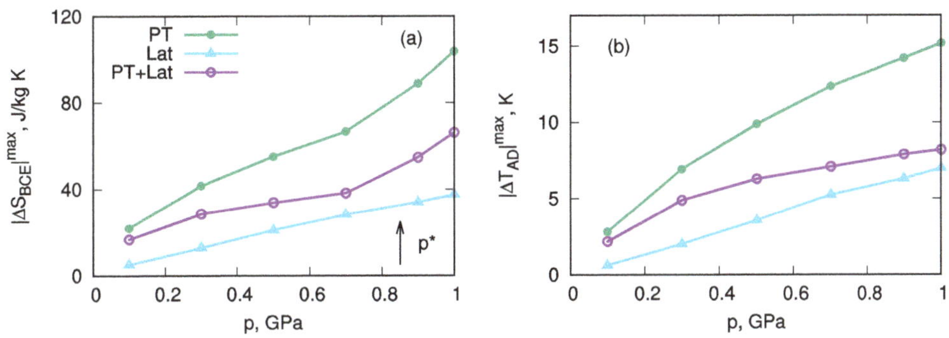

Figure 4.18. Pressure dependences of maximum values of extensive (a) and intensive (b) BCE in $(NH_4)_2NbOF_5$.

to increase again at $p > 0.7$–0.9 GPa. Due to the different signs of the thermal expansion coefficient of the crystal lattice ($\beta > 0$) and the coefficient of anomalous expansion associated with phase transitions in $(NH_4)_2NbOF_5$, the change in lattice entropy under pressure leads to a significant decrease in BCE (figure 4.17(c, d) and figure 4.18).

4.5 Conclusion

We have considered a number of fluorides and oxyfluorides that undergo individual and/or sequential phase transitions, predominantly of the order–disorder type, and demonstrate significant conventional and inverse barocaloric effects. Due to the small hysteresis of the phase transition temperature, the giant values of the isothermal entropy change and the adiabatic temperature change turned out to be reversible at sufficiently low pressures. The parameters of barocaloric effects near first-order phase transitions in some fluorides and oxyfluorides at a pressure of 0.2 GPa are presented in table 4.2.

The change in chemical pressure associated with the anion substitution and/or the central atom plays a significant role in the formation of both characteristics of the phase transitions (table 4.1) and barocaloric parameters (table 4.2).

To show the role of the cationic substation, we added to the table 4.2 oxyfluoride $(NH_4)_3TiOF_5$ not considered above, related to Rb_2KTiOF_5 and at $p = 0$ also undergoing one phase transition, characterized, on the one hand, by a very small baric coefficient, $dT/dp = 6.3$ K GPa^{-1}, but, on the other hand, a giant change in entropy $\Delta S_0 = 18.1$ J $(mol\ K)^{-1}$ [40, 67].

However, for $p \geqslant 0.196$ GPa, corresponding to the triple point pressure, $(NH_4)_3TiOF_5$ undergoes two successive transformations, each of which exhibits a very high sensitivity to the hydrostatic pressure: $dT_1/dp = 223$ K GPa^{-1}, $dT_2/dp = -176$ K GPa^{-1}. Due to this point, despite the division of the total entropy into two parts, $\Delta S_1 = 5.0$ J $(mol\ K)^{-1}$ and $\Delta S_2 = 12.8$ J $(mol\ K)^{-1}$, the barocaloric efficiency of ammonium oxyfluoride at T_2 turned out to be high even at a fairly low pressure, $p \approx 0.2$ GPa, and is characterized by parameters close to those found for Rb_2KTiOF_5 (table 4.2).

Table 4.2. Barocaloric parameters of some fluorides and oxyfluorides at pressure change 0.2 GPa.

Compound	T_0, K	ΔS_{BCE}^{max}, J kg K^{-1}	ΔT_{AD}^{max}, K	CRP	RC, kJ kg^{-1}	p_{rev}, GPa
Rb_2KTiOF_5	215	−40	12	0.95	0.7	0.056
$(NH_4)_2SnF_6$	110	57	−11	0.65	1.1	0.003
Rb_2KFeF_6	198	−47	17	1.47	1.2	0.035
$(NH_4)_2NbOF_5$	258 219	27	−4	0.27	0.5	0.013
$(NH_4)_2MoO_2F_4$	271 180	−45	8	0.49	1.0	0.009
$(NH_4)_2WO_2F_4$	201.5 161	−30	2.5	0.96	0.2	0.104
$(NH_4)_3TiOF_5$ [40]		−20 60	4.4 −11.5			

In this regard, in our opinion, it is promising to study solid solutions, for example, $(NH_4)_{3-x}(Rb_2K)_x$ $TiOF_5$, among which, quite likely, one can find a composition with phase transition characteristics that contribute to the implementation of optimal barocaloric parameters at even lower pressure.

In a number of fluorides and oxyfluorides under consideration, which have a triple point on the T–p phase diagram, the transition from the usual BCE to the reverse effect is observed in a narrow temperature range, in which successive structural transformations with different signs of dT/dp occur, accompanied by large changes in entropy and temperature under pressure. Materials of this kind can be considered as promising solid-state refrigerants in combined thermodynamic cycles, similar to cycles on materials with a pronounced anisotropy of the piezocaloric effect [68].

The large thermal expansion of the crystal lattice of complex oxyfluorides as well as fluorides has led in a number of crystals, characterized by $dT/dp > 0$, to an additional significant contribution comparable to the effect associated with the phase transition to the conventional BCE (figures 4.13 and 4.15). On the other hand, in the case of the inverse effect associated with $dT/dp < 0$ and, as a rule, $\beta_{Lat} > 0$, taking into account the lattice contribution reduces the total BCE. At the same time, at temperatures significantly higher than the temperature of the structural transformation in oxyfluorides, BCE may turn out to be even higher than the anomalous one observed in the region T_0 (figure 4.17).

Recently, an averaged estimate of a number of BCE parameters was carried out for some families of materials of different physical nature [34]. In comparison with these data, fluorides and oxyfluorides undergoing order–disorder transformations, accompanied by high sensitivity to hydrostatic pressure, show high barocaloric efficiency and characterized by quite satisfactory barocaloric parameters. As a result, they can be considered as promising solid-state refrigerants in the cooling devices of the future. These compounds are somewhat inferior in absolute barocaloric characteristics to some compounds with organic cations. However, due to both relatively low pressures and a small temperature hysteresis leading to a high reversibility of the barocaloric effect, they are superior to polymers and plastic crystals.

References

[1] Mañosa L, Planes A and Acet M 2013 *J. Mater. Chem.* A **1** 4925–36

[2] Kitanovski A, Plaznik U, Tomc U and Poredoš A 2015 *Int. J. Refrig.* **57** 288–98

[3] Lorusso G *et al* 2013 *Adv. Mater.* **25** 4653–6

[4] Michaelis N, Welsch F, Kirsch S M, Schmidt M, Seelecke S and Schütze A 2019 *Int. J. Refrig.* **100** 167–74

[5] Kitanovski A 2020 *Adv. Energy Mater* **10** 1903741

[6] Gschneidner K A Jr, Pecharsky V K and Tsokol A O 2005 *Rep. Prog. Phys.* **68** 1479–539

[7] Franco V, Blázquez J, Ingale B and Conde A 2012 *Annu. Rev. Mater. Res.* **42** 305–42

[8] Smith A, Bahl C R, Bj/ork R, Engelbrecht K, Nielsen K K and Pryds N 2012 *Adv. Energy Mater.* **2** 1288–318

[9] Zhong W, Au C T and Du Y W 2013 *Chin. Phys.* B **22** 057501

[10] Planes A, Mañosa L and Acet M 2009 *J. Phys.: Condens. Matter* **21** 233201

[11] Brück E 2008 *Handbook of Magnetic Materials* (Amsterdam: Elsevier)

[12] Scott J 2011 *Annu. Rev. Mater. Res.* **41** 229–40

[13] Valant M 2012 *Prog. Mater. Sci.* **57** 980–1009

[14] Tishin A, Spichkin Y, Zverev V and Egolf P 2016 *Int. J. Refrig.* **68** 177–86

[15] Zverev V, Pyatakov A, Shtil A and Tishin A 2018 *J. Magn. Magn. Mater.* **459** 182–6

[16] Greco A, Aprea C, Maiorino A and Masselli C 2019 *AIP Conf. Proc.* **2191** 020091

[17] March 2014 Energy savings potential and RD&D opportunities for non-vapor compression HVAC technologies *Report of the U.S. Department of Energy*

[18] Aznar A, Lloveras P, Barrio M, Negrier P, Planes A, Mañosa L, Mathur N D, Moya X and Tamarit J L 2020 *J. Mater. Chem.* A **8** 639–47

[19] Bermúdez-García J M, Yáñez-Vilar S, García-Fernández A, Sánchez-Andújar M, Castro-García S, López-Beceiro J, Artiaga R, Dilshad M, Moya X and Señarís-Rodríguez M A 2018 *J. Mater. Chem.* C **6** 9867–74

[20] Bermúdez-García J M, Sánchez-Andújar M and Señarís-Rodríguez M A 2017 *J. Phys. Chem. Lett.* **8** 4419–23

[21] Li B *et al* 2019 *Nature* **567** 506–10

[22] Bermúdez-García J M, Sánchez-Andújar M, Castro-García S, López-Beceiro J, Artiaga R and Señarís-Rodríguez M A 2017 *Nat. Commun.* **8** 15715

[23] Ouyang G, Pan C, Wolf S, Mohapatra P, Takeuchi I and Cui J 2020 *Appl. Phys. Lett.* **116** 251901

[24] Zarkevich N A, Johnson D D and Pecharsky V K 2017 *J. Phys. D: Appl. Phys.* **51** 024002

[25] Gorev M, Bogdanov E and Flerov I 2017 *J. Phys. D: Appl. Phys.* **50** 384002

[26] Gorev M, Bogdanov E, Flerov I and Laptash N 2010 *J. Phys.: Condens. Matter* **22** 185901

[27] Hou H *et al* 2019 *Science* **366** 1116–21

[28] Pu Y, Zhang Q, Li R, Chen M, Du X and Zhou S 2019 *Appl. Phys. Lett.* **115** 223901

[29] Zhang G *et al* 2020 *Appl. Phys. Lett.* **116** 023902

[30] Bradeško A, Juričić D, Santo Zarnik M, Malič B, Kutnjak Z and Rojac T 2016 *Appl. Phys. Lett.* **109** 143508

[31] Hanrahan B, Easa J, Payne A, Espinal Y, Alpay S P, Kareem H, O'Brien C and Smith A 2020 *Cell Rep. Phys. Sci.* **1** 100075
[32] Liu Y, Wei J, Janolin P E, Infante I C, Kreisel J, Lou X and Dkhil B 2014 *Phys. Rev.* B **90** 104107
[33] Mañosa L and Planes A 2017 *Adv. Mater.* **29** 1603607
[34] Lloveras P and Tamarit J L 2021 *MRS Energy Sustain* **8** 3–15
[35] Cazorla C 2019 *Appl. Phys. Rev.* **6** 041316
[36] Wei Z Y, Sun W, Shen Q, Shen Y, Zhang Y F, Liu E K and Liu J 2019 *Appl. Phys. Lett.* **114** 101903
[37] Xiao F, Li Z, Chen H, Li Z, Huang K, Jin X and Fukuda T 2020 *Materialia* **9** 100547
[38] Gui W, Qu Y, Cao Y, Zhao Y, Liu C, Zhou Q, Chen J and Liu Y 2022 *J. Mater. Res. Technol.* **19** 4998–5007
[39] Pirc R, Kutnjak Z, Blinc R and Zhang Q M 2011 *Appl. Phys. Lett.* **98** 021909
[40] Gorev M, Bogdanov E and Flerov I 2017 *Scr. Mater* **139** 53–7
[41] Flerov I, Gorev M, Bogdanov E and Laptash N 2016 *Ferroelectrics* **500** 153–63
[42] Strässle T, Furrer A, Hossain Z and Geibel C 2003 *Phys. Rev.* B **67** 054407
[43] Gorev M, Flerov I, Bogdanov E, Voronov V and Laptash N 2010 *Phys. Solid State* **52** 377–83
[44] Flerov I N, Kartashev A V, Gorev M V, Bogdanov E V, Mel'nikova S V, Molokeev M S, Pogoreltsev E I and Laptash N M 2016 *J. Fluor. Chem.* **183** 1–9
[45] Flerov I, Gorev M, Tressaud A and Laptash N 2011 *Crystallogr. Rep.* **56** 9–17
[46] Aznar A, Negrier P, Planes A, Mañosa L, Stern-Taulats E, Moya X, Barrio M, Tamarit J L and Lloveras P 2021 *Appl. Mater. Today* **23** 101023
[47] Leblanc M, Maisonneuve V and Tressaud A 2015 *Chem. Rev.* **115** 1191–254
[48] Tressaud A 2010 *Functionalized Inorganic Fluorides* (Chichester: Wiley-Blackwell)
[49] Tressaud A and Poeppelmeier K (ed) 2016 *Photonic and Electronic Properties of Fluoride Materials* (Amsterdam: Elsevier)
[50] Aleksandrov K S, Anistratov A T, Beznosikov B V and Fedoseeva N V 1981 *Phase Transitions of Halide ABX_3 Compounds* (Novosibirsk: Nauka)
[51] Aleksandrov K S and Beznosikov B V 2004 *Perovskites. Present and Future* (SB RAS: Novosibirsk)
[52] Flerov I, Gorev M, Aleksandrov K, Tressaud A, Grannec J and Couzi M 1998 *Mater. Sci. Eng. R Rep.* **24** 81–151
[53] Gautier R, Gautier R, Chang K B and Poeppelmeier K R 2015 *Inorg. Chem.* **54** 1712–9
[54] Udovenko A A and Laptash N M 2008 *Acta Cryst.* **64** 645–51
[55] Udovenko A A, Vasiliev A D and Laptash N M 2010 *Acta Cryst.* **B66** 34–9
[56] Fokina V D, Flerov I N, Molokeev M S, Pogorel'tsev E I, Bogdanov E V, Krylov A S, Bovina A F, Voronov V N and Laptash N M 2008 *Phys. Solid State* **50** 2175–83
[57] Gorev M V, Bogdanov E V, Flerov I N, Voronov V N and Laptash N M 2010 *Ferroelectrics* **397** 76–80
[58] Fokina V D, Bogdanov E V, Gorev M V, Molokeev M S, Pogorel'tsev E I, Flerov I N and Laptash N M 2010 *Phys. Solid State* **52** 781–8
[59] Fokina V D, Bogdanov E V, Pogorel'tsev E I, Bondarev V S, Flerov I N and Laptash N M 2010 *Phys. Solid State* **52** 158–66
[60] Gorev M, Bogdanov E, Flerov I, Kocharova A and Laptash N 2010 *Phys. Solid State* **52** 167–75

[61] Flerov I N, Gorev M V, Tressaud A and Grannec J 1998 *Ferroelectrics* **217** 21–33

[62] Javier Zúñiga F, Tressaud A and Darriet J 2006 *J. Solid State Chem.* **179** 3607–14

[63] Flerov I N, Fokina V D, Gorev M V, Bogdanov E V, Molokeev M S, Bovina A F and Kocharova A G 2007 *Phys. Solid State* **49** 1149–56

[64] Bogdanov E V, Pogorel'tsev E I, Mel'nikova S V, Gorev M V, Flerov I N, Molokeev M S, Kartashev A V, Kocharova A G and Laptash N M 2013 *Phys. Solid State* **55** 409–18

[65] Bogdanov E V, Mel'nikova S V, Pogoreltsev E I, Molokeev M S and Flerov I N 2016 *Solid State Sci.* **61** 155–60

[66] Udovenko A A and Laptash N M 2008 *Acta Cryst.* **B64** 527–33

[67] Flerov I N, Gorev M V, Fokina V D, Bovina A F and Laptash N M 2004 *Phys. Solid State* **46** 915–21

[68] Mikhaleva E A, Gorev M V, Molokeev M S, Kartashev A V and Flerov I N 2020 *J. Alloys Compd.* **839** 155085

Chapter 5

Spin crossover complexes for barocaloric cooling

Karl G Sandeman and Malcolm A Halcrow

Molecules, molecular crystals and inorganic compounds which possess a spin crossover (SCO) (or 'spin transition') at temperature T_{SCO} have been studied extensively for the past 50 years. Several applications ranging from sensing to display technology have been suggested for the material class. The extension of research in SCO materials to include the barocaloric effect (BCE) is relatively recent and is founded on two known factors: first, the possibility of large relative volume changes at the spin transition and second, the magnitude of the pressure sensitivity of the spin crossover temperature. The latter feature yields the possibility of large adiabatic temperature changes driven by applied pressure. This chapter is organized as follows: in section 5.1, we introduce the phenomenon of spin crossover and define terms relevant to the chapter. In section 5.2 we present a brief review of the properties of SCO materials under pressure that were known prior to BCE-focused studies. The barocaloric effects that have been explored thus far in SCO materials are summarized and discussed in section 5.3. Finally, we outline some areas for further exploration in section 5.4.

5.1 Introduction to spin crossover

5.1.1 History and fundamentals

The phenomenon of spin crossover was first observed almost 100 years ago [1]. It can be described as a change in the spin state of a material, from low spin (LS) to high spin (HS), triggered by a change in temperature or some other, applied, stimulus. Spin crossover (or a spin transition[1]) occurs when the crystal field (ligand field) is strong enough to break the degeneracy of the electron levels in an ion but

[1] In this chapter, we will use 'spin crossover' and 'spin transition' interchangeably since both terms occur in the literature.

doi:10.1088/978-0-7503-4690-0ch5

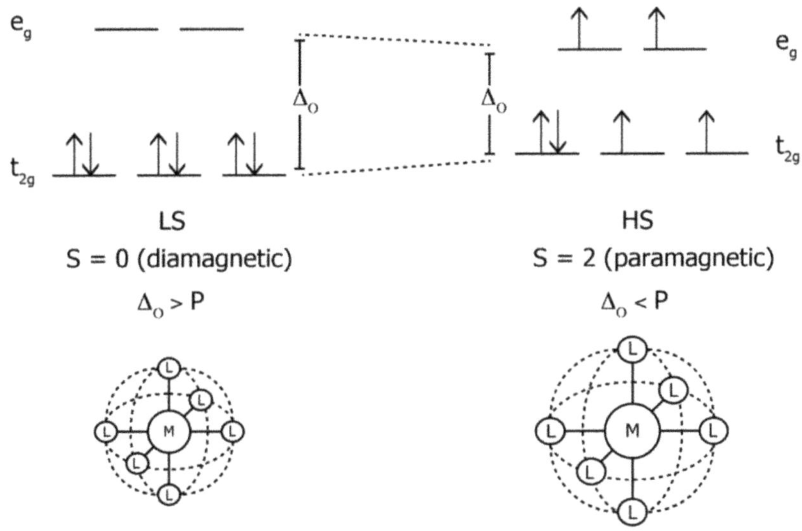

Figure 5.1. A schematic of the formation of a low spin (LS) or high spin (HS) state in the case of a magnetic ion ('M') with 6 d-electrons which is octahedrally coordinated to ligands labelled 'L'. The change of spin state can occur as a result of increasing temperature or via a change in the strength of the crystal field (Δ_0). Reproduced with permission from Vallone *et al* [9]. Copyright 2019 Wiley-VCH Verlag GmbH & Co. KGaA, Weinheim.

not so strong as to make population of the highest energy levels energetically unfeasible. The original discovery of spin transitions is credited to the work of Cambi and Szegö, whose 1931 article 'Über die magnetische Susceptibilität der komplexen Verbindungen' ('On the magnetic susceptibility of complex compounds') explored the room temperature magnetic susceptibility of a large number of Ni, Co, Fe and Mn dithiocarbamate complexes $T(S_2C{\cdot}NR'R'')_3$ (T=Ni, Co, Fe, Mn), finding highly temperature-dependent magnetization states (and therefore, magnetic moments) [1]. Their work in the 1930s resulted in a several articles on this topic [1–7]. As pointed out by Gütlich [8], however, those early observations were attributed to the existence of magnetic isomers, rather than a spin transition as we understand it today.

The spin states which occur on either side of the spin crossover, and the external conditions required to access either state, depend on the type of ligand field and the number of available electrons. Figure 5.1 illustrates the concept of spin crossover for a d^6 ion subject to an octahedral crystal field which splits the d orbitals as shown. The crystal field splitting between the $e_{.g.}$ and t_{2g} orbitals is denoted as Δ_0. At low temperature when Δ_0 is greater than the spin-pairing energy, P, the electrons will occupy the t_{2g} orbitals in preference to excitation across the crystal field gap. This LS state breaks Hund's First Rule. However, at higher temperature, and/or if the magnitude of Δ_0 is reduced to below the spin-pairing energy, for example by an external stimulus, excitation becomes possible, and a high spin state is formed.

Table 5.1. Spin states in systems exhibiting SCO and examples of ions which form them, for various numbers of d valence electrons. The * beside the LS state in the d^4 case is to note that in some (non-octahedral) coordination geometries, the LS state is in fact $S = 0$.

Number of d electrons	S {LS}	S {HS}	Example ions
4	1*	2	Cr(II), Mn(III)
5	$\frac{1}{2}$	$\frac{5}{2}$	Fe(III), Mn(II)
6	0	2	Fe(II), Co(III)
7	$\frac{1}{2}$	$\frac{3}{2}$	Co(II)

If Δ_0 is too large, spin crossover is not observed, and only an LS state forms. If Δ_0 is very small, an HS state forms in accordance with Hund's First Rule. Accordingly, there is a relatively narrow selection of magnetic ions which tend to form spin crossover compounds. A discussion of this selectivity can be found in an essay by Gütlich [8]. In summary, for the example of octahedrally-coordinated complexes, both HS and LS configurations can occur in transition metal ions with 4–7 3d valence electrons, making possible the thermally-driven spin state switching that is associated with spin crossover.

In the example shown in figure 5.1, the LS state has total spin $S = 0$ and the HS state has $S = 2$, but we see that if there are more or fewer valence electrons, a transition with a finite change of spin would still be possible. In table 5.1 we list the spin transitions that are possible for transition metals with 4–7 3d valence electrons in octahedrally-coordinated environments, along with examples of relevant ions where such LS ↔HS transitions may be found.

In the 90 years since Cambi and Szegö's work on Fe(III) complexes, the range of metal ions in which SCO is found has extended to all of the examples in table 5.1 and to a range of material types, of which octahedrally-coordinated molecular complexes are the most dominant group. Within that group, there has been a further emphasis on Fe(II) complexes featuring nitrogen donor ligands [10]. It is not the purpose of this chapter to review the SCO literature or the families of SCO compounds; in that regard, we refer the reader to several very comprehensive treatments elsewhere [8, 10–21]. However, it is worth mentioning here the sheer range of compound types and geometries in which spin crossover may be found, as this lends weight to the consideration of SCO transitions as a rich seam for exploration of barocaloric effects. As noted by Halcrow [10], spin crossover can be found in material types which include molecular complexes [22]; coordination polymers with 1D, 2D or 3D geometries [23]; metal organic frameworks (MOFs) [24, 25]; coordination cages and other supramolecular assemblies [26]; organometallic compounds [27, 28]; dinuclear [29], trinuclear [30] and larger metal-metal bonded materials such as metal cluster compounds [31]; and, inorganic solids such as the cobaltites [32–34].

5.1.2 Classification of spin transitions

From a caloric point of view, the sharpness and thermal hysteresis at the SCO transition is of great interest. This is because, as with all caloric materials, larger caloric effects are normally associated with sharp phase transitions, while reversibility (and a lack of thermal hysteresis) is desirable if a single stimulus, such as pressure, is to be used in the construction of a cooling cycle [35, 36]. To this point, SCO transitions exhibit a wide range of transition types, as depicted in figure 5.2. The degree to which the metal ions interact is known as the cooperativity of the material, and its effect is akin to that of magneto-elastic coupling in magnetocaloric materials. Namely, materials with smaller cooperativity exhibit gradual SCO transitions while those with large cooperativity exhibit first-order SCO transitions with accompanying thermal hysteresis—the difference between the transition temperature on heating and on cooling—which can be as large as over 100 Kelvin [37]. In addition to complete, single transitions, double-step and incomplete transitions are also possible [38].

5.1.3 Volume change at a spin transition

A crucial aspect of the barocaloric potential of SCOs is the change of volume at the spin transition. Its origin is particularly clear to visualize in the case of those materials which have Fe–ligand bonds (as opposed to inorganic solids). The origin of this volume change was discussed by Ewald *et al* in 1964 in their examination of

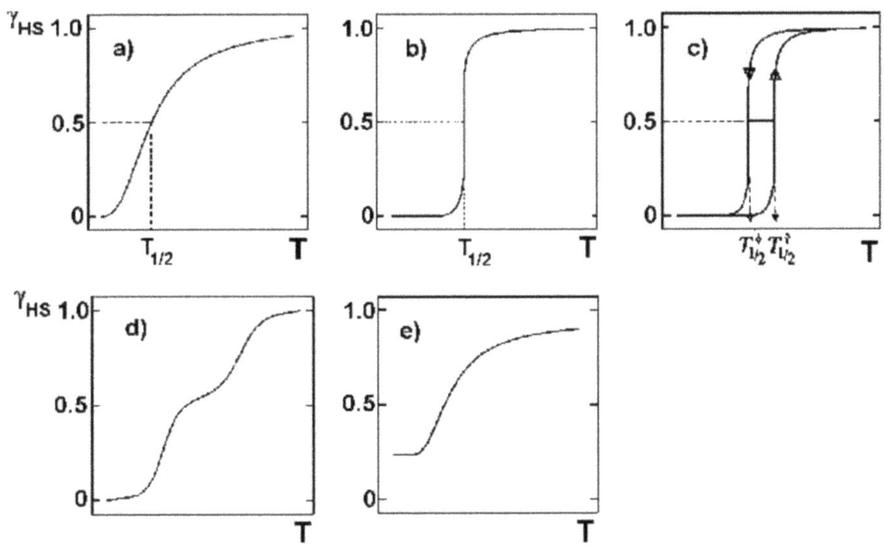

Figure 5.2. There are several types of SCO transition, often categorized in the following ways: (a) gradual, (b) steep or abrupt, (c) (abrupt) with hysteresis, (d) step-wise, (e) incomplete. In each case, the relative proportion of the sample in the HS state is given by $\gamma_{HS}(T)$. The temperature at which the HS fraction is 50% is denoted as $T_{1/2}$. This figure was originally published in reference [38] and is reproduced with permission from [39]. Copyright © 2003 Wiley-VCH Verlag GmbH & Co. KGaA, Weinheim.

several iron(III) dialkyldithiocarbamates [40] on the basis of earlier more general work on the influence of the crystal field on ionic radii by van Santen and van Wieringen [41]. The latter had pointed out that the transfer of an electron from a bonding t_{2g} orbital to an antibonding e_g orbital results in an increase of Fe–ligand distance. This change of Fe–ligand distance also decreases the crystal field splitting, Δ_0, as shown in figure 5.1.

The change of volume at a spin transition is thus dependent on the electronic configuration of both the LS state and the HS state. The largest volume changes are generally found in Fe(II) complexes with N-donor ligands, which accounts for their popularity with SCO researchers [42]. As will become evident from what follows, the change of volume is significant in contributing to the barocaloric effects of SCO materials. The Fe–ligand distance can change by up to ~10%, with corresponding unit cell volume changes typically of the order of 2%–4% in Fe(II) complexes [42]. However, in cases of polymeric chain SCO materials such as the ([Fe(Htrz)$_2$(trz)] (BF$_4$)]), unit cell volume changes can be as high as 11% [43, 44] and have even attracted the idea of the use of SCO compounds embedded in polymer matrices for actuation [45].

5.1.4 Caloric properties

The caloric properties of an SCO transition were first explored without applied pressure, typically via heat capacity measurements under ambient conditions. In 1974, Sorai and Seki used heat capacity measurements to probe the abrupt phase spin transition in [Fe(phen)$_2$(NCS)$_2$] (phen = 1,10-phenanthroline; T_{SCO} = 176K) and [Fe(phen)$_2$(NCSe)$_2$] (T_{SCO} = 231K) for the first time [46]. In both cases, they found latent heats of around ΔS=50 J K^{-1} mol^{-1} and that the magnetic contribution (~13 J K^{-1} mol^{-1}) was only a small fraction of the total ΔS. In the 1985 work of Kaji and Sorai [47], the heat capacity of [Fe(2-pic)$_3$]Cl$_2$·EtOH (pic = 2-{amino-methyl}pyridine) was studied in order to investigate the nature of the apparent two-step transition (type (d), in figure 5.2) in this compound. Similar conclusions to those of Sorai and Sekai regarding the size of the latent heat and the relatively small percentage magnetic contribution were obtained. Given the normal density of spin crossover complexes, these (unoptimized) entropy change values corresponded to 100 J K^{-1} kg^{-1} or 0.2 J K^{-1} cm^{-3}, values which can be described as 'giant' caloric effects if they could be released by an applied stimulus such as pressure.

Given the caloric potential of SCO compounds, and the opportunity afforded by applied pressure to couple to the volume change that is present in so many material examples, barocaloric studies of SCO compounds have, so far, fallen into two categories:

(1) An examination of SCO compounds for which the pressure dependence of the SCO transition temperature is already known but not explored for barocaloric purposes (see section 5.2). In such studies, literature data on a compound is typically used to postulate or model the expected BCE. The SCO transition might be gradual or sharp.

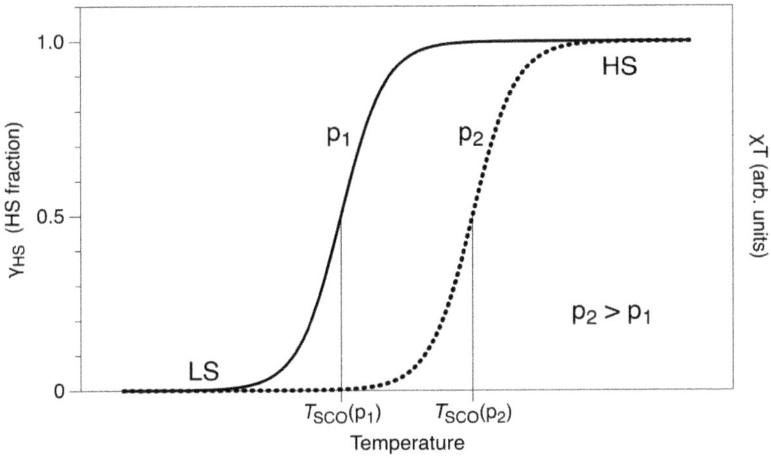

Figure 5.3. A schematic of the typical LS state-stabilizing effect of hydrostatic pressure on a SCO compound. In this example, p_1 and p_2 are two applied pressures, where $p_2 > p_1$. The fraction of the HS state is given by γ. An alternative measure of the change of spin state is the product of magnetic susceptibility, χ and temperature. The transition from LS to HS is normally accompanied by an increase in volume, in which case the above schematic is relevant. The spin crossover transition temperature, T_{SCO}, therefore increases with applied pressure, resulting in a conventional barocaloric effect. The above schematic is of a continuous spin transition, and so no thermal hysteresis is depicted. In practice, the pressure dependence of thermal hysteresis at a first-order spin transition has been a subject of considerable research [48].

(2) An examination of SCO compounds in which the effect of applied pressure is being studied for the first time, because of BCE-driven interest. Due to the focus on the BCE, most materials in this category have sharp transitions near room temperature (type (c) in figure 5.2), with varying levels of thermal hysteresis (see section 5.3).

In section 5.3, we will consider both BCE research themes, but before doing so, we next examine the effect of hydrostatic pressure on SCO compounds.

5.2 Effect of hydrostatic pressure on SCO compounds

We start this section with a general statement about the expected variation of T_{SCO} with pressure, since it is in the vicinity of T_{SCO} that the largest BCEs are expected. As described in section 5.3, the LS state of a SCO compound is typically of lower volume than the HS state. In such cases, the effect of hydrostatic pressure will be to stabilize the LS state, as shown in figure 5.3 for a generic spin crossover transition. Accordingly, SCO compounds are conventional barocalorics, in that they possess a positive ΔT_{ad} when pressure is applied adiabatically, and a negative isothermal entropy change, $\Delta S(T, \Delta p)$ when pressure is applied isothermally.

5.2.1 Application of pressure to materials with spin transitions below room temperature

Until there was interest in the barocaloric properties of SCO compounds around their spin transitions, the investigation of the effect of pressure on a spin transition

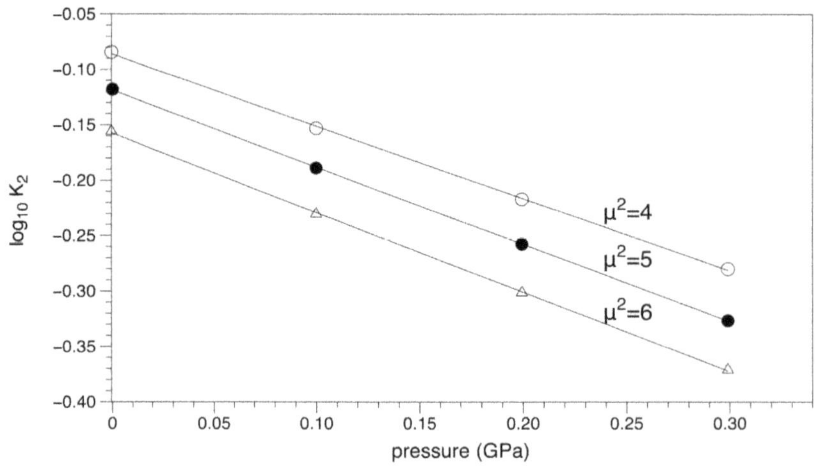

Figure 5.4. Variation of the ratio of the room temperature equilibrium constant, $\frac{\gamma_{HS}}{1-\gamma_{HS}}$, expressed as K_2, in [Fe($S_2C{\cdot}Nn\text{-}Bu_2)_3$] under pressures of up to 0.3 GPa (adapted from reference [40] with permission from the Royal Society of Chemistry; the original data have been replotted here). The data are extracted from magnetic susceptibility measured as a function of applied pressure. The drop in the equilibrium constant with pressure indicates a rise in the LS population. The three different lines fits to the experimental data correspond to different values of the effective magnetic moment of the LS state, μ, postulated by the authors of reference [40].

was undertaken for fundamental research purposes. The first studies of the effect of pressure on an SCO compound were by Ewald *et al* in 1964 [40] and in 1969 [49] and by Fisher and Drickamer in 1971 [50]. Ewald *et al* observed the magnetic moments of several iron(III) dialkyldithiocarbamates, $Fe(S_2C{\cdot}NR_2)_3$, using the Gouy method as a function of temperature and applied pressure and discussed their findings in the context of Cambi's original work [40]. A pressure of 0.3 GPa was applied to an *n*-butyl complex, [Fe($S_2C{\cdot}Nn\text{-}Bu_2)_3$] which exhibited a broad spin transition at around 120 K at ambient pressure. The authors found that the balance of the HS state and the LS state at room temperature was duly altered by applied pressure, with the LS population increasing—see figure 5.4.

If pressure is applied at constant temperature, then $\frac{\Delta V}{RT} = \frac{\partial(\ln K_{eq})}{\partial p}$ where $K_{eq} = \frac{\gamma_{HS}}{1-\gamma_{HS}}$ and ΔV is the difference of volume between the HS and LS states. Ewald *et al* used this relation to infer the volume change of [Fe($S_2C{\cdot}Nn\text{-}Bu_2)_3$] from their magnetic susceptibility measurements; further investigations of ferric dithiocarbamate complexes [$Fe(S_2CNR_2)_3$] were published in 1969 [49].

The subsequent work of Fisher and Drickamer [50] was on one of the most studied mononuclear phenanthroline complexes of the form [$Fe(phen)_2(X)_2$] including those with ligands X=NCS and NCSe (isothiocyanate and isoselenocyanate, respectively). This appears to be the first study of a Fe(II) SCO complex under pressure. The authors used Mössbauer spectroscopy (isomer shifts) and optical absorption to examine the effects of pressure. In figure 5.5, their results, an observation of the percentage LS state at room temperature in [$Fe(phen)_2(X)_2$]

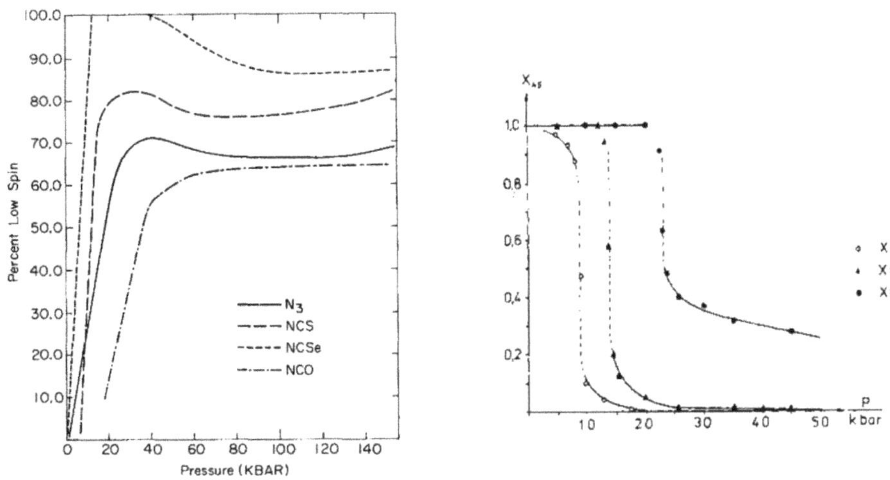

Figure 5.5. Left: Effect of pressure on several [Fe(phen)$_2$(X)$_2$] complexes at room temperature, as determined by Fisher and Drickamer [50] and reproduced from [50] with the permission of AIP Publishing. The initial rapid rise in the percentage of LS state is expected; the downturn at higher pressures was not reproduced in subsequent studies [52, 53] and, as conjectured by Adams [52], may have been due to a lack of hydrostatic conditions caused by the solid nature of the boron/lithium hydride pressure medium used by Fisher and Drickamer [54]. Right: Data from 3 of the same [Fe(phen)$_2$(X)$_2$] complexes, measured under pressure by Pebler [53]. These data show the expected monotonic decrease in the HS fraction as pressure is increased. This subfigure is reproduced with permission from [53], copyright 1983 American Chemical Society.

with X=N_3, NCS, NCSe and NCO, are reproduced. Since T_{SCO} at atmospheric pressure in the X = NCS and X = NCSe compounds is 174 K and 232 K, respectively [51], the observation of the LS state at room temperature constitutes a T_{SCO} shift of 50–100 K in 20 kbar (2 MPa), depending on the compound. From those early data, the barocaloric potential of SCO compounds was evident, even if the transition temperatures of these particular material examples were below room temperature and therefore not immediately applicable to all cooling applications.

For a comprehensive review of the effect of pressure on spin crossover (and valence tautomeric) material systems studied up to 2004/2005, we refer the reader to references [48, 55]. For a review of the effect of pressure on spin crossover compounds up to 2018, we refer the reader to reference [56]. In table 5.2, we list studies of SCO compounds under pressure from 1971 to 2000, with a focus on the pressure-driven change in T_{SCO} and the magnitude of the maximum applied pressure in each study. We note that, up to the mid-1990s, many of the pressure data were taken using a high-pressure cell designed for Mössbauer experiments. This pressure cell suffered from non-hydrostatic conditions—as seen the left panel of figure 5.5—as well as requiring isotopic changes to be made to the (relatively small) samples [48]. All studies in table 5.2 were not motivated by the barocaloric effect, since interest in BCEs began around 2016. Before moving on to barocaloric properties studied since that time, we here highlight just a few examples of compounds studied up to 2016, which have received subsequent interest from the barocaloric community.

Table 5.2. Examples of the first 30 years of studies of SCO compounds under pressure, from 1971 to 2000, where we here focus on the pressures applied and the change of T_{SCO} with applied pressure. Before 2018, these studies were not motivated by the barocaloric effect, and so heat capacity data under pressure were typically not supplied. In some cases, the apparent $\frac{dT_{SCO}}{dp}$ has been extracted from the data in the relevant article. In the cases marked with an asterisk (*), some approximation has been required, for example, linearization of the available data and therefore the values of $\frac{dT_{SCO}}{dp}$ extrapolated are only a guide. In those studies indicated with a **, the lack of hydrostatic pressure conditions affected the observed $\frac{dT_{SCO}}{dp}$. Lastly, we note that $\frac{dT_{SCO}}{dp} < 0$ for Nb_6I_{11}; we include it here because it serves as a reminder that the discussion of the sign of $\Delta V = V_{HS} - V_{LS} > 0$ discussed in the text is not true for every spin transition. Ligand abbreviations are explained below.[a]

Reference	Year	Compound(s)	Maximm pressure (GPa)	T_{SCO} (K) in ambient pressure	$\frac{dT_{SCO}}{dp}$ (K GPa^{-1})
[40, 49]	1964, 1969	Several Fe(III) $[Fe(S_2CNR_2)_3]$ compounds	0.3	~120 K ($[Fe(S_2C{\cdot}N\{n\text{-}Bu\}_2)_3]$)	
[65]	1971	Several Fe(II) $[Fe(phen)_2(X)_2]$ compounds	1.4	Not stated in this article.	**
[52]	1982	$[Fe(phen)_2I_2]$ $[Fe(phen)_2(NCS)_2]$ $[Fe(phen)_2(NCSe)_2]$ $[Fe(phen)_2(N_3)_2]$	2.8 1.4 22 4.1	Isothermal data presented as a function of pressure	Not quantified due to sparse pressure data
[53]	1983	$[Fe(phen)_2(NCS)_2]$ $[Fe(phen)_2(NCSe)_2]$ $[Fe(phen)_2(N_3)_2]$	4.5 1.8 4.5	182 235 — (HS only)	90 (est.)* 80 (est.)*
[57]	1983	$[Fe_xZn_{1-x}(2\text{-pic})_3]Cl_2{\cdot}EtOH$ ($x = 0.5, 1.0$)	0.015	~120 K (x=1.0)	300*
[66]	1984	Nb_6I_{11}	15	274 [31]	negative
[67]	1985	$[Fe(phen)_2(NCS)_2]$ $[Fe(bipy)_2(NCS)_2]$	0.98 0.98	177.5 ~220	216 (avg.) negative

(*Continued*)

Table 5.2. (*Continued*)

Reference	Year	Compound(s)	Maximm pressure (GPa)	T_{SCO} (K) in ambient pressure	$\frac{dT_{SCO}}{dp}$ (K GPa^{-1})
[68]	1985	$[Fe(phy)_2](BF_4)_2$	0.24	279	~0**
		$[Fe(phy)_2](ClO_4)_2$		245	~0**
[38]	1990	$[Fe(2\text{-}pic)_3]Cl_2{\cdot}EtOH$	0.1350	120 K	150
		$[Fe(2\text{-}pic\text{-}ND_2)_3]Cl_2{\cdot}EtOD$	0.1200	135 K	167
[69]	1993	$[Fe(phen)_2(NCS)_2]$	1.0	176 K	195 (293 K at 0.6 GPa)
[70]	1994	$[Fe(bt)_2(NCS)_2]$	0.19	183 K	0**
[63]	1997	$[Fe(ptz)_6](PF_6)_2$	0.1	74 K	280
[71]	1998	$Fe(PM\text{-}L)_2(NCS)_2$;	1.0	Incomplete crossover	*
		L=A	1.08	189 K	~130
		L=Aza	1.0	(HS only)	—
		L=DMA	1.26	~170 K	~66
		L=BiA	0.26	~210 K	negative
		L=PEA			
[61]	1998	$[Fe(hyetrz)_3]A_2{\cdot}3H_2O$ (A^- = 3-nitrophenylsulfonate)	0.89	105	191 (275 K at 0.89 GPa)
[62]	2000	$[Fe(hyptrz)_3]$(4-chlorobenzenesulfonate)$_2{\cdot}H_2O$	0.59	185	236 (324 K at 0.59 GPa)

Notes: [a] Ligand abbreviations: bipy = 2,2′-bipyridine; bt = bis(2,2′-bithiazoline); hyetrz = 4-(2′-hydroxyethyl)-1,2,4-triazole; hyptrz = 4-(3′-hydroxypropyl)-1,2,4-triazole; phen = 1,10-phenanthroline; phy = 1,10-phenanthroline-2-carbaldehyde phenylhydrazone; pic = 2-picolylamine, 2-(aminomethyl)pyridine; PM-A = *N*-(2′-pyridylmethylene)aniline; PM-Aza = *N*-(2′-pyridylmethylene)-4-(phenylazo)aniline; PM-BiA = *N*-(2-pyridylmethylene)aminobiphenyl; PM-DMA *N*-(2′-pyridylmethylene)-2,6-dimethylaniline; PM-PEA = *N*-(2′-pyridylmethylene)-4-(phenylethynyl)aniline; ptz = 1-{*n*-propyl}tetrazole.

5.2.1.1 [$Fe(phen)_2(NCS)_2$]

This is perhaps the most studied of all SCO compounds, and, as can be seen from table 5.2, several authors have probed its response to applied pressure. As already mentioned, calorimetric data and ambient pressure magnetic susceptibility established the presence of a first-order spin transition at 174 K [51]. Table 5.2 reveals that, in common with other SCO complexes, the pressure variation of the spin transition is over 200 K GPa^{-1}.

5.2.1.2 [$Fe(2\text{-}pic)_3$]Cl_2·EtOH

Two early studies, in 1983 and in 1990, examined the effect of pressure on [Fe(2-pic)$_3$]Cl_2·EtOH and related complexes. The first of these, by Meissner *et al*, was at very low applied pressure (0.015 GPa) on [Fe_xZn_{1-x}(2-pic)$_3$]Cl_2·EtOH ($x = 0.5, 1.0$) [57]. The $x = 1.0$ material is notable because it was an early example of a 'two-step' transition, with a small plateau in the temperature-dependent magnetization at the transition temperature of 120 K [57, 58]. Temperature-dependent Mössbauer effect data were taken on both compounds in ambient pressure and under 0.015 GPa. In the case of the $x = 1.0$ compound, these were correlated with temperature-dependent x-ray diffraction data, in order to test the elastic model of Spiering *et al* [59], which could predict the HS population away from the plateau region.

The second study, by Köhler *et al*, involved the application of a larger pressures to [Fe(2-pic)$_3$]Cl_2·EtOH and [Fe(2-pic-ND_2)$_3$]Cl_2·EtOD [60] in a Mössbauer effect experiment. The two-step transition of the former compound in ambient pressure was seen to become a single gradual transition as pressure was applied. Due to the sparse nature of the pressure points (ambient, 0.6 GPa and 1.35 GPa), the exact pressure at which the phase transition changes could not accurately be identified. An applied pressure of up to 1.2 GPa was applied to the deuterated compound; in all applied pressures the spin transition was found to be gradual.

5.2.1.3 Polymeric SCO compounds including [$Fe(hyptrz)_3$](4-chlorobenzenesulfonate)$_2$ H_2O

In 1998, a study of the first-order one-dimensional polymeric SCO compound [Fe(hyetrz)$_3$]A_2·3H_2O (hyetrz = 4-(2′-hydroxyethyl)-1,2,4-triazole; $A^- =$ 3-nitrophenylsulfonate) was performed [61]. From magnetic susceptibility measurements up to 0.89 GPa, there was a shift in the spin transition temperature from 105 K at ambient pressure to 275 K under 0.89 GPa. The hysteresis width of around 10 K was preserved in this applied pressure.

In 2000, Garcia *et al* performed magnetic susceptibility experiments under hydrostatic pressure on another one-dimensional polymeric SCO compound, [Fe(hyptrz)$_3$](4-chlorobenzenesulfonate)$_2$·H_2O (hyptrz = 4-(4′-hydroxypropyl)-1,2,4-triazole), which has a first-order spin transition at 185 K in ambient conditions. This transition was raised to 324 K in a pressure of 0.59 GPa. The thermal hysteresis at the spin transition was diminished in a pressure of 0.41 GPa but increased again at the highest pressures, thus confirming that the first-order nature of the transition was retained under pressure. Given the one-dimensional nature of the polymeric chains, the authors of the study concluded that intrachain interactions were responsible for the preserved thermal hysteresis and first-order nature of the spin transition [62].

5.2.1.4 [Fe(ptz)$_6$](PF$_6$)$_2$

Jeftić *et al* [63] studied single crystals of [Fe(ptz)$_6$](PF$_6$)$_2$ (ptz = 1-propyltetrazole) via optical absorption, at three different pressures of up to 0.1 GPa. HS fraction data were obtained by comparing the integrated area under the $^1A_1 \rightarrow {}^1T_1$ absorption band to the same area at absolute zero. From a thermodynamic analysis based on a mean field model, the authors interpreted the relatively sharp spin transition as being near-critical. This transition broadened with applied pressure, with a non-linear variation of $T_{SCO}(p)$.

5.2.1.5 [CrI$_2$(depe)$_2$]

Ksenofontov *et al* carried out magnetic susceptibility measurements on *trans*-bis[1,2-bis-(diethylphosphino)ethane]di-iodochromium(II) ([CrI$_2$(depe)$_2$]) in pressures of up to 0.55 GPa [64]. This material has a sharp spin transition at 169 K at ambient pressure and is noteworthy in the context of this chapter because it is a rare example of a Cr(II) complex studied under applied pressure. The authors did not find any thermal hysteresis, and the transition broadened with applied pressure, with a highly non-linear variation of $T_{SCO}(p)$. The latter was attributed to the existence of a threshold pressure, below which only the iodide ion volumes are altered, and above which T_{SCO} starts to change significantly from its ambient pressure value, because the Cr–P distances between the Cr(II) ion and the donor P atoms begin to be affected [64].

5.2.2 Modeling of SCO compounds

The effect of pressure on SCO compounds can be captured by models which can be divided into three types: classical thermodynamic treatments, as reviewed in reference [56], microscopic (quantum mechanical) approaches as reviewed in reference [72] and mechanoelastic approaches, as reviewed in reference [73]. (We leave any discussion of *ab initio* modeling to the outlook section of this chapter.) Here were summarize only the key characteristics of the thermodynamic and quantum mechanical approaches since these have been most used in subsequent barocaloric studies to date.

5.2.2.1 Thermodynamic treatment

Semi-classical models of spin transitions express the free energy of the LS/HS system as a phase mixture, to take into account those transitions which are first order and in which there is thermal hysteresis. The free energy was expressed by Slichter and Drickamer in 1972 as [74]:

$$G = n_{LS}G_{LS} + n_{HS}G_{HS} - TS_{mix} + \Gamma n_{LS}n_{HS}.$$

Here, the molar fractions of the LS and HS states are n_{LS} and n_{HS}, respectively, while the free energies of those states are G_{LS} and G_{HS}. The entropy of mixing is given by

S_{mix} and Γ is a phenomenological parameter which indicates the degree of interaction between Fe centers (a cooperativity measure). By minimizing the free energy with respect to, for example, the LS state fraction, the population of LS or HS states can be mapped as a function of temperature. Slichter and Drickamer also included a pressure term in their 1972 work which enabled a first, thermodynamic treatment of the effect of pressure on spin transitions:

$$\Delta G(T, p) = \Delta G(T, p = 0) + p\Delta V + \frac{p^2}{2}\Delta\left(\frac{V(0)}{B}\right).$$

Here, the change of volume, ΔV is defined as $\Delta V = V_{HS} - V_{LS}$ is normally positive for the metal complexes most examined since. The bulk modulus is allowed to change from the LS to the HS state. As the interaction strength, Γ, is increased, the model of Slichter and Drickamer produces a change from a second-order thermally-induced spin transition to a first-order one. With a positive value of ΔV, the model also produces a positive $\frac{dT_{SCO}}{dp}$, along with a pressure-induced reduction of the thermal hysteresis at the spin transition, and, eventually, a pressure-induced critical point [56].

The above model allowed Slichter and Drickamer to introduce an equation of state, using a method analogous to the evaluation of the magnetization of a ferromagnet as a function of applied field and temperature [74]. The equation of state relates the HS fraction, n_{LS} to the essential thermodynamic parameters at the transition and the interaction parameter, Γ:

$$T = \frac{\Delta H - \Gamma(1 - 2n_{HS})}{R\ln(\frac{1-n_{HS}}{n_{HS}}) + \Delta S}$$

This model has been used extensively in the research literature to, for example, extract the interaction parameter from available thermodynamic data [48, 75]. We will refer to the use of the thermodynamic model in barocaloric studies after 2016 in section 5.3, where relevant.

5.2.2.2 Microscopic (quantum mechanical) models

As outlined by Gudyma and Ivashko [76], early microscopic models of spin crossover systems started with the viewpoint that the high spin fraction is a viable order parameter. In 1971, Wajnflasz [77] and Wajnflasz and Pick [78] used a (ferromagnetic) Ising model of fictitious spins, $\sigma = \pm 1$, associated with each Fe(II) ion. Applying a molecular field approximation (MFA), they found a first-order transition between LS and HS—because of the differing degeneracies of the LS and HS states—at an effective transition point was that at which the average spin, $\langle\sigma\rangle = 0$. However, their analysis was disputed in 1972 by Bari and Sivardière [79] because of a lack of dependence of the spin crossover temperature on the spin–spin interaction strength, J. Bari and Sivardière instead employed a coupling of the lattice distortion to the magnetic spins, and obtained a first-order transition in an extension of a thermodynamic model that was originally developed in 1964 by

Chesnut [80] to account for the temperature-dependent excitation of tetracyanoquinodimethan-ion-radical salts from their singlet ground state to a triplet excited state. Chesnut's model employed a free energy expansion in powers of excitation density that contained a quadratic 'effective particle–particle interaction' term, and within Chesnut's article was the suggestion that this effective interaction could have, as its origin, a coupling of the magnetic subsystem to a lattice distortion.

The timing of the evolution of such ideas in magneto-elastic coupling is significant; the Bean–Rodbell model of a first-order ferromagnetic ordering transition was published just two years prior to Chesnut's work [81] and so it seems that the development of models of spin transitions ran in parallel with the development of models of the Curie transitions that have been utilized in magnetocaloric research [82]. Following the work of Bari and Sivardière, in 1974 Kambara sought to make explicit the origin of the lattice–magnetism coupling, and suggested a Jahn–Teller coupling arising from the coupling of lattice distortion to the degenerate, HS state [83]. In 1977, Zimmerman and König introduced spin–orbit coupling and a dependence of the Debye temperature on the HS state occupancy, thereby suggesting the importance of atom–phonon coupling [84]. In what has since been regarding as a precursor of so-called 'elastic' models [73], atom–phonon coupling (APC) was explicitly introduced to the Hamiltonian in a one-dimensional SCO model by Nasser [85], who also showed that a two-step transition could be accounted for in this framework [86]. Further iterations of the APC model have been presented by Gîndulescu *et al* [87, 88].

Subsequent modelling efforts of bulk SCO properties have, for example, extended microscopic models of mononuclear spin transitions to give a better account of the experimentally observed temperature dependence of magnetic properties. As described by Pavlik and Linares in their 2018 review [72], simple microscopic models have LS–HS population curves which are limited in their form. Extensions to the microscopic model have therefore included: solvent disorder effects [89], bi-step transitions such as shown in figure 5.2(d) [90, 91], and a redefinition of the Ising model for binuclear species [92].

5.3 Barocaloric effects in spin crossover compounds

5.3.1 General thermodynamics and modeling

The giant barocaloric potential of SCO compounds was first suggested by Sandeman in 2016 [93] via an analysis of ambient pressure calorimetric data from the literature and the known pressure dependence of the spin transition temperature for several SCO systems with sub-room temperature values of T_{SCO}.

Table 5.3, reproduced from that paper, contains predictions of the maximum (irreversible) adiabatic temperature change, ΔT_{ad}, along with the pressures at which such values of ΔT_{ad} might be found for five compounds. Sandeman suggested that near-room temperature T_{SCO} should be explored for their barocaloric properties, while noting that thermal hysteresis at the spin transition would need to be minimized in order to maximize the reversible BCE.

Shortly afterwards, von Ranke *et al* published the first of a series of studies in which a microscopic, quantum mechanical model has been used to fit experimental

Table 5.3. Caloric potential of several SCO compounds, estimated from literature data using available isothermal entropy data, formula weights, and a typical specific heat capacity of 2000 J kg^{-1} K^{-1}. This table was first published in reference [93] and is reproduced with the permission of AIP Publishing. We see that the adiabatic temperature change available in modest applied pressures is substantial. The magnitude of the isothermal entropy change, ΔS is shown. Ligand abbreviations are explained below.[a]

Compound	T_{SCO} (K)	$\|\Delta S\|$ (J K^{-1} mol^{-1})	$\|\Delta S\|$ (J K^{-1} kg^{-1})	$\frac{dT_{SCO}}{dp}$ (K GPa^{-1})	Maximum ΔT_{ad} (K)	Pressure required (GPa)	References for T_{SCO}, ΔS, $\frac{dT_{SCO}}{dp}$
$[Fe(PM\text{-}BiA)_2(NCS)_2]$	170	58	84	66	8.4	0.12	[71, 94]
$[Fe(phen)_2(NCS)_2]$	180	49	92	220	8.3	0.04	[64, 95]
$[Fe(H_2Bpz_2)_2(bipy)]$	160	48	95	188	7.6	0.04	[96–98]
$[Fe(pmd)_2\{Cu(CN)_2\}_2]$	140	36	80.5	380 (avg.)	5.6	0.013	[99]
$[Fe(pmea)_2(NCS)_2]$	184	60 (calc.)	44	146	4	0.03	[100]

Notes: [a] Ligand abbreviations: bipy = 2,2′-bipyridine; H_2Bpz_2 = dihydrido-bis(pyrazol-1-yl)borate; phen = 1,10-phenanthroline; PM-BiA = *N*-(2-pyridylmethylene) aminobiphenyl; pmd = pyrimidine; pmea = bis(pyrid-2-ylmethyl)(2-{pyrid-2-yl}ethyl)amine.

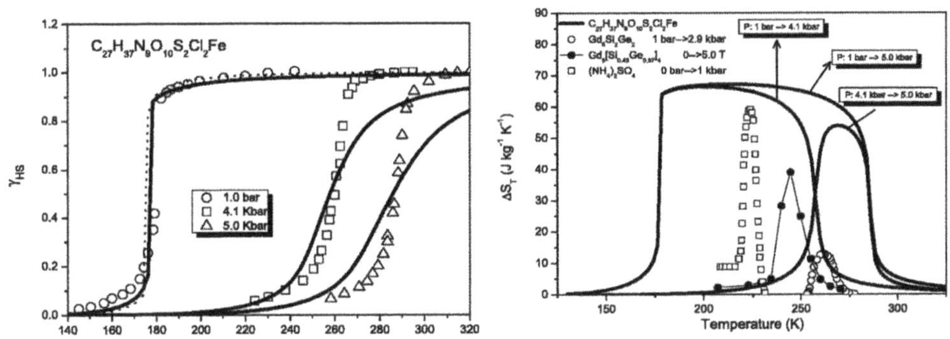

Figure 5.6. Left: A fit of a microscopic Ising-like model (solid lines) to the known temperature- and pressure-variation of the HS fraction in the 1D polymeric SCO compound $[Fe(hyptrz)_3](4\text{-chlorophenylsulfonate})_2{\cdot}H_2O$ (open symbols). Right: based on this fit, von Ranke *et al* were able to predict giant BCE (ΔS) in this material [101]. In addition, the authors compared the predicted BCE with the magnetocaloric effect in Gd–Si–Ge and the BCE in ammonium sulfate. The experimental data used in this modelling study (left subfigure) were obtained from reference [62]. Both figures are reproduced from von Ranke *et al* [101] with the permission of AIP Publishing.

data from the literature, and thereby predict the barocaloric effect [101]. In the first study [101], a microscopic Ising model was used to fit the known pressure dependence of the high spin fraction in the one-dimensional polymeric SCO compound $[Fe(hyptrz)_3](4\text{-chlorophenylsulfonate})_2{\cdot}H_2O$ as a function of temperature and pressure. The experimental data in that first BCE modelling work were obtained from reference [62] and are described in section 5.2 of this chapter. In a method reminiscent of the early work of Bari and Sivardière [79], the model employed by von Ranke *et al* used two order parameters: the fictitious spin and the volumetric deformation. These two order parameters were coupled through a Taylor expansion of the spin coupling parameter. Thermodynamic predictions were made possible by use of the mean field approximation (MFA). Figure 5.6 shows these first BCE-driven modeling results of an SCO compound: the fitting of the HS fraction, γ_{HS}, as a function of temperature and applied pressure, and the resulting $\Delta S(T, p)$ curves, for several different changes of pressure. The authors noted the strong pressure variation of the ligand field parameter with pressure, indicative of the strong effect of pressure on the Fe–N bond length.

A follow-up study by von Ranke and co-workers extended modeling of the same material ($[Fe(hyptrz)_3](4\text{-chlorophenylsulfonate})_2{\cdot}H_2O$), to include the adiabatic temperature change and refrigerant capacity for a change of applied pressure of 0.09 GPa, from 0.41 GPa to 0.5 GPa [102]. These pressure values were determined by the requirement to fit to the same experimental data in reference [62], which are relatively sparse in pressure. The applied pressure brings the spin transition close to room temperature. For the investigated pressure change, from 0.41 to 0.5 GPa, the models established an expected adiabatic temperature change of 10 K.

An APC-type microscopic model of experimental data from Jeftić *et al* [63] on [Fe $(ptz)_6](PF_6)_2$(ptz = 1-propyltetrazole) was presented by von Ranke *et al* in 2018 [103]. This model was the first to assess the BCE in a SCO complex in terms of both ΔS and ΔT_{ad} from zero pressure to some finite pressure; the experimental data obtained in reference [63] on $[Fe(ptz)_6](PF_6)_2$ allowed for this. At ambient pressure, the material has a sharp—but not first-order—spin transition at 74 K. Applied pressure increases the spin transition temperature at a rate of 280 K GPa^{-1} in the 0–0.1 GPa pressure range [63]. The authors of reference [103] interpolated their microscopic model out to applied pressures of 0.2 GPa, finding that, for a change in applied pressure from ~0 to 0.1 GPa, $\Delta S = -46$ J K^{-1} kg^{-1} and $\Delta T = 27$ K, which can be classed as giant BCE values. Reference [103] also proposed a thermodynamic Maxwell-type relation specific to the calculation of BCEs in SCO compounds:

$$\left(\frac{\partial S}{\partial p}\right)_T = -\Delta V_{LH} \times \left(\frac{\partial \gamma_{HS}}{\partial T}\right)_p,$$

integration of which can be used to estimate the pressure-induced isothermal entropy change:

$$\Delta S(T, \Delta p) = -\frac{\Delta V_{LH}}{\Delta T}\left[\int_{p_1}^{p} \gamma_{HS}(T + \Delta T)dp - \int_{p_1}^{p_2} \gamma_{HS}(T)dp\right].$$

Here, ΔV_{LH} is the volume difference between the LS and HS states, and γ_{HS} is the HS state fraction.

Further modelling work of the BCE in SCOs by von Ranke and co-workers has included a 2021 study of the thermal expansion coefficient and the expected BCE in $\{Fe[H_2B(pz)_2]_2(bipy)\}$ (bipy = 2,2′-bipyridine) [104], based on magnetic susceptibility data published by Galet *et al* in 2006 [96]. The model used was of the same APC-type as in reference [103]. A 2022 study of the same model examined the impact of the crystalline field on the order of the spin transition and any associated BCEs [105]. The authors further proposed that the size of the volume change, ΔV_{LH}, at a spin transition may be calculated from integration of barocaloric entropy change curves:

$$|\Delta V_{LH}| = \frac{1}{\Delta p}\int_0^{\infty} \Delta S(T, \Delta p)dT.$$

5.3.2 Experimental studies

The first direct experimental evidence of a giant BCE near room temperature in a spin crossover material was obtained by Vallone *et al* [9], in a study of the mononuclear Fe(II) complex, $[FeL_2][BF_4]_2$ [L = 2,6-di(pyrazol-1-yl)pyridine]. This material has a first-order spin transition at 260 K (the average of the heating and cooling values) and a thermal hysteresis of 4 K. It had been the subject of considerable structural investigation at ambient pressure by Halcrow and co-workers [106–109]. The BCE study by Vallone *et al* combined evidence from neutron diffraction under pressures of up to 0.12 GPa, magnetometry under

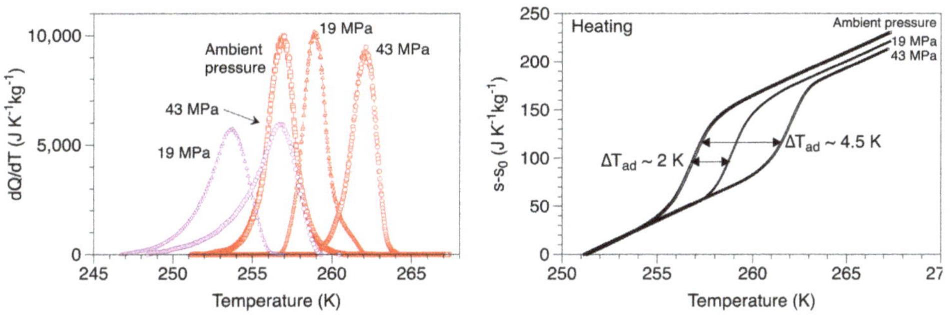

Figure 5.7. Calorimetry data on partially deuterated $[FeL_2][BF_4]_2$, showing how the spin crossover transition shifts in 19 MPa and 43 MPa on heating (red) and cooling (blue). Data in ambient conditions are shown in heating, only. Calculated entropy versus temperature showing that the maximum irreversible adiabatic temperature change in 19 MPa is 2 K and in 43 MPa is about 4.5 K. Entropy is plotted as $s–s_0$, where s_0 is the entropy at ambient pressure and 251 K, away from the phase transition. In the construction of the entropy curves, a fixed background heat capacity of 3R per mole-atom has been added. Figure reproduced with permission from [9]. Copyright 2019 Wiley-VCH Verlag GmbH & Co. KGaA, Weinheim.

pressures of up to 0.49 GPa and calorimetry under pressures of up to 0.043 GPa. Figure 5.7 shows the calorimetry data (left) and the resulting entropy in pressures of 0.019 GPa and 0.043 GPa, measured relative to the entropy, s_0 measured at ambient pressure and 251 K.

From calorimetry data, an irreversible ΔT_{ad} value of over 4 K in a pressure change of 0–0.043 GPa was extracted. To put this result in another caloric context, a temperature change of 2 K is of the same order as a 0–0.7 Tesla magnetic field change acting on the magnetocaloric effect (MCE) material, Gd, at room temperature. Similarly, the 0–1 Tesla MCE in Gd (ΔT_{ad}~3 K) would equate to a pressure of change of 0–300 bar (0–0.030 GPa) here. The entropy change found on heating was pressure dependent, ranging from 86 J K^{-1} kg^{-1} at ambient pressure to 68 J K^{-1} kg^{-1} at 0.043 GPa.

Of note is the presence of a 4 K thermal hysteresis which reduces the *reversible* effect in this material to zero if the pressure applied is less than about 0.04 GPa. From the calorimetry data in figure 5.7 (left) taken on heating and cooling, the heating curve under ambient pressure overlaps with the cooling curve taken under the maximum pressure of the calorimetry experiment, 0.043 GPa. In other words, the thermal hysteresis is only closed by a pressure of 0.043 GPa. This is to be expected, as the authors found that $\frac{dT_{\mathrm{SCO}}}{dp}$ was of the order of 100 K GPa^{-1} and so a thermal hysteresis of 4 K at the spin transition should require a pressure of around 0.04 GPa for an isobaric calorimetry curve taken on cooling to be shifted enough in temperature so as to overlap with the ambient pressure heating curve. What this means in practice is that the reversible BCE in this compound is zero until at least ~0.04 GPa is applied. A similar statement can be made about any first-order caloric material, for example the reversible component of the MCE in MnAs is relatively small until a magnetic field equivalent to the thermal hysteresis is applied [110].

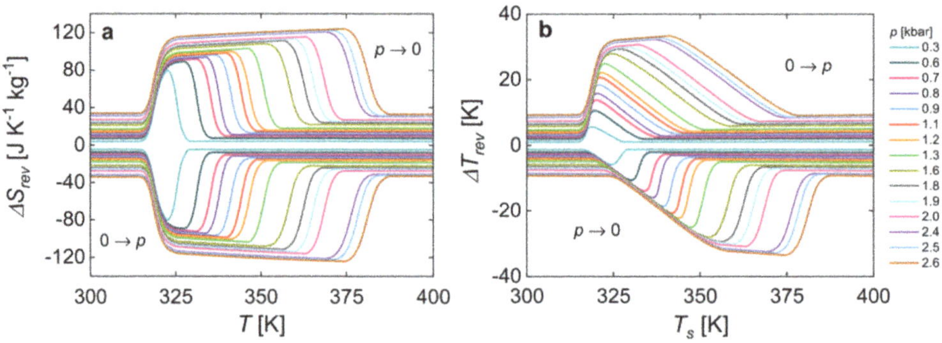

Figure 5.8. The reversible pressure-driven entropy change and adiabatic temperature change for the trinuclear Fe(II) complex, $Fe_3(bntrz)_6(tcnset)_6$ (FBT). Data are shown on compression and decompression at the pressures indicated in the legend. In addition to the BCE, which grows significantly with applied pressure, the strong variation of spin transition temperature with pressure ($\frac{dT_{SCO}}{dp}$ = 250K GPa^{-1}) can be seen. This figure is reproduced with permission from reference [111]. Copyright -2021 Wiley-VCH Verlag GmbH & Co. KGaA, Weinheim.

This first experimental study demonstrated the potential of SCO compounds but also how imperative it is to reduce thermal hysteresis as much as possible in order to maximize the use of low pressures. Accordingly, the two subsequent experimental studies of BCEs in spin crossover materials have examined BCEs in materials with lower thermal hysteresis at the spin transition.

In a 2021 study of the trinuclear Fe(II) complex, $Fe_3(bntrz)_6(tcnset)_6$ (FBT), where bntrz = 4-(benzyl)-1,2,4-triazole and tcnset = 1,1,3,3-tetracyano-2-thioethylepropenide, Romanini *et al* observed a giant BCE via calorimetry in pressures of up to 0.26 GPa [111]. The study combined calorimetry with x-ray diffraction under the same maximum pressure. Importantly, FBT has a first-order transition (T_{SCO} = 318K, with $\frac{dT_{SCO}}{dp}$ = 250K GPa^{-1}) with a thermal hysteresis of only 2 K which remained relatively pressure-independent. This enabled a giant reversible BCE to be obtained: ΔT_{ad} = 6 K and ΔS=80 J K^{-1} kg^{-1} for compression to 0.033 GPa and ΔT_{ad} = 35 K and ΔS=120 J K^{-1} kg^{-1} for compression to 0.26 GPa. The reversible pressure-driven entropy change and adiabatic temperature change for FBT obtained in reference [111] are shown in figure 5.8.

The most recent experimental exploration of the BCE in a spin crossover compound at the time of writing was in 2022 by Seo *et al* [112] and was of a mononuclear Fe(II) complex, $Fe[HB(tz)_3]_2$ which has a first-order spin transition at 332 K and yet lower thermal hysteresis than FBT, at only 0.2 K. The structure and spin transition of the material had been examined in previous studies [113–117], with thermal hysteresis values of around 2 K observed [117]. Prior to the work of Seo *et al*, attention had been given to a number of features and/or potential uses of the material: the heating rate-dependence of the thermal hysteresis at the spin transition, the possibility of depositing thin films for smart device applications [118], the use of the material in a tunable Ag/SCO/Ag multilayer Fabry–Perot cavity [119], and the mechanical properties of the material on either side of its spin transition [120]. The extremely low thermal hysteresis value found by Seo *et al* [112] was determined by finding the onset of the spin transition in the differential scanning calorimetry (DSC) as opposed to using peak calorimetry values. Such a low thermal hysteresis places the material at the near-tricritical ideal for a caloric compound [35].

Seo *et al* combined x-ray diffraction in pressures up to 0.03 GPa with calorimetry up to 0.015 GPa, focusing on the barocaloric performance in low applied pressures. From the calorimetry data, the reversible ΔS was obtained from the overlap of the compression-driven and decompression-driven isothermal entropy change curves, as shown in figure 5.9(c), and the pressure hysteresis was found to be consistent with the low thermal hysteresis of 0.2 K. From figure 5.9(c), it is clear that a large proportion of the available entropy change of around 99 J K^{-1} kg^{-1} is available at a very low applied pressure of only 0.004 GPa. This is a result of the low thermal hysterisis and the

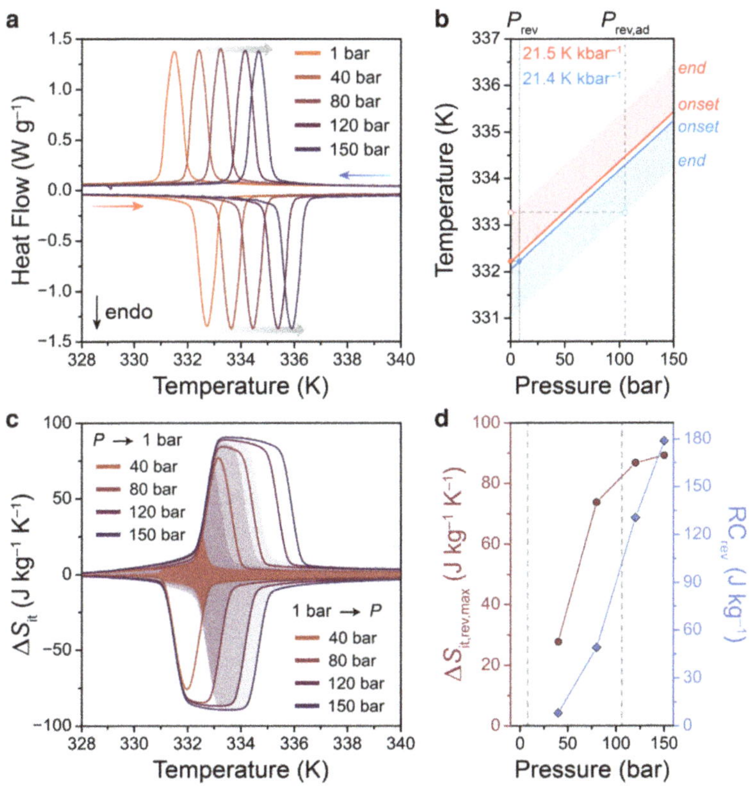

Figure 5.9. The pressure sensitivity of the spin transition and the associated barocaloric effect in $Fe[HB(tz)_3]_2$, as determined by Seo *et al* [112]. In (a), the effect of up to 0.015 GPa of applied pressure is shown on the calorimetry signal associated with the spin transition. The resulting phase diagram of the LS/HS states is shown in (b). From the calorimetry data, the isothermal entropy change is determined and shown in (c). The reversible component is shown as the shaded overlap between compression and decompression curves. In (d), the reversible isothermal entropy change and the refrigerant capacity, *RC*, are shown as a function of applied pressure. This figure is reproduced with permission from reference [112]. Copyright 2022 American Chemical Society.

pressure dependence of the spin transition, which in this case is $\frac{dT_{SCO}}{dp} = 215\ \mathrm{K\ GPa^{-1}}$—see figure 5.9(a). Accordingly, the pressure above which reversible BCEs may be expected is only 0.2K/(215K/GPa)≈ 0.001 GPa. The actual reversible pressure is shown as P_{rev} on the phase diagram in figure 5.9(b). The authors of the study were able to calculate that the dissipation due to thermal hysteresis accounts for only 0.7% of the available thermal energy. Reversible entropy change values and refrigerant capacity are given in figure 5.9(d).

5.4 Perspectives for future development

From the few experimental studies of BCE on SCOs that exist, it is already clear that very large, reversible BCEs are possible in this material class. The research field is at only an early stage in terms of optimization of the total BCE. For example, the range of structure types selected for BCE analysis thus far has been relatively limited, as compared to the variety of structures in which SCO is found (outlined towards the start of the chapter). Similarly, the temperature range at which this particular BCE may be found—and optimized—is ripe for exploration, given that the crystal field, and hence spin transition temperature, can be highly influenced by the choice of ligand [38]. Chemical doping can be used to tune the spin transition temperature predictably and, in favourable cases, can be used to reduce or remove hysteresis from first-order SCO materials [121–123].

In this context, it may be argued that—as with other caloric materials—the development and optimization of the BCE in spin transition materials could benefit from the increased use of *ab initio* predictions across all forms of material discovery [124]. Here, as with the development of other caloric materials, there are some difficulties which must be, and are being, overcome. The difficulties associated with predicting the transition temperatures of (ferro-) and anti-ferromagnetic materials for magnetocaloric effect development are well known. In the case of barocaloric effects, structure prediction at finite temperatures is most important. And here, the prediction of spin transitions has a particular difficulty because of the size of exchange correlations [125, 126]. It has been noted elsewhere that even the predicted existence of a spin gap can be dependent on the exchange functional used in a density functional theory (DFT) calculation [127, 128]. Therefore, recent advances in density-corrected DFT [128], and in the parameterization of a Hubbard U parameter to use in tandem with an appropriate DFT functional [129], are among the most promising with regard to the development of BCE in SCO 'by design'.

In conclusion, this interdisciplinary research field, which can harness the latest developments in ligand chemistry, caloric measurement, and *ab initio* modelling, is just in its infancy. It may also contribute to, and further motivate, future applications of SCO materials involving thermal energy management [130] or photo-induction of the spin transition [131].

References

[1] Cambi L and Szegö L 1931 Über die magnetische Susceptibilität der komplexen Verbindungen *Berichte der Dtsch. Chem. Gesellschaft (A B Ser.)* **64** 2591–8

[2] Cambi L, Szegö L and Cagnasso A 1932 The magnetic susceptibility of complexes. IV. Ferric *N*,*N*-dipropyldithiocarbamates *Atti dell'Accademia Naz. Dei Lincei* **13** 266–71
[3] Cambi L, Szegö L and Cagnasso A 1932 The magnetic susceptibility of complexes. V. Iron dibutyldithio-carbamates *Atti dell'Accademia Naz. Dei Lincei* **13** 329–35
[4] Cambi L and Cagnasso A 1931 Iron dithiocarbamates and nitroso dithiocarbamates *Atti dell'Accademia Naz. Dei Lincei* **13** 809–13
[5] Cambi L and Szegö L 1933 Über die magnetische Susceptibilität der komplexen Verbindungen (II. Mitteil.) *Berichte der Dtsch. Chem. Gesellschaft (A B Ser.)* **66** 656–61
[6] Cambi L and Cagnasso A 1933 The structure and magnetic susceptibility of complex ferrous compounds *Gazz. Chim. Ital.* **63** 767–78
[7] Cambi L and Malatesta L 1937 Magnetismus und Polymorphie innerer Komplexsalze: Eisensalze der Dithiocarbamidsäuren *Berichte der Dtsch. Chem. Gesellschaft (A B Ser.)* **70** 2067–78
[8] Gütlich P 2013 Spin crossover—quo vadis? *Eur. J. Inorg. Chem.* **2013** 581–91
[9] Vallone S P, Tantillo A N, dos Santos A M, Molaison J J, Kulmaczewski R, Chapoy A, Ahmadi P, Halcrow M A and Sandeman K G 2019 Giant barocaloric effect at the spin crossover transition of a molecular crystal *Adv. Mater.* **31** 1807334
[10] Halcrow M A 2011 Structure:function relationships in molecular spin-crossover complexes *Chem. Soc. Rev.* **40** 4119–42
[11] Kumar K S and Ruben M 2017 Emerging trends in spin crossover (SCO) based functional materials and devices *Coord. Chem. Rev.* **346** 176–205
[12] Kumar K S and Ruben M 2021 Sublimable spin-crossover complexes: from spin-state switching to molecular devices *Angew. Chem. Int. Ed.* **60** 7502–21
[13] Kipgen L, Bernien M, Tuczek F and Kuch W 2021 Spin-crossover molecules on surfaces: from isolated molecules to Ultrathin films *Adv. Mater.* **33** 2008141
[14] Gaspar A B, Ksenofontov V, Seredyuk M and Gütlich P 2005 Multifunctionality in spin crossover materials *Coord. Chem. Rev.* **249** 2661–76
[15] Bousseksou A, Molnár G, Salmon L and Nicolazzi W 2011 Molecular spin crossover phenomenon: recent achievements and prospects *Chem. Soc. Rev.* **40** 3313–35
[16] Murray K S 2013 The development of spin-crossover research *Spin-Crossover Materials: Properties and Applications* (New York: Wiley) 1st edn pp 1–54
[17] Brooker S 2015 Spin crossover with thermal hysteresis: practicalities and lessons learnt *Chem. Soc. Rev.* **44** 2880–92
[18] Harding D J, Harding P and Phonsri W 2016 Spin crossover in iron(III) complexes *Coord. Chem. Rev.* **313** 38–61
[19] Molnár G, Rat S, Salmon L, Nicolazzi W and Bousseksou A 2018 Spin crossover nanomaterials: from fundamental concepts to devices *Adv. Mater.* **30** 1703862
[20] Coronado E 2019 Molecular magnetism: from chemical design to spin control in molecules, materials and devices *Nat. Rev. Mater.* **5** 87–104
[21] Kumar K S, Bayeh Y, Gebretsadik T, Elemo F, Gebrezgiabher M, Thomas M and Ruben M 2019 Spin-crossover in iron(II)–Schiff base complexes *Dalt. Trans.* **48** 15321–37
[22] Gütlich P, Garcia Y and Goodwin H A 2000 Spin crossover phenomena in Fe(II) complexes *Chem. Soc. Rev.* **29** 419–27
[23] Muñoz M C and Real J A 2013 Polymeric spin-crossover materials *Spin-Crossover Materials: Properties and Applications* (New York: Wiley) 1st edn pp 214–44

[24] Bigdeli F, Lollar C T, Morsali A and Zhou H 2020 Switching in metal–organic frameworks *Angew. Chem. Int. Ed.* **59** 4652–69

[25] Southon P D, Liu L, Fellows E A, Price D J, Halder G J, Chapman K W, Moubaraki B, Murray K S, Létard J-F and Kepert C J 2009 Dynamic interplay between spin-crossover and host–guest function in a nanoporous metal–organic framework material *J. Am. Chem. Soc.* **131** 10998–1009

[26] Huang W, Ma X, Sato O and Wu D 2021 Controlling dynamic magnetic properties of coordination clusters via switchable electronic configuration *Chem. Soc. Rev.* **50** 6832–70

[27] Sitzmann H 2001 Maximum spin cyclopentadienyl complexes of 3d transition metals *Coord. Chem. Rev.* **214** 287–327

[28] Meredith M B, Crisp J A, Brady E D, Hanusa T P, Yee G T, Pink M, Brennessel W W and Young V G 2008 Tunable spin-crossover behavior in Polymethylated bis(indenyl)chromium(II) complexes: the significance of benzo-ring substitution *Organometallics* **27** 5464–73

[29] Angaridis P, Cotton F A, Murillo C A, Villagrán D and Wang X 2005 Structural and magnetic evidence concerning spin crossover in formamidinate compounds with Ru_2^{5+} cores *J. Am. Chem. Soc.* **127** 5008–9

[30] Clérac R, Cotton F A, Daniels L M, Dunbar K R, Kirschbaum K, Murillo C A, Pinkerton A A, Schultz A J and Wang X 2000 Linear tricobalt compounds with di(2-pyridyl)amide (dpa) ligands: temperature dependence of the structural and magnetic properties of symmetrical and unsymmetrical forms of $Co_3(dpa)_4Cl_2$ in the solid state *J. Am. Chem. Soc.* **122** 6226–36

[31] Imoto H and Simon A 1982 Structural study of the spin-crossover transition in the cluster compounds niobium iodide (Nb_6I_{11}) and hydrogen niobium iodide (HNb_6I_{11}) *Inorg. Chem.* **21** 308–19

[32] Rao C N R, Cheetham A K and Thirumurugan A 2008 Hybrid inorganic–organic materials: a new family in condensed matter physics *J. Phys. Condens. Matter* **20** 083202

[33] Rotter M, Wang Z S, Boothroyd A T, Prabhakaran D, Tanaka A and Doerr M 2014 Mechanism of spin crossover in $LaCoO_3$ resolved by shape magnetostriction in pulsed magnetic fields *Sci. Rep.* **4** 7003

[34] Okimoto Y, Saitoh T, Kobayashi Y and Ishihara S (ed) (ed.) 2021 *Spin-Crossover Cobaltite* (Singapore: Springer)

[35] Takeuchi I and Sandeman K G 2015 Solid-state cooling with caloric materials *Phys. Today* **68** 48–54

[36] Gottschall T, Gràcia-Condal A, Fries M, Taubel A, Pfeuffer L, Mañosa L, Planes A, Skokov K P and Gutfleisch O 2018 A multicaloric cooling cycle that exploits thermal hysteresis *Nat. Mater.* **17** 929–34

[37] Seredyuk M, Znovjyak K, Valverde-Muñoz F J, da Silva I, Muñoz M C, Moroz Y S and Real J A 2022 105 K wWide room temperature spin transition memory due to a supramolecular latch mechanism *J. Am. Chem. Soc.* **144** 14297–309

[38] Gütlich P, Hauser A and Spiering H 1994 Thermal and optical switching of iron(II) complexes *Angew. Chem. Int. Ed. Engl.* **33** 2024–54

[39] Gütlich P, Garcia Y and Spiering H 2003 Spin transition phenomena *Magnetism: Molecules to Materials* (Germany: Wiley-VCH) vol 34 pp 271–344

[40] Ewald A H, Martin R L, Ross I G and White A H 1964 Anomalous behaviour at the $^6A_1$–$^2T_2$ crossover in iron (III) complexes *Proc. R. Soc. London. Ser. A. Math. Phys. Sci.* **280** 235–57

[41] van Santen J H and van Wieringen J S 1952 Some remarks on the ionic radii of iron-group elements. The influence of crystalline field *Recl. Des Trav. Chim. Des Pays-Bas* **71** 420–30
[42] Gütlich P, Gaspar A B and Garcia Y 2013 Spin state switching in iron coordination compounds *Beilstein J. Org. Chem.* **9** 342–91
[43] Urakawa A, Van Beek W, Monrabal-Capilla M, Galán-Mascarós J R, Palin L and Milanesio M 2011 Combined, modulation enhanced X-ray powder diffraction and Raman spectroscopic study of structural transitions in the spin crossover material $[Fe(Htrz)_2(trz)](BF_4)$ *J. Phys. Chem.* C **115** 1323–9
[44] Grosjean A, Négrier P, Bordet P, Etrillard C, Mondieig D, Pechev S, Lebraud E, Létard J F and Guionneau P 2013 Crystal structures and spin crossover in the polymeric material $[Fe(Htrz)_2(trz)](BF_4)$ including coherent-domain size reduction effects *Eur. J. Inorg. Chem.* **2** 796–802
[45] Piedrahita-Bello M, Angulo-Cervera J E, Enriquez-Cabrera A, Molnár G, Tondu B, Salmon L and Bousseksou A 2021 Colossal expansion and fast motion in spin-crossover@polymer actuators *Mater. Horizons* **8** 3055–62
[46] Sorai M and Seki S 1974 Phonon coupled cooperative low-spin $^1A_1 \rightleftarrows$high-spin $^5T_2$ transition in $[Fe(phen)_2(NCS)_2]$ and $[Fe(phen)_2(NCSe)_2]$ crystals *J. Phys. Chem. Solids* **35** 555–70
[47] Kaji K and Sorai M 1985 Heat capacity and dual spin-transitions in the crossover system $[Fe(2\text{-}pic)_3]Cl_2{\cdot}EtOH$ *Thermochim. Acta* **88** 185–90
[48] Gütlich P, Ksenofontov V and Gaspar A 2005 Pressure effect studies on spin crossover systems *Coord. Chem. Rev.* **249** 1811–29
[49] Ewald A H, Martin R L, Sinn E and White A H 1969 Electronic equilibrium between the $^6A_1$ and $^2T_2$ states in iron(III) dithio chelates *Inorg. Chem.* **8** 1837–46
[50] Fisher D C and Drickamer H G 1971 Effect of pressure on the spin state of iron in ferrous phenanthroline compounds *J. Chem. Phys.* **54** 4825–37
[51] König E and Madeja K 1967 $^5T_2$–$^1A_1$ equilibriums in some iron(II)-bis(1,10-phenanthroline) complexes *Inorg. Chem.* **6** 48–55
[52] Adams D M, Long G J and Williams A D 1982 Spectroscopy at very high pressures. 36. An infrared study of spin-state equilibriums in some iron(II) complexes *Inorg. Chem.* **21** 1049–53
[53] Pebler J 1983 Reinvestigation of the iron-57 Moessbauer effect at high pressures in some iron(II) bis(phenanthroline) complexes *Inorg. Chem.* **22** 4125–8
[54] Debrunner P, Vaughan R W, Champion A R, Cohen J, Moyzis J and Drickamer H G 1966 Versatile high pressure Mössbauer apparatus *Rev. Sci. Instrum.* **37** 1310–5
[55] Ksenofontov V, Gaspar A B and Gütlich P 2004 Pressure effect studies on spin crossover and valence tautomeric systems *Spin Crossover in Transition Metal Compounds III* (Berlin: Springer) pp 23–64
[56] Gaspar A B, Molnár G, Rotaru A and Shepherd H J 2018 Pressure effect investigations on spin-crossover coordination compounds *C. R. Chim.* **21** 1095–120
[57] Meissner E, Köppen H, Spiering H and Gütlich P 1983 The effect of low pressure on a high-spin—low-spin transition *Chem. Phys. Lett.* **95** 163–6
[58] Köppen H, Müller E and Köhler C 1982 Unusual spin-transition anomaly in the crossover system $[Fe(2\text{-}pic)_3]$ Cl_2 EtOH *Chem. Phys.* **91** 348–52
[59] Spiering H, Meissner E, Köppen H, Müller E W and Gütlich P 1982 The effect of the lattice expansion on high spin $\rightleftharpoons$ low spin transitions *Chem. Phys.* **68** 65–71

[60] Köhler C P, Jakobi R, Meissner E, Wiehl L, Spiering H and Gütlich P 1990 Nature of the phase transition in spin crossover compounds *J. Phys. Chem. Solids* **51** 239–47
[61] Garcia Y, Van Koningsbruggen P J, Lapouyade R, Fournès L, Rabardel L, Kahn O, Ksenofontov V, Levchenko G and Gütlich P 1998 Influences of temperature, pressure, and lattice solvents on the spin transition regime of the polymeric compound $[Fe(hyetrz)_3]A_2{\cdot}3H_2O$ (hyetrz = 4-(2′-hydroxyethyl)-1,2,4-triazole and A^- = 3-nitrophenylsulfonate) *Chem. Mater.* **10** 2426–33
[62] Garcia Y, Ksenofontov V, Levchenko G and Gütlich P 2000 Pressure effect on a novel spin transition polymeric chain compound *J. Mater. Chem.* **10** 2274–6
[63] Jeftić J, Hinek R, Capelli S C and Hauser A 1997 Cooperativity in the iron(II) spin-crossover compound $[Fe(ptz)_6](PF_6)_2$ under the influence of external pressure (ptz = 1-*n*-propyltetrazole) *Inorg. Chem.* **36** 3080–7
[64] Ksenofontov V, Gaspar A B, Levchenko G, Fitzsimmons B and Gütlich P 2004 Pressure effect on spin crossover in $[Fe(phen)_2(NCS)_2]$ and $[CrI_2(depe)_2]$ *J. Phys. Chem.* B **108** 7723–7
[65] Fisher D C 1971 Effect of pressure on the spin state of iron in ferrous phenanthroline compounds *J. Chem. Phys.* **54** 4825–37
[66] Wu M K, Chu C W, Disalvo F J and Tarascon J M 1984 High pressure study of the cluster compound Nb_6I_{11} *Mater. Res. Soc. Symp. Proc.* 125–7
[67] Usha S, Srinivasan R and Rao C N R 1985 High-pressure magnetic susceptibility studies of spin-state transition in Fe(II) complexes *Chem. Phys.* **100** 447–55
[68] König E, Ritter G, Waigel J and Goodwin H A 1985 The effect of pressure on the thermal hysteresis of the first-order spin transition in bis(1,10-phenanthroline-2-carbaldehyde phenylhydrazone) iron (II) complexes *J. Chem. Phys.* **83** 3055–61
[69] Granier T, Gallois B, Gaultier J, Real J A and Zarembowitch J 1993 High-pressure single-crystal X-ray diffraction study of two spin-crossover iron(II) complexes: $Fe(Phen)_2(NCS)_2$ and $Fe(Btz)_2(NCS)_2$ *Inorg. Chem.* **32** 5305–12
[70] König E, Ritter G, Gruensteudel H, Dengler J and Nelson J 1994 Effect of a simultaneous change of temperature and pressure on the spin-state transition in bis(thiocyanato)bis(2,2′-bi-2-thiazoline)iron(II) *Inorg. Chem.* **33** 837–9
[71] Ksenofontov V, Levchenko G, Spiering H, Gütlich P, Létard J F, Bouhedja Y and Kahn O 1998 Spin crossover behavior under pressure of $Fe(PM\text{-}L)_2(NCS)_2$ compounds with substituted 2′-pyridylmethylene 4-anilino ligands *Chem. Phys. Lett.* **294** 545–53
[72] Pavlik J and Linares J 2018 Microscopic models of spin crossover *C. R. Chim.* **21** 1170–8
[73] Enachescu C and Nicolazzi W 2018 Elastic models, lattice dynamics and finite size effects in molecular spin crossover systems *C. R. Chim.* **21** 1179–95
[74] Slichter C P and Drickamer H G 1972 Pressure-induced electronic changes in compounds of iron *J. Chem. Phys.* **56** 2142–60
[75] Levchenko G, Gaspar A B, Bukin G, Berezhnaya L and Real J A 2018 Pressure effect studies on the spin transition of microporous 3D polymer $[Fe(pz)Pt(CN)_4]$ *Inorg. Chem.* **57** 8458–64
[76] Gudyma I V and Ivashko V V 2016 Spin-crossover molecular solids beyond rigid crystal approximation *Nanoscale Res. Lett.* **11** 196
[77] Wajnflasz J 1970 Etude de la transition ‘low spin’–‘high spin’ dans les complexes octaédriques d’ion de transition *Phys. Status Solidi* **40** 537–45

[78] Wajnflasz J and Pick R 1971 Transitions 'low spin'–'high spin' Dans Les Complexes de Fe^{2+} *Le J. Phys. Colloq.* **32** C1–91-C1–92
[79] Bari R A and Sivardière J 1972 Low-spin–high-spin transitions in transition-metal-ion compounds *Phys. Rev.* B **5** 4466–71
[80] Chesnut D B 1964 Simple model for some dense magnetic exciton systems *J. Chem. Phys.* **40** 405–11
[81] Bean C and Rodbell D 1962 Magnetic disorder as a first-order phase transformation *Phys. Rev.* **126** 104–15
[82] Basso V 2011 The magnetocaloric effect at the first-order magneto-elastic phase transition *J. Phys. Condens. Matter* **23** 226004
[83] Kambara T 1979 Theory of high-spin⇄low-spin transitions in transition metal compounds induced by the Jahn–Teller effect *J. Chem. Phys.* **70** 4199–206
[84] Zimmermann R and König E 1977 A model for high-spin/low-spin transitions in solids including the effect of lattice vibrations *J. Phys. Chem. Solids* **38** 779–88
[85] Nasser J A 2001 First order high-spin/low-spin phase transition induced by acoustic-phonons *Eur. Phys. J.* B **21** 3–10
[86] Nasser J A, Boukheddaden K and Linares J 2004 Two-step spin conversion and other effects in the atom–phonon coupling model *Eur. Phys. J.* B **39** 219–27
[87] Gîndulescu A, Rotaru A, Linares J, Dimian M and Nasser J 2010 Excited metastables electronic spin states in spin crossover compounds studies by atom–phonon coupling model: gradual and two-step transition cases *J. Appl. Phys.* **107** 09A959
[88] Gîndulescu A, Rotaru A, Linares J, Dimian M and Nasser J 2011 Analysis of phase transitions in spin-crossover compounds by using atom–phonon coupling model *J. Phys. Conf. Ser.* **268** 012007
[89] Chernyshov D, Klinduhov N, Törnroos K W, Hostettler M, Vangdal B and Bürgi H B 2007 Coupling between spin conversion and solvent disorder in spin crossover solids *Phys. Rev.* B **76** 014406
[90] Boussekou A, Nasser J, Linares J, Boukheddaden K and Varret F 1992 Ising-like model for the two-step spin-crossover *J. Phys.* I **2** 1381–403
[91] Boussekou A, Varret F and Nasser J 1993 Ising-like model for the two step spin-crossover of binuclear molecules *J. Phys.* I **3** 1463–73
[92] Pavlik J, Nicolazzi W, Molnár G, Boča R and Bousseksou A 2013 Coupled magnetic interactions and the Ising-like model for spin crossover in binuclear compounds *Eur. Phys. J.* B **86** 292
[93] Sandeman K G 2016 Research update: the mechanocaloric potential of spin crossover compounds *APL Mater.* **4** 111102
[94] Létard J F, Guionneau P, Rabardel L, Howard J A K, Goeta A E, Chasseau D and Kahn O 1998 Structural, magnetic, and photomagnetic studies of a mononuclear iron(II) derivative exhibiting an exceptionally abrupt spin transition. Light-induced thermal hysteresis phenomenon *Inorg. Chem.* **37** 4432–41
[95] Levchenko G G, Bukin G V, Gaspar A B and Real J A 2009 The pressure-induced spin transition in the $Fe(phen)_2(NCS)_2$ model compound *Russ. J. Phys. Chem.* A **83** 951–4
[96] Galet A, Gaspar A B, Agusti G, Muñoz M C, Levchenko G and Real J A 2006 Pressure effect investigations on the spin crossover systems $\{Fe[H_2B(pz)_2]_2(bipy)\}$ and $\{Fe[H_2B(pz)_2]_2(phen)\}$ *Eur. J. Inorg. Chem.* **2006** 3571–3

[97] Real J A, Muñoz M C, Faus J and Solans X 1997 Spin crossover in novel dihydrobis(1-pyrazolyl)borate [$H_2B(pz)_2$]-containing iron(II) complexes. Synthesis, X-ray structure, and magnetic properties of [FeL{$H_2B(pz)_2$}$_2$] (L = 1,10-phenanthroline and 2,2′-bipyridine) *Inorg. Chem.* **36** 3008–13
[98] Moliner N *et alet al* 2002 Thermal and optical switching of molecular spin states in the {[FeL[$H_2B(pz)_2$]$_2$]} spin-crossover system (L = bpy, phen) *J. Phys. Chem.* B **106** 4276–83
[99] Agustí G, Thompson A L, Gaspar A B, Muñoz M C, Goeta A E, Rodríguez-Velamazán J A, Castro M, Burriel R and Real J A 2008 Thermal, pressure and light induced spin transition in the two-dimensional coordination polymer [Fe(pmd)$_2$[Cu(CN)$_2$]$_2$] *Dalton Trans.* **5** 642–9
[100] Li B, Wei R J, Tao J, Huang R B and Zheng L S 2010 Pressure effects on a spin-crossover monomeric compound [Fe(pmea)(SCN)$_2$] (pmea = bis[(2-pyridyl)methyl]-2-(2-pyridyl)ethylamine *Inorg. Chem.* **49** 745–51
[101] von Ranke P J 2017 A microscopic refrigeration process triggered through spin-crossover mechanism *Appl. Phys. Lett.* **110** 181909
[102] von Ranke P J, Alho B P, Ribas R M, Nobrega E P, Caldas A, de Sousa V S R, Colaço M V, Marques L F, Rocco D L and Ribeiro P O 2018 Colossal refrigerant capacity in [Fe (hyptrz)$_3$]$A_2 \cdot H_2O$ around the freezing temperature of water *Phys. Rev.* B **98** 224408
[103] von Ranke P J, Alho B P and Ribeiro P O 2018 First indirect experimental evidence and theoretical discussion of giant refrigeration capacity through the reversible pressure induced spin-crossover phase transition *J. Alloys Compd.* **749** 556–60
[104] von Ranke P J, Alho B P, da Silva P H S, Ribas R M, Nobrega E P, de Sousa V S R, Carvalho A M G and Ribeiro P O 2021 Correlation between anomalous thermal expansion coefficient and barocaloric effect: application to spin crossover systems *Solid State Commun.* **336** 114427
[105] von Ranke P J, Alho B P, Ribas R M, Nobrega E P, de Sousa V S R and Ribeiro P O 2022 Theoretical investigation of entropic barocaloric effect in spin-crossover systems *J. Magn. Magn. Mater.* **564** 170121
[106] Holland J M, Kilner C A, Thornton-Pett M, Halcrow M A, McAllister J A and Lu Z 2001 An unusual abrupt thermal spin-state transition in [FeL_2][BF_4]$_2$ [L = 2,6-di(pyrazol-1-yl) pyridine] *Chem. Commun.* **6** 577–8
[107] Barrett S A, Kilner C A and Halcrow M A 2011 Spin-crossover in [Fe(3-bpp)$_2$][BF_4]$_2$ in different solvents—a dramatic stabilisation of the low-spin state in water *Dalt. Trans.* **40** 12021–4
[108] Halcrow M A 2014 Recent advances in the synthesis and applications of 2,6-dipyrazolylpyridine derivatives and their complexes *New J. Chem.* **38** 1868–82
[109] Cook L J K, Kulmaczewski R, Cespedes O and Halcrow M A 2016 Different spin-state behaviors in isostructural solvates of a molecular iron(II) complex *Chem. Eur. J.* **22** 1789–99
[110] Tocado L, Palacios E and Burriel R 2006 Adiabatic measurement of the giant magnetocaloric effect in MnAs *J. Therm. Anal. Calorim.* **84** 213–7
[111] Romanini M *et alet al* 2021 Giant and reversible barocaloric effect in trinuclear spin-crossover complex $Fe_3(bntrz)_6(tcnset)_6$ *Adv. Mater.* **33** 2008076
[112] Seo J, Braun J D, Dev V M and Mason J A 2022 Driving barocaloric effects in a molecular spin-crossover complex at low pressures *J. Am. Chem. Soc.* **144** 6493–503

[113] Janiak C 1994 Binary [hydrotris(1,2,4-triazolyl)borate]metal complexes [M{HB $(C_2H_2N_3)_3\}_2$] with M = Fe, Co, Cu, Zn: synthesis, characterization, magnetochemistry, and X-ray structure of [Cu{HB$(C_2H_2N_3)_3\}_2]\cdot 4CH_3OH$ and K[HB$(C_2H_2N_3)_3]\cdot 2H_2O$ *Chem. Ber.* **127** 1379–85

[114] Janiak C, Scharmann T G, Green J C, Parkin R P G, Kolm M J, Riedel E, Mickler W, Elguero J, Claramunt R M and Sanz D 1996 Effects of nitrogen substitution in poly (pyrazolyl)borato ligands: from orbital energy levels to C–H···O hydrogen bonding *Chem. Eur. J.* **2** 992–1000

[115] Janiak C, Scharmann T G and Bra T 1998 Zeitschrift fu È r anorganische und allgemeine chemie spin crossover studies on bis {hydro-tris(1, 2, 4-triazolyl)borato}iron(II) *Z. Anorg. Allg. Chem.* **624** 769–74

[116] van Koningsbruggen P J and Miller J S 2008 Observation of a 331 K (58 °C) spin transition for bis[hydrotris(1,2,4-triazol-1-yl)borate]iron(II) by variable temperature infrared spectroscopy and magnetic susceptibility measurements *Solid State Sci.* **10** 1804–6

[117] Rat S, Ridier K, Vendier L, Molnár G, Salmon L and Bousseksou A 2017 Solvatomorphism and structural-spin crossover property relationship in bis[hydrotris (1,2,4-triazol-1-yl)borate]iron(II) *CrystEngComm* **19** 3271–80

[118] Bovo G 2016 Room temperature spin crossover in molecular and polymeric materials *PhD Thesis* University of London

[119] Zhang Y *et alet al* 2020 A molecular spin-crossover film allows for wavelength tuning of the resonance of a Fabry–Perot cavity *J. Mater. Chem.* C **8** 8007–11

[120] Mikolasek M *et alet al* 2018 Complete set of elastic moduli of a spin-crossover solid: spin-state dependence and mechanical actuation *J. Am. Chem. Soc.* **140** 8970–9

[121] Kahn O, Sommier L and Codjovi E 1997 Spin transition molecular alloys: an attempt of fine tuning of the transition temperatures *Chem. Mater.* **9** 3199–205

[122] Tayagaki T, Galet A, Molnár G, Carmen Muñoz M, Zwick A, Tanaka K, Real J A and Bousseksou A 2005 Metal dilution effects on the spin-crossover properties of the three-dimensional coordination polymer Fe(pyrazine)[Pt$(CN)_4$] *J. Phys. Chem.* B **109** 14859–67

[123] Baldé C, Desplanches C, François Létard J and Chastanet G 2017 Effects of metal dilution on the spin-crossover behavior and light induced bistability of iron(II) in [$Fe_xNi_{1-x}(bpp)_2$] $(NCSe)_2$ *Polyhedron* **123** 138–44

[124] Oganov A R, Pickard C J, Zhu Q and Needs R J 2019 Structure prediction drives materials discovery *Nat. Rev. Mater.* **4** 331–48

[125] Chen J, Millis A J and Marianetti C A 2015 Density functional plus dynamical mean-field theory of the spin-crossover molecule Fe$(phen)_2(NCS)_2$ *Phys. Rev.* B **91** 241111

[126] Römer A, Hasecke L, Blöchl P and Mata R A 2020 A review of density functional models for the description of Fe(II) spin-crossover complexes *Molecules* **25** 1–16

[127] Droghetti A, Alfè D and Sanvito S 2012 Assessment of density functional theory for iron(II) molecules across the spin-crossover transition *J. Chem. Phys.* **137** 124303

[128] Song S, Kim M C, Sim E, Benali A, Heinonen O and Burke K 2018 Benchmarks and reliable DFT results for spin gaps of small ligand Fe(II) complexes *J. Chem. Theory Comput.* **14** 2304–11

[129] Vela S, Fumanal M, Cirera J and Ribas-Arino J 2020 Thermal spin crossover in Fe(II) and Fe(III). Accurate spin state energetics at the solid state *Phys. Chem. Chem. Phys.* **22** 4938–45

[130] Ridier K, Zhang Y, Piedrahita-Bello M, Quintero C M, Salmon L, Molnár G, Bergaud C and Bousseksou A 2020 Heat capacity and thermal damping properties of spin-crossover molecules: a new look at an old topic *Adv. Mater.* **32** 2000987

[131] Ridier K, Nicolazzi W, Salmon L, Molnár G and Bousseksou A 2022 Sequential activation of molecular and macroscopic spin-state switching within the hysteretic region following pulsed light excitation *Adv. Mater.* **34** 2105468

Chapter 6

Hybrid organic–inorganic materials: introducing soft barocaloric materials for low-pressure applications

Javier García-Ben, Ignacio Delgado-Ferreiro, Jorge Salgado-Beceiro and Juan Manuel Bermúdez-García

6.1 Introduction

6.1.1 General definition and classification of hybrid organic–inorganic materials

Over the last few years, hybrid organic–inorganic materials have been presented as a deep pool to dive into the search of new barocaloric compounds, which can present large thermal changes under the application of very low pressures.

The term 'hybrid organic–inorganic material', often shortened just as 'hybrid material', covers any compound formed by an intimate mixture of different inorganic and organic components, usually interpenetrated on scales of below 1 μm [1, 2]. Depending on the field of study, hybrid materials can refer to either composites or single-phase materials [3]. In this chapter, and in the field of barocaloric materials, we defined hybrid materials as single-phase solid-state compounds that integrate organic and inorganic moieties, also known as 'building-blocks', in the same crystal structure, see figure 6.1 [4].

In general, there are three main building-blocks that can configure any given hybrid material: (i) inorganic metal cations acting as coordination centres or nodes; (ii) organic or inorganic bridge anions acting as ligands to the metal cations through strong chemical bonds; and (iii) organic molecules or ions occupying empty spaces left by the previous two strongly-bound moieties, and interacting with them through weak chemical bonds. This latter component can be absent, although the coexistence of both organic and inorganic units in the crystal structure must be maintained for a material to be defined as a hybrid compound.

Accordingly, hybrid materials are normally classified following two main criteria: (i) as a function of the coordination dimensionality into 3D, 2D, 1D and 0D

doi:10.1088/978-0-7503-4690-0ch6

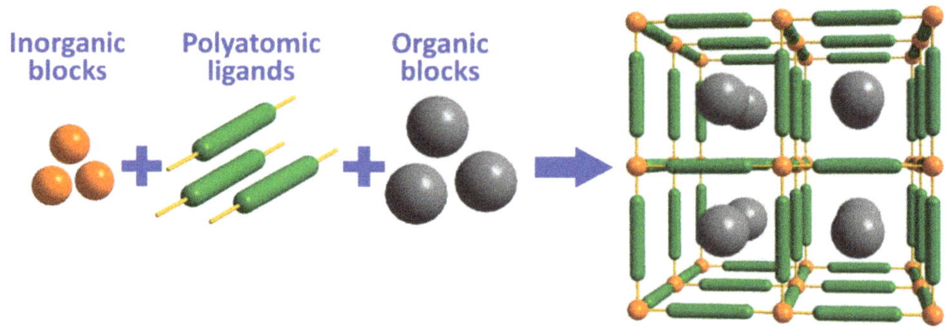

Figure 6.1. Schematic of the building-blocks and general structure for hybrid organic–inorganic materials.

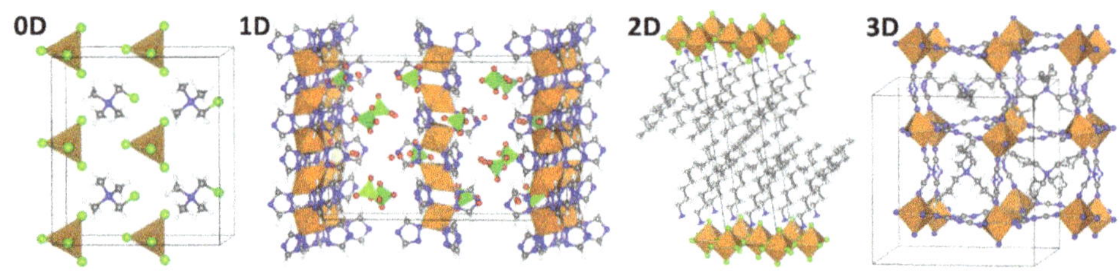

Figure 6.2. Selection of hybrid organic–inorganic materials with different coordination dimenstionality: 0D—$[(CH_3)_3(ClCH_2)]MCl_4$ [7], 1D—$[Cu(hyetrz)_3](ClO_4)_2{\bullet}3H_2O$ (hyetrz = 4-(2-hydroxyethyl)-1,2,4-triazole) [8], 2D—$[CH_3(CH_2)_9NH_3]MCl_4$[9], and 3D—$[(CH_3CH_2CH_2)_4N][M(dca)_3]$ [10]. *Note:* blue = N, grey = C, red = O, white = H, green = Cl, black lines = unit cell. The coordination polyhedra of the metal cations are represented in orange colour. Ligands of the 1D-compound have been simplified for facilitating visualization.

materials [5]; and (ii) as a function of the porosity into dense (non-porous) materials and porous materials [6].

Regarding the first classification criterion, the coordination dimensionality refers to the directions towards where the strong covalent/ionic bonds between the different building-blocks is extended in the structure.

In that aspect, hybrid materials that do not show any strong chemical interaction between the organic and inorganic moieties are named 0D or ionic/molecular hybrid materials, see figure 6.2. Here, the different moieties are isolated organic and inorganic ions or molecules that only interact through weak hydrogen bonds, Van der Waals, π–π, and/or other electrostatic interactions. This is the case, for example, with alkylammonium tetrahalometallates of general formula $[A][MX_4]$, where A are mid-size alkylammonium cations, M are normally trivalent transition metal cations, and X are different halide anions [7]. Those materials are formed by $[MX_4]$ tetrahedra alternating with organic cations, where both units interact through weak electrostatic forces and/or hydrogen bonds, see figure 6.2.

When increasing the dimensionality, 1D hybrid compounds exhibit isolated columns or chains of inorganic metal cations linked by organic (and sometimes also inorganic) ligands, which are interconnected through weak bonds and can accommodate other organic or inorganic species between the chains. This hybrid structure is very common, for example, in spin crossover materials of general formula $M(L)(ClO_4).xH_2O$ (M = transition metal cations, L = different organic ligands) [11]. Those particular compounds present –[M–L–M]– one-dimensional chains where perchlorate anions (ClO_4^-) and water molecules (H_2O) are located within chains and interlinked through hydrogen bonds and electrostatic forces, see figure 6.2.

In the next dimensional step, 2D hybrid compounds are formed by two-dimensional layers of inorganic metal cations linked through inorganic (or even organic) anions, where larger organic molecules/ions are place between layers acting as separators that interact through weak chemical bonds. That would be the case, for example, of 2D hybrid perovskites—also known as layered perovskites—of general formula A_2BX_4, where A are normally long-chain alkylammonium cations, B are divalent metal cations and X are generally different halide anions [12]. Here, the 2D layers are formed by $[BX_6]$ octahedra sharing corners and interleaved with the organic cations, see figure 6.2.

The higher coordination dimensionality corresponds to 3D hybrid materials. They present strong covalent/ionic bonds in the three directions of space, forming frameworks of metal cations linked by bridge ligands (normally organic anions), see figure 6.2. These frameworks present cavities that can often allocate additional organic species, which can slightly distort and adapt the framework to their size and shape. Depending on the size of those cavities, which directly affects the porosity of the compounds, 3D hybrid frameworks can be sub-classified into porous or dense (non-porous) materials, as previously described [6].

Porous hybrid materials are also defined as metal–organic frameworks (MOFs), where the organic ligands linking the metallic nodes play an important role in the topology and in the size and shape of the cavities and pores, see figure 6.3. This family of compounds, reported for the first time in the 1990s [13], can present a wide range of pore size from micropores (<2 nm) to macropores (>50 nm) [14], where guest species can freely enter, diffuse through, and leave the material without compromising the integrity of the structure. Actually, those MOFs can present an exceptionally large amount of empty space with above 90% of free volume [14], which makes them suitable for many different applications in the fields of gas adsorption, storage and separation [15], water harvesting in extremely dry environments [16], or evaporative refrigeration [17], for instance.

In contrast, dense hybrid framework materials show negligible pores and voids and, in general, the species located in their cavities cannot freely diffuse or leave without collapsing their structure [6]. In these materials, the interactions between the framework and the guest species in the cavities tend to be stronger than in MOFs. In turn, this intimate relation can give rise to functional properties, such as optolectronic, ferroic and multiferroic properties [6, 19], where the molecules in the cavities play an important role. In these materials, and similarly to MOFs, the selection of the different building-blocks can also lead to different topologies. However, probably, the most studied hybrid materials with dense 3D frameworks are those presenting perovskite-like structure of general formula ABX_3 (A = mid-size alkylammonium cations, B = divalent transition metal cations, X = halide anions

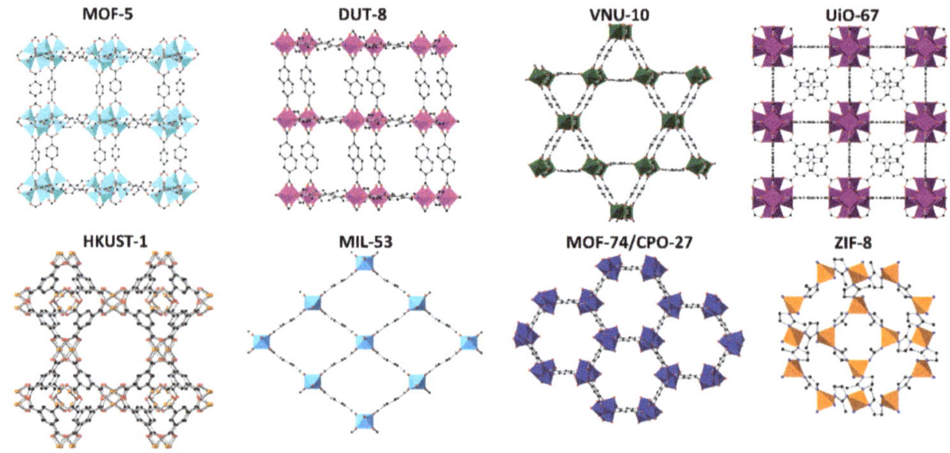

Figure 6.3. Selection of different MOFs topologies depending on the selected building-blocks. Crystal structures obtained from CCDC [18]. Hydrogen atoms and guest molecules are omitted for facilitating visualization. Note: MOFs names often refer to their institution of origin (DUT = Dresden University of Technology, VNU = Vietnam National University, UiO = Universitetet i Oslo, HKUST = Hong Kong University of Science and Technology, MIL = Material of Institute Lavoisier, CPO = Coordination Polymer of Oslo) or their specific composition (ZIF = zeolite imidazole framework).

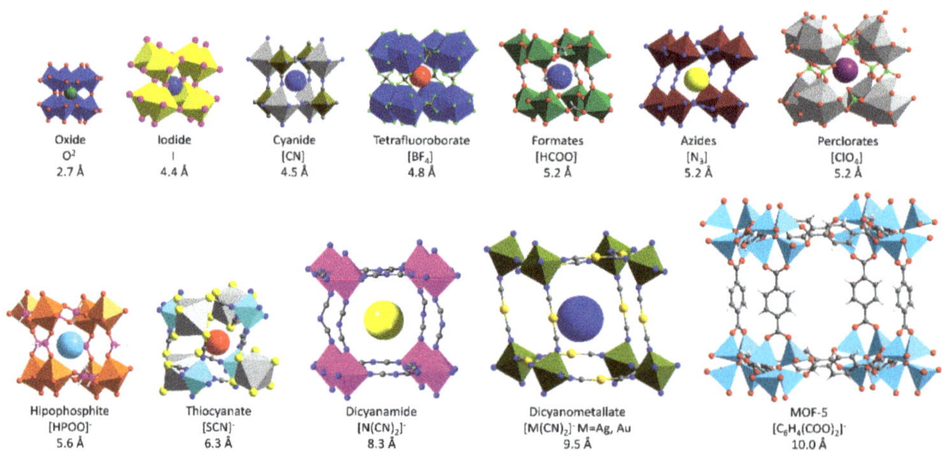

Figure 6.4. Representation of crystal structures for dense hybrid perovskites with different X ligands in comparison with fully inorganic perovskites and porous hybrid MOF-5. Figure reproduced from reference [6] (https://creativecommons.org/licenses/by-nc-nd/4.0/).

or polyatomic bidentate bridge ligands) [6, 19]. In these structures, the X ligands are connecting $[BX_6]$ octahedra through their corners forming pseudo-cubooctahedral cavities than can allocate different mid-size A cations.

Thus, the use of different X ligands will provide cavities of different size, which will allow for the accommodation of different A cations. Figure 6.4 shows a schematic of different crystal structures for dense hybrid perovskites in comparison with a prototypical fully inorganic oxide perovskite and with the porous hybrid

MOF-5, one of the MOFs with smaller pore size. Here, it can be seen that by increasing the length of the [–M–L–M–] bridge ligands in hybrid perovskites, from halide anions to dicyanometallate anions, the pore size can increase from 4.4 Å to 9.5 Å, which could be used to allocate bigger organic cations.

6.1.2 Physico-chemical and mechanical properties of interest for barocaloric effects

In a similar manner to any other material, hybrid organic–inorganic materials should present three main requirements to be considered as a potential barocaloric candidate [4]:

(1) Large and reversible entropy changes related to solid–solid phase transitions, which are especially significant in first-order ferroic transitions. In this case, entropy changes associated with thermally-induced phase transitions (ΔS_t) can be used to estimate the maximum barocaloric effect attainable (ΔS_{it}), considering that thermally- and pressure-induced transitions could be equivalent or very similar. Fortunately, as already mentioned, hybrid materials have been widely explored as ferroic materials [6, 19, 20], and there is plenty of literature containing preliminary calorimetric studies of a vast number of hybrid materials, which already pose a great advantage in the quest of identifying new barocaloric candidates.

(2) A large barocaloric coefficient, defined as the pressure dependence of the transition temperature ($\mathrm{d}T_t/\mathrm{d}p$). This parameter is highly affected by the flexibility and pressure-responsiveness of a given material, which is especially large in the case of hybrid materials [6, 21], as will be discussed below. A large barocaloric coefficient is desired to enlarge the temperature span (T_{span})—or operating temperature range—of a material by applying a lower pressure [22]. It is important to pay special attention to the sign of the barocaloric coefficient, which can be positive or negative and give rise to the so-called conventional and inverse barocaloric effects, respectively [23]. This sign will indicate whether the temperature span will extend towards higher temperatures (conventional) or lower temperatures (inverse) from the initial temperature of transition. In addition to that, it is worth noting that a low thermal hysteresis ($|T_{t(h)} - T_{t(c)}|$) will also decrease the required operating pressure and increase the efficiency of a given device [24].

(3) An adequate transition temperature, T_t, which matches well with that of the desired application and the suitable climate zone of the world where the barocaloric material will be used. For instance, the operating conditions for commercial fridges are classified into three main climate classes, namely, normal (N) for climate temperatures of 16 °C–32 °C, subnormal (SN) for 10 °C–32 °C, subtropical (ST) for 18 °C–38 °C, and Tropical (T) for 18 °C–43 °C. In the same way, air conditioners are classified into class 1 (for moderate climates of 18 °C–43 °C), class 2 (for chill climates of 10 °C–35 °C) and class 3 for (hot climates of 21 °C–52 °C). Meanwhile, in the case of larger industrial applications, power plants and other specific applications, these operating temperatures can expand from −40 °C up to 400 °C [25]. Here, the wide chemical versatility of hybrid materials plays an important

design role, since it is possible to chemically tune the transition temperature by simple strategies such as, for example, forming solid solutions, creating hydrogen bonds, or incorporating π···π interactions [6].

The already described chemical and structural versatility will offer a wide variability and tunability of physical properties of such compounds. For example, it has been reported that the different combinations of different building-blocks can allow for a large number of the resulting degrees of freedom in hybrid materials [26]. These degrees of freedom would be directly related to the origin of caloric effects. In that regard, hybrid materials—especially those with 3D frameworks—present new 'unconventional' degrees of freedom that are not presented by the traditional inorganic materials, which enlarge the elements to consider when designing new barocaloric hybrids. The degrees of freedom that we can find in hybrid organic materials can be classified in four main types [26], see figure 6.5:

(1) Rotational: refers to the cooperative rotation of the metal–ligand coordination polyhedra, such as in the case of hybrid perovskites, where neighbour octahedra can rotate in opposite directions (conventional octahedra tilting) or in the same direction (unconvenctional octahedra tilting).
(2) Translational: can be understood as collective displacement of building-blocks out of their initial position. This can be, for example, the cooperative displacement of [–metal–ligand–metal–] columns ('columnar shifts') along a given direction in the case of some hybrid perovskites or MOFs (such as DUT-8(Ni)), or the 'wine-rack' displacement in the case of MIL-53.
(3) Orientational: relates to the general orientation of a whole building-block along a given direction, which can be observed in both ligands (i.e. imidazole or dicyanamide anions) and molecules or ions (i.e. dimethylammonium or methylammonium cations) interacting through weak bonds with the hybrid framework. This is commonly found in order–disorder transitions where at low temperature the building-blocks can be oriented towards one direction, while at high temperature they become disordered and lose such orientation.
(4) Conformational: differently from the general orientational degrees of freedom, the conformational ones refer to the capacity of the building-blocks of

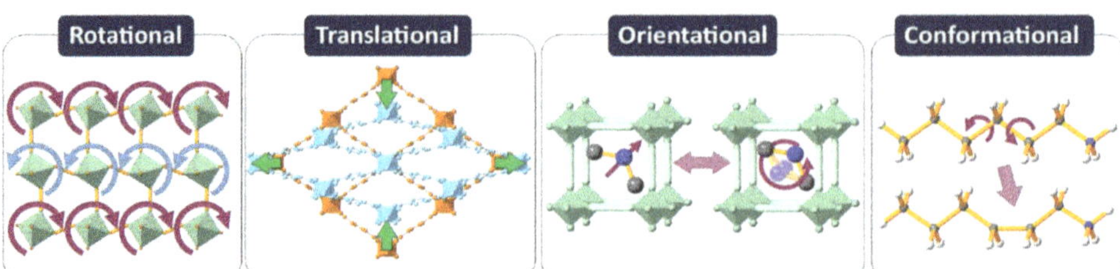

Figure 6.5. Selected examples of the main types of degrees of freedom found in hybrid organic–inorganic frameworks: rotational unconventional octahedra tilting, translational 'wine-rack' displacement, orientational order–disorder transition, and conformational modification of dihedral angles.

changing between different local conformations, such as torsion or dihedral angles of a given molecule, anion or cation.

In the pursue of caloric effects, these degrees of freedom can be easily modulated through solid–solid first-order phase transitions induced by the application of an external stimuli (magnetic field, electric field and/or pressure), giving rise to large stimuli-induced enthalpy and entropy changes [21, 26].

Regarding the specific stimulus of applied pressure, the length of the selected organic/inorganic bridge ligands will have a direct impact in the barocaloric response. The length of the ligand will affect the specific volume, density or mechanical properties of the selected compounds [6, 27–31]. In that manner, hybrid materials with larger ligands will present a larger unit cell volume and, in general, larger empty space and lower density. At the same time, in principle, larger ligands will also confer larger softness and flexibility (elasticity and/or compressibility) to the material. As a matter of example, figure 6.6 shows a comparison of those physicochemical properties for different 3D materials, including traditional inorganic oxide-perovskites, dense (non-porous) hybrid perovskites and porous MOFs.

As can be observed in figure 6.6, MOFs and hybrid perovskites with larger bridge ligands (such as dicyanamide anions, $N(CN)_2^-$) exhibit a larger unit cell volume and a lower density, elastic modulus and hardness. These characteristics, in principle, will make those materials more responsive towards the application of external pressure.

Actually, barocaloric hybrid materials were initially proposed as a low-pressure alternative [32] to the barocaloric materials of that time, mainly metallic alloys and inorganic perovskites or ionic salts [33–36], which required operating pressures of above 1000 bar (0.1 GPa). It should be noted that this value of 0.1 GPa was already considered relatively low-pressure, since the pressure used in extreme conditions science was often of several hundreds of GPa [37]. In order to better picture the magnitude of the mentioned pressure values, figure 6.7 illustrates some of the

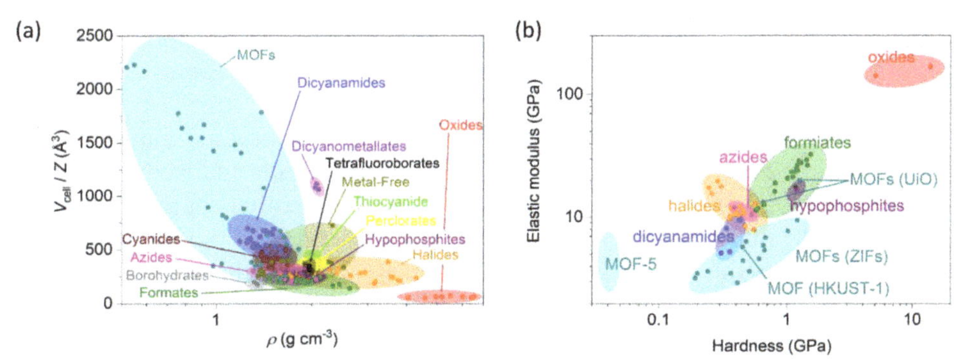

Figure 6.6. Selected physical properties affected by the length of bridge ligands in different 3D materials, including oxide-perovskites, dense hybrid perovskites and porous MOFs. (a) Unit cell volume *versus* density, and (b) elastic modulus *versus* hardness. Figures reproduced from reference [6] (https://creativecommons.org/licenses/by-nc-nd/4.0/).

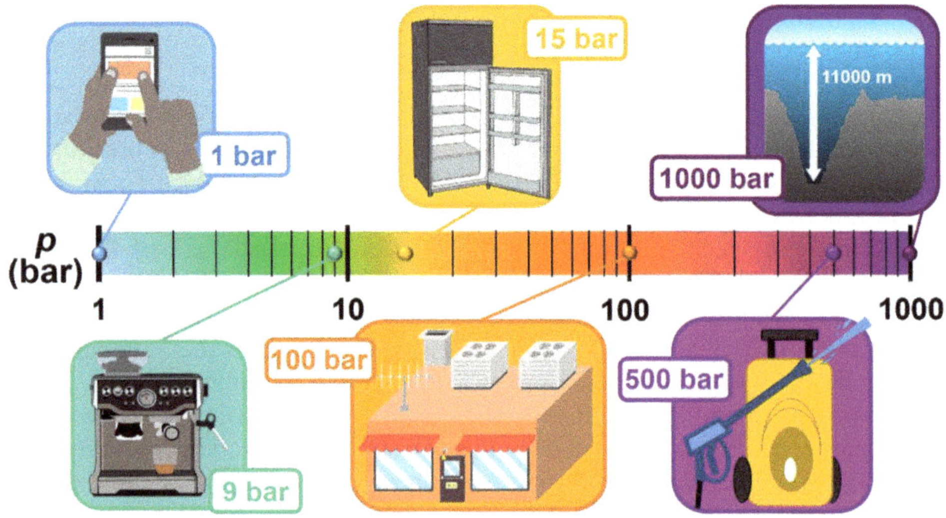

Figure 6.7. Representation of approximated pressure magnitudes that we encounter in our quotidian life. Figure designed including open-access images from free-stock sites (no attribution required).

quotidian pressures that we can find around, from human-generated pressure when manipulating objects (in the scale ~1 bar) to that of some devices that we use in our daily life, such as the pressure of a coffee machine (~9 bar). Here, it can be seen that 1000 bar is equivalent to the pressure exerted by the water column on to the deepest point of the Earth's oceans—the Mariana trench-, which is at around 11 000 m below the sea level [38]. Therefore, 1000 bar is still a very large value when referring to daily life cooling technologies, such as fridges, freezers or air conditioning devices. Those quotidian technologies normally operate under few tens of bar, or around 100 bar in the case of CO_2 (R744) cooling systems used in large stores and/or malls. For barocaloric materials to be efficiently implemented in cooling technologies, it has been recently estimated that they should be able to operate below 300 bar [39], which is slightly below the magnitude that we can reach using a high-pressure water jet washer. And beyond that, barocaloric materials have been even proposed for future cooling technologies that could be activated with the small forces of our body motion, such as our footsteps [40].

In this context, as already discussed, hybrid organic–inorganic materials might have all the ingredients to present very large caloric effects induced by very low pressures in the range of commercial needs. Therefore, in this chapter we compile and compare the main advances on barocaloric effects on hybrid organic–inorganic materials made in the recent few years. Here, we start with the first studies of barocaloric properties in dense (non-porous) hybrid materials, specifically in 3D hybrid perovskites, and we show the evolution towards new barocaloric hybrid materials of different dimensionalities, such as 2D hybrid perovskites and 0D hybrid ionic plastic crystals. It should be noted that 1D barocaloric hybrid materials are spin crossover compounds that, given the particular nature of the spin crossover transitions, will be discussed in a separate chapter of this book. Finally, we will also discuss a recent discovery of new pressure-induced caloric effects found in porous MOFs, which show some important resemblances with the barocaloric effects found in dense hybrid materials.

6.2 Exploring first barocaloric insights in 3D hybrid perovskites

In 2017, the earliest reports on hybrid organic–inorganic materials displaying barocaloric effects were made on the family of 3D hybrid perovskites [4, 32]. The first experimental studies were published for the [TPrA][Mn(dca)$_3$] (TPrA = $(CH_3CH_2CH_2)_4N^+$, dca = $N(CN)_2^-$) hybrid perovskite [32]. This specific compound can be categorized into the sub-family of dicyanamide-perovskites, since hybrid perovskites are often classified as a function of the X ligand present in the ABX_3 chemical formula.

Additionally, this initial publication already anticipated that the described barocaloric perovskite would not be an isolated example, since other dense hybrids materials and porous MOFs already presented the basic ingredients required to be potential barocaloric materials [32]. In fact, a following publication analysed in detail reported calorimetric and structural data spread in the literature, and identified 16 additional barocaloric candidates belonging to four different sub-families of 3D hybrid perovskites, namely, dicyanamide-, azide-, formate- and halide-perovskites, see figure 6.8 [4].

As can be seen in figure 6.9, the hybrid perovskites identified as potential barocaloric materials exhibited very large entropy changes (ΔS_t), from ~15 J K^{-1} kg^{-1} to ~110 J K^{-1} kg^{-1}, associated with thermally-induced phase transitions, and in a wide range of transition temperatures from 140 to 360 K. Furthermore, the barocaloric coefficients indirectly estimated by using the Clausius–Clapeyron equation, $dT_t/dp = \Delta v/\Delta S_t$ (Δv = volume change across the phase transition obtained from structural data), ranged from 2 to 23 K kbar^{-1}.

From this initial assessment, it can be seen that the family of halide- and formate-perovskites present large thermal changes (ΔS_t) with very different transition temperatures (T_t), which offer a new catalogue of barocaloric materials that could work on different operating temperatures. However, these two families would exhibit a relatively limited response towards pressure, as evidenced by the

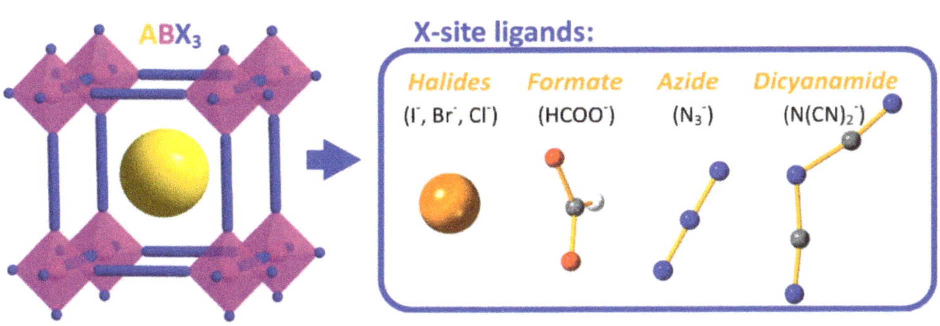

Figure 6.8. General structure of different selected sub-families of hybrid perovskites based on the X ligand used in the crystal structure: halide-, formate-, azide-, and dicyanamide-, perovskites.

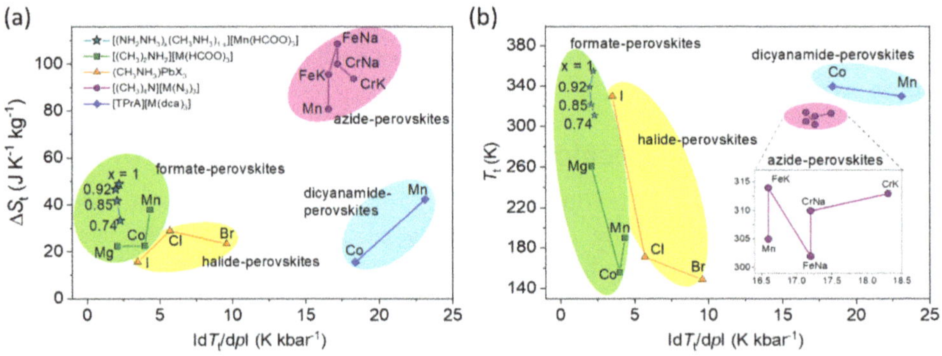

Figure 6.9. Barocaloric coefficients *versus* (a) entropy change of the thermally-induced phase transition, and (b) transition temperature for different 3D hybrid perovskites. Figure adapted with permission from reference [4], copyright (2017) American Chemical Society.

barocaloric coefficient, dT_t/dp. Meanwhile, azide- and dicyanamide-perovskites present very large barocaloric coefficients due to the larger pressure response of such materials. Nevertheless, and different from the previous hybrid compounds, they could mainly operate in warm climates, as shown by their transition temperatures slightly above room temperature. Particularly, azide-perovskites exhibit very large thermal changes in the range of 100 J K^{-1} kg^{-1}, which could anticipate very large barocaloric effects.

In recent years, further and more detailed experimental studies have confirmed the potentiality of the different sub-families of 3D hybrid perovskites as barocaloric materials, as will be described in the following sections.

Note: We would like to bring to the reader's attention that, to date, all experimental barocaloric studies in 3D hybrid perovskites (as well as in 0D hybrid materials and porous MOFs) have been performed in powder polycrystalline samples. This differs from 2D perovskites, where studies on polycrystalline powders and single crystals were made, as will be explained in the corresponding section.

6.2.1 Dicyanamide-perovskites

To date, there are only two publications that report experimental barocaloric studies on dicyanamide-perovskites, specifically in the [TPrA][M(dca)$_3$] (TPrA = $(CH_3CH_2CH_2)_4N^+$; M = Mn^{2+}, Cd^{2+}; dca = $N(CN)_2^-$) compounds [32]. These hybrid perovskites present a crystal structure formed by a $[M(dca)_3]^-$ extended frameworks displaying pseudo-cubic cavities that allocate $TPrA^+$ cations, see figure 6.10.

The Mn-compound presents a single thermally-induced solid–solid phase transition at T_t ~ 330 K (as observed by differential scanning calorimetry (DSC)) between the low-temperature tetragonal $P\bar{4}2_1c$ polymorph and the high-temperature tetragonal $I4mcm$ polymorph [10], see figure 6.7. Here, the thermal changes registered for the transitions are as large as $|\Delta H_t|$ ~ 14 kJ kg and $|\Delta S_t|$ ~ 42.5 J K^{-1} kg^{-1} [32]. In addition, the thermal hysteresis is remarkably low, with values of ~0.9 using the onset of the peaks or ~3.0 using the peaks' maximum.

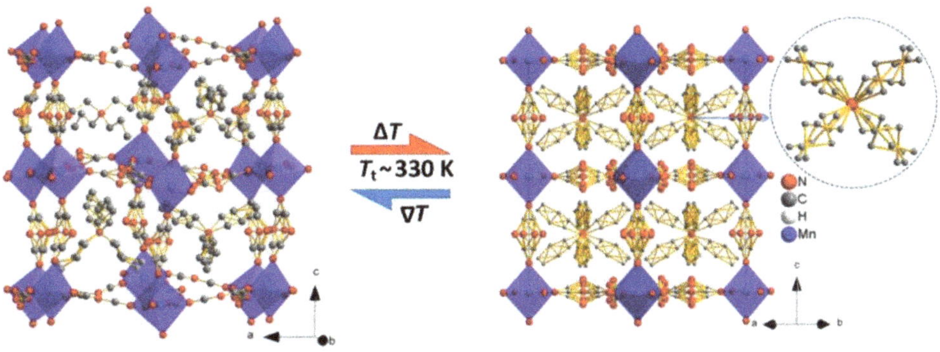

Figure 6.10. Crystal structures of the low-temperature and high-temperature polymorphs of the [TPrA][Mn (dca)$_3$] hybrid perovskite. Inset: detail of the configurational disorder in the C atoms of the TPrA cations. Figure adapted with permission from reference [10], copyright (2015) American Chemical Society.

In contrast, the analogous Cd-compound displays multiple phase transitions, although only the transition occurring at $T_t \sim 385$ K showed a sufficient latent heat of interest for barocaloric properties. This transition takes place between the low-temperature orthorhombic *I*bam polymorph and the high-temperature tetragonal *I*4mcm polymorph [41, 42]. In this case, the thermal changes are noticeably smaller than for the Mn-compound, with values of $|\Delta H_t| \sim 6.2$ kJ kg^{-1} and of $|\Delta S_t| \sim 16.2$ J K^{-1} kg^{-1}. However, the thermal hysteresis is equally low, ranging from ~1.4 (onset) to ~3.0 K (peak's maximum).

Trying to elucidate the origin of thermal changes of such structural transitions, the analysis of the crystal structures of [TPrA][M(dca)$_3$] materials revealed the coexistence of several types of degrees of freedom that are modulated across the transition. For example, these materials exhibit displacement of TPrA$^+$ cations out of their crystallographic centered positions, translational columnar shifts, and octahedra tilting, among others [6, 43, 44]. Nevertheless, the entropy change was mainly attributed to configurational order–disorder processes between the two polymorphs of the materials, where the C-atoms of the TPrA and the N- and C-atoms of the dicyanamide largely increase their structural disorder, see figure 6.7. According to crystallographic data and the equation $\Delta S_{config.} = R \ln(N)$ (R = ideal gas constant, N = ratio of configurations between both polymorphs), the configurational contribution seemed to account for the totality of the entropy change observed by DSC [32].

In view of the notable thermal properties found in both compounds at atmospheric pressure, these materials were further studied by variable-temperature and variable pressure differential scanning calorimetry (VT- and VP-DSC) to explore the potential barocaloric effects.

The VT-DSC analysis at different isobaric conditions revealed that the transition temperatures of the Mn- and Cd-compounds shift towards higher temperatures when increasing the pressure, even for pressures as low as 68.9 bar and up to 1000 bar in the case of the Mn-material, see figure 6.11 [32, 41]. Remarkably, both materials exhibit very large barocaloric coefficients of $dT_t/dp_{(Mn)} \sim 23.1$ K kbar^{-1}

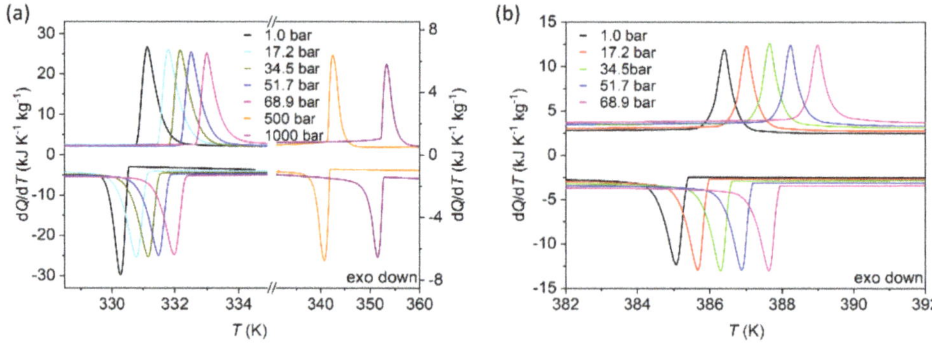

Figure 6.11. VT-DSC curves at different pressures for the (a) [TPrA][Mn(dca)$_3$] and (b) [TPrA][Cd(dca)$_3$] hybrid perovskites. Note: low-pressure (up to 68.9 bar) and high-pressure (up to 1000 bar) DSC measurement from panel (a) were made in different instruments. Note: figure (a) recreated using data from reference [32] (http://creativecommons.org/licenses/by/4.0/) and figure (b) adapted with permission from reference [41] with permission from the Royal Society of Chemistry.

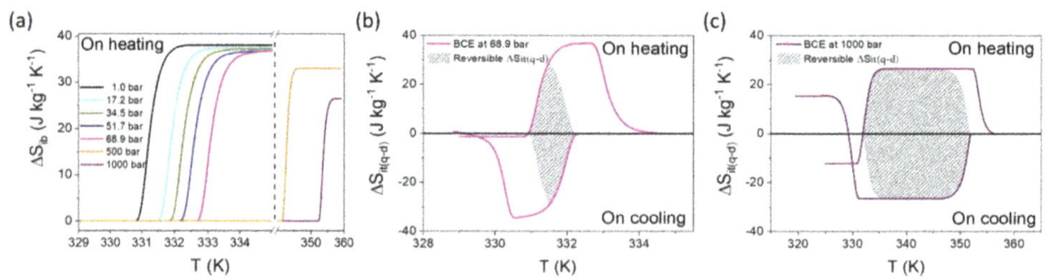

Figure 6.12. Quasi-direct barocaloric studies illustrated for the case of [TPrA][Mn(dca)$_3$]: (a) Isobaric entropy change (ΔS_{ib}) as a function of temperature at different pressures from 1 to 1000 bar. (b and c) Isothermal entropy change calculated by quasi-direct methods ($\Delta S_{it(q\text{-}d)}$) on heating and cooling for selected pressure of 68.9 bar (b) and 1000 bar (c). Figure adapted from reference [32] (http://creativecommons.org/licenses/by/4.0/).

and $dT_t/dp_{(Cd)} \sim 38.2$ K kbar^{-1}, which still range among the largest pressure responsiveness found in barocaloric materials [45]. This large pressure responsiveness, was fully in agreement with Clausius–Clapeyron calculations, $dT_t/dp = \Delta v/\Delta S_t$, where the Mn- and Cd-compounds display a volume change of $\Delta v_{(Mn)} \sim 10.0 \times 10^{-6}$ m^3 kg^{-1} and $\Delta v_{(Cd)} \sim 5.3 \times 10^{-6}$ m^3 kg^{-1}, according to crystallographic data [32, 41].

Furthermore, by using the so-called quasi-direct methods [23], the barocaloric effects (in terms of isothermal entropy change, ΔS_{it}) were calculated from these VT-DSC data, see figure 6.12.

In the range of the low operating pressures of 68.9 bar (close to that of commercial cooling systems), the [TPrA][Mn(dca)$_3$] material exhibits a reversible barocaloric effects of $|\Delta S_{it(q\text{-}d)}|_{rev} \sim 30.5$ J K^{-1} kg^{-1}—see figures 6.12(a) and (b)—with an initially estimated adiabatic temperature change of $|\Delta T_{ad}|_{rev} = (T/C_p) \times |\Delta S_{it(q\text{-}d)}|_{rev} \sim 4.1$ K [32]. Meanwhile, in the Cd-analogue compound this effect decreases down to $|\Delta S_{it(q\text{-}d)}| \sim 11.5$ J K^{-1} kg^{-1}.

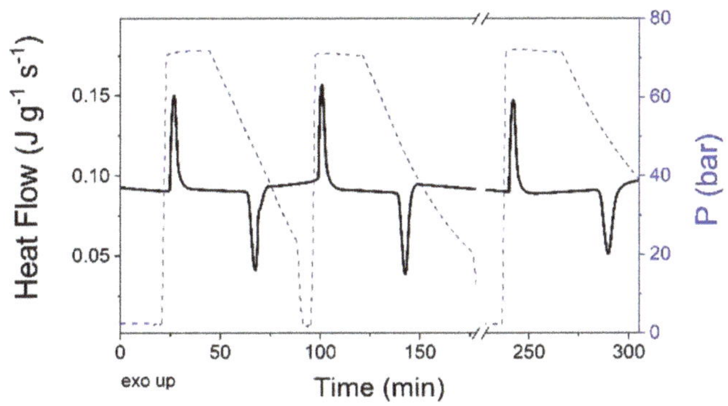

Figure 6.13. VP-DSC curve under isothermal conditions at ~331.4 K for several pressurization and depressurization cycles of the [TPrA][Mn(dca)$_3$] compound over time. Figure reproduced from reference [32] (http://creativecommons.org/licenses/by/4.0/).

This can be attributed to a larger structural ordering when applying pressure in the case of the Mn-compound. The values of both thermal changes can be considered large, in comparison with barocaloric materials existing at the time of the discovery [45]. However, the temperature span (operating temperature range) under this low pressure range was limited to a few degrees.

Advantageously, the reported data demonstrate that the temperature span can be enlarged up to ~20 K in the case of the Mn-compound, at the expense of applying larger pressures of up to 1000 bar, see figure 6.12(c). Unfortunately, the application of such large pressure also results into a ~30% decrease of the entropy change, $\Delta S_{\mathrm{it(q\text{-}d)}}$, in comparison with lower pressures, see figure 6.12(a). Therefore, depending on the desired application, there should be a compromise between the amount of barocaloric effect, the applied pressure, and the temperature span.

Furthermore, isothermal VP-DSC studies were also performed in the Mn-compound to further study barocaloric effects by direct methods [23], which still remains a scarce technique even nowadays. These data confirmed a reversible barocaloric effect of $|\Delta S_{\mathrm{it(d)}}| \sim 35.1$ J K^{-1} kg^{-1} at ~331 K under the application of ~70 bar (figure 6.13), which is slightly larger than the estimated by quasi-direct methods [32].

In summary, these findings demonstrated the enormous potential of hybrid materials to exhibit very large barocaloric effects under easily accessible pressures. And, although important efforts must be done before any commercial implementation can take place, the catalogue of barocaloric hybrid materials has been enlarged over the years including new examples with even more enhanced barocaloric characteristics.

6.2.2 Formate-perovskites

In 2018, the formate-perovskite [$(CH_3)_2NH_2$][$Mg(HCOO)_3$] was remarkably described as the first hybrid material that presented inverse barocaloric effects and that, in addition, appeared slightly below room temperature [46].

The $[(CH_3)_2NH_2][Mg(HCOO)_3]$ compound exhibits a thermally-induced phase transition at $T_t \sim 264$ K ($|T_{t(h)} - T_{t(c)}| \sim 2$ K) with thermal changes of $|\Delta H_t| \sim 14.5$ kJ kg^{-1} and $|\Delta S_t| \sim 55$ J K^{-1} kg^{-1}, considering also the pre-transitional effects [46]. Assuming that the entropy change arises from configurational disorder and, thus, applying the equation $\Delta S_{config.} = R \ln(N)$, the authors obtained a ratio of $N \sim 3.9$. This is in agreement with a three-fold disordering of the structure across the transition, as evidence by the crystallographic studies, see figure 6.1 [46, 47].

According to such crystallographic data, below the transition temperature, the compound exhibits a monoclinic space group *Cc* structure, where $[(CH_3)_2NH_2]^+$ cations are perfectly ordered inside the pseudo-cuboctahedral cavities of the perovskite. Actually, those cations are linked to the $[Mg(HCOO)_3]^-$ framework through hydrogen bonds between the H atoms of the $-NH_2$ groups of the cations and the O-atoms of the formate ligands, see figure 6.14. Meanwhile, when increasing the temperature above $T_t \sim 264$ K, the crystal structure transforms into a trigonal $R\bar{3}c$ structure, and the $[(CH_3)_2NH_2]^+$ cations get disordered into three equivalent crystallographic positions (figure 6.14), which would explain the observed entropy change [46].

Upon the phase transition, the high-temperature polymorph exhibits a lower volume than the low-temperature phase. This is attributed to a dynamic formation and breakage of hydrogen bonds between the cations and the framework in the high-temperature phase, which results into a larger cation-framework attraction that decreases the unit cell volume [46]. In turn, this material exhibits a negative volume change of $\Delta v \sim -2.28 \times 10^{-6}$ m^3 kg^{-1}, which is characteristic of inverse barocaloric materials, see figure 6.15(a). However, it should be noted that this volume change is lower in magnitude than the positive volume changes of the dicyamanide-perovskites ($\Delta v_{(Mn)} \sim 10.0 \times 10^{-6}$ m^3 kg^{-1}, and $\Delta v_{(Cd)} \sim 5.3 \times 10^{-6}$ m^3 kg^{-1}), probably due to the shorter formate ligand placed in the X-site of the perovskite.

The barocaloric effects in this formate-perovskite were studied by quasi-direct methods using variable pressure differential thermal analysis (VP-DTA) at different

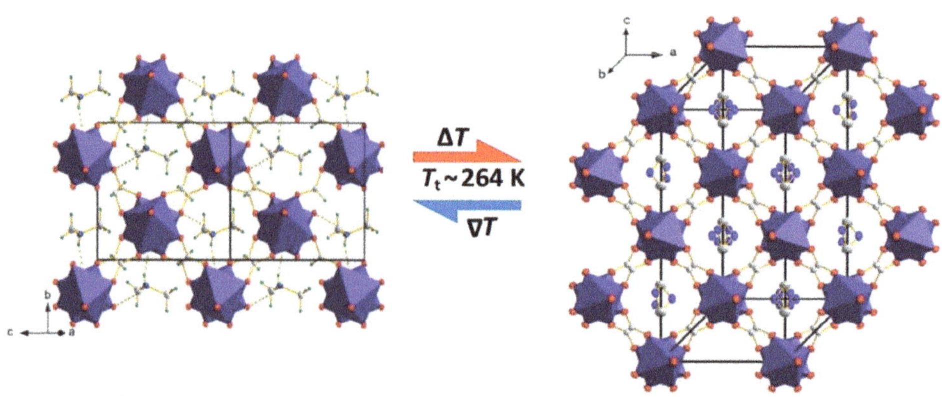

Figure 6.14. Crystal structures of the low temperature and high temperature (ambient) polymorphs of the $[CH_3NH_2][Mg(HCOO)_3]$ hybrid perovskite. Note: H-bonds in the low temperature polymorph are indicted as dashed yellow lines. Figures adapted from reference [46] (http://creativecommons.org/licenses/by/4.0/).

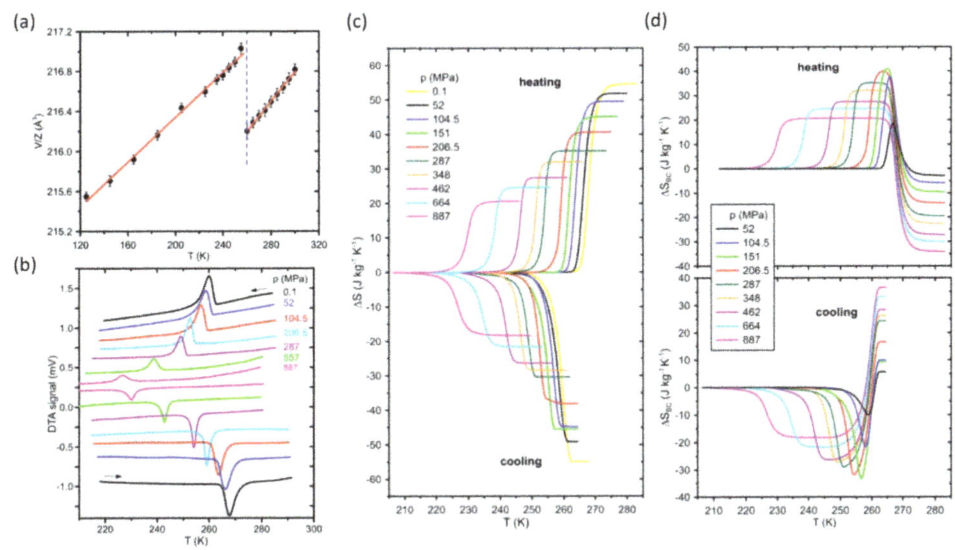

Figure 6.15. Selection of barocaloric parameters for the $[CH_3NH_2][Mg(HCOO)_3]$ perovskite: (a) Specific volume change as a function of temperature; (b) VT-DTA curves at different pressures of up to ~9000 bar (887 MPa); (c) Isobaric entropy change as a function of temperature at different pressures from 1 to ~9000 bar. (d) Isothermal entropy change calculated by quasi-direct methods on heating and cooling for the different applied pressures. Figures reproduced from reference [46] (http://creativecommons.org/licenses/by/4.0/).

relatively high pressures from 1 to ~9000 bar, see figure 6.15. These studies revealed that, effectively, the transition temperature shifts towards lower temperatures when increasing the pressure, as expected for an inverse barocaloric material, see figure 6.15(b) [46]. Remarkably, this material exhibits large barocaloric effects even for for pressures as large as ~9000 bar. However, with a barocaloric coefficient of $dT_t/dp \sim -4.6$ K kbar^{-1}, this material displays a lower pressure-responsiveness than dicyanamide-perovskites. This is probably due to the less flexible nature of formate-perovskites, see previous figure 6.6. For that reason, even if the value of $|\Delta S_{it(q\text{-}d)}|$ can be as large as 41.2 J K^{-1} kg^{-1} (and $|\Delta T_{ad}| = (T/C_p) \times |\Delta S_{it(q\text{-}d)}| \sim 8.1$ K), this material requires very large pressures (of over 2000 bar) to present reversible barocaloric effects [46], see figure 6.15(b). Furthermore, for such large pressures, there is a progressive decrease of the $\Delta S_{it(q\text{-}d)}$ of ~50% at ~9000 bar [46]—see figures 6.15(c) and (d)—, in a similar way to what happens with the [TPrA][Mn $(dca)_3$] perovskite that showed a $\Delta S_{it(q\text{-}d)}$ reduction of ~30% when increasing the pressure up to 1000 bar [32].

6.2.3 Azide-perovskites

The last example, to date, of barocaloric 3D hybrid perovskites is the azide-perovskite with formula $[(CH_3)_4N][Mn(N_3)_3]$, reported in the year 2020 [48]. The length of azide ligands is placed in between dicyanamide and formate ligands, which would anticipate an intermediate pressure responsiveness of azide-perovskites in comparison with the previous dicyanamide- and formate-perovskites. In addition,

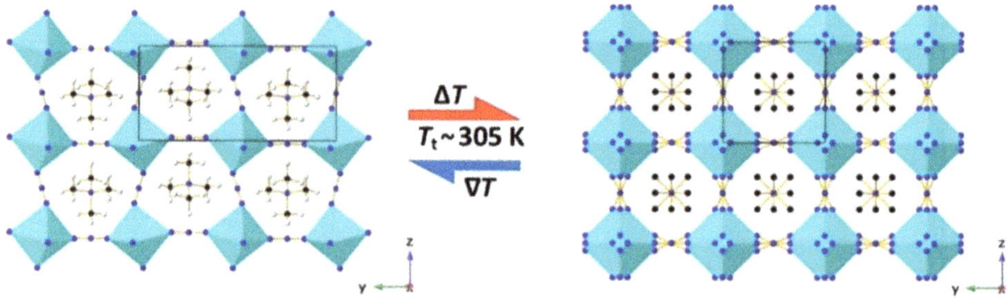

Figure 6.16. Crystal structures of the low temperature ($P2_1$/m) and high temperature ($Pm\bar{3}m$) polymorphs of the $[(CH_3)_4N][Mn(N_3)_3]$ hybrid perovskite. Note: black solid lines indicate the unit cell. Crystallographic data obtained from reference [49].

azide-perovskites are known for showing very large configurational disorder when increasing the temperature, even larger than that of dicyanamide-perovskites [4].

In the particular case of the $[(CH_3)_4N][Mn(N_3)_3]$, it exhibits a thermally-induced phase transition very close to room temperature ($T_t \sim 305$ K, $|T_{t(h)} - T_{t(c)}| \sim 7$ K) between a monoclinic $P2_1$/m polymorph and a cubic $Pm\bar{3}m$ polymorph, see figure 6.16.

The thermally-driven phase transition give rise to large thermal changes of $|\Delta H_t| \sim 25$ kJ kg^{-1} and $|\Delta S_t| \sim 80$ J K^{-1} kg^{-1}, which are one of the largest values observed in 3D hybrid perovskites so far. Such large values can be attributed to a large configurational disorder of the C-atoms of the $[(CH_3)_4N]^+$ cations and the N-atoms of the $[N_3]^-$ ligands. At low temperature these ions are well ordered and located in a single crystallographic position. Meanwhile, when increasing the temperature each C-atom of the $[(CH_3)_4N]^+$ cations and the two terminal N-atoms of the $[N_3]^-$ anions are disordered into four different crystallographic positions each one. As observed by single-crystal x-ray diffraction, the number of crystallographic configurations between both polymorphs give rise to a value of $N \sim 12$, see figure 6.16 [48, 49]. This in in full agreement with the ratio of $N \sim 11$ calculated from the equation $\Delta S_{config.} = R \ln(N)$, which confirms that the thermal changes are, again, originated by order–disorder processes similarly to the other barocaloric perovskites.

Delving into the structural characterization, the phase transition of the $[(CH_3)_4N][Mn(N_3)_3]$ was found to exhibit a positive volume change of $\Delta v \sim 9.4 \times 10^{-6}$ m^3 kg^{-1} similar to the [TPrA][Mn(dca)$_3$] perovskite, see figure 6.17(a). This value anticipated a conventional barocaloric behaviour with a pressure responsiveness of $dT_t/dp \sim 12$ K kbar^{-1}, as calculated by Clausius–Clapeyron and further confirmed by VT-DSC, see figure 6.17(b) [48]. In fact, further calorimetric studies of $[(CH_3)_4N][Mn(N_3)_3]$ revealed a barocaloric effect as large as $|\Delta S_{it}| \sim 70$ J K^{-1} kg^{-1} using both quasi-direct and direct methods, figures 6.17(c) and (d) [48].

Those calorimetric studies highlighted that the pressure-responsiveness of the azide-perovskite lies between that of dicyanamide- and formate-perovskites, while the barocaloric effect is almost double that in these latter [32, 41, 46]. Nevertheless, due to the also larger thermal hysteresis, the pressure required to induce a reversible

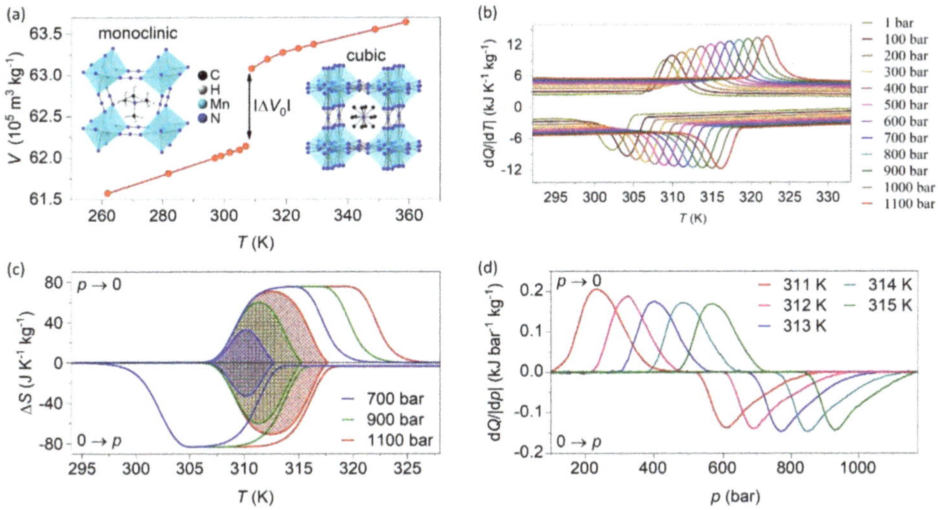

Figure 6.17. Selection of barocaloric parameters for the $[(CH_3)_4N][Mn(N_3)_3]$ perovskite: (a) Specific volume change as a function of temperature (Inset: prototypical pseudo-cubes of the different polymorphs); (b) VT-DSC curves at different pressures of up to 1100 bar; (c) Isothermal entropy change calculated by quasi-direct methods on heating and cooling for different applied pressures (shadowed areas represent the reversible effects); (d) VP-DSC curves at different temperatures from 311 to 315 K. Figures reproduced from reference [48] (https://creativecommons.org/licenses/by-nc/3.0/).

effect is still superior to 900 bar, although such large pressures do not seem to decrease the total entropy change in contrast with the other 3D perovskites.

6.2.4 Conclusions

In summary, we can extract the following main conclusions for barocaloric 3D hybrid perovskites:

- They have exhibited very large barocaloric effects ($|\Delta S_{it}| \sim$ [11.5–70] kJ K^{-1} kg^{-1}) in a wide range of ambient temperatures ($T_{span} \sim$ [220 K–390] K) and operating pressures ($p \sim$ [70–9000] bar), parameters that could be even improved in future members of hybrid barocalorics.
- The length of the ligand is directly related with the pressure-responsiveness and the barocaloric coefficient of the material, which increases from formate- to azide- and to dicyanamide-perovskites.
- It is possible to obtain high configurational entropy changes with relatively short ligands, such as azide ligands, which will provide very large barocaloric effects.
- The application of high pressures seems to decrease the total barocaloric effect, except for the $[(CH_3)_4N][Mn(N_3)_3]$ azide-perovskite.
- The presence of dynamic hydrogen bonds between the A cations and the framework can lead towards negative volume changes and—in turn—to inverse barocaloric effects, as observed in the $[(CH_3)_2NH_2][Mg(HCOO)_3]$ formate-perovskite.

6.3 Enhancing barocaloric performance in 2D hybrid perovskites

Looking for new barocaloric hybrid materials that could present larger entropy changes under the application of lower pressures, 2D hybrid perovskites have recently arisen as very powerful candidates. This family of hybrid compounds has been widely studied for thermal energy storage applications since the late 70s, due to their very large latent heat near ambient temperature associated with thermally-induced phase transitions [50].

Even more recently, since 2021, these layered hybrid perovskites have been reported to display colossal barocaloric effects under very low applied pressures. Specifically, three of these compounds have been studied so far, namely, $[C_{10}H_{21}NH_3]_2MnCl_4$, $[C_9H_{19}NH_3]_2MnCl_4$, and $[C_9H_{19}NH_3]_2CuBr_4$, see figures 6.18(a) and (b) [9, 51, 52]. In addition, over 50 additional 2D perovskites with general formula $[CnH_2n+1NH_3]MX_4$ ($n = [7–16]$; $M = Mn^{2+}, Cu^{2+}, Cd^{2+}, Pb^{2+}$; $X = Cl^-, Br^-, I^-$) have been also proposed as potential barocaloric materials in view of their large entropy change associated with a thermally-induced solid–solid phase transition and occurring near room temperature, see figure 6.18(c) [51].

From the structural point of view, these compounds are formed by 2D layers of $[MX_6]$ octahedra sharing equatorial corners and that are separated by two layers of alkylammonium chains, see figures 6.18(a) and (b). These alternating layers follow an *inorganic–organic–organic–inorganic* scheme in a sandwich-like structure. Here,

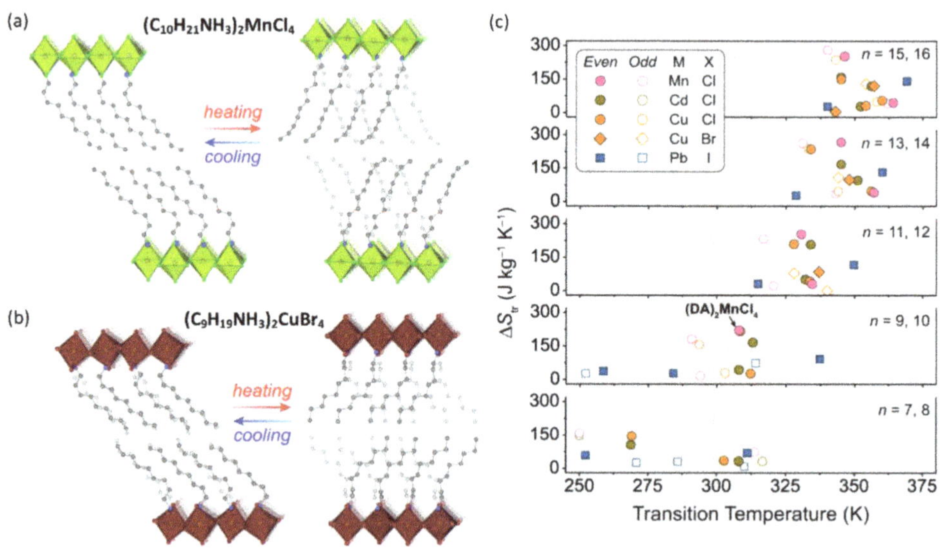

Figure 6.18. Crystal structures of the low temperature and high temperature polymorphs of the barocaloric materials $[C_{10}H_{21}NH_3]_2MnCl_4$ (a) and $[C_9H_{19}NH_3]_2CuBr_4$ (b), respectively. (c) Thermally-induced phase transition entropy *versus* transition temperature for different 2D hybrid perovskites of general formula $[C_nH_{2n+1}NH_3]_2MX_4$ ($n = [7–16]$; $M = Mn^{2+}, Cd^{2+}, Cu^{2+}, Pb^{2+}$; $X = Cl^-, Br^-, I^-$). Note: $(DA)_2MnCl_4$ in panel (c) refers to the $[C_{10}H_{21}NH_3]_2MnCl_4$ compound. Figures adapted from reference [51] (http://creativecommons.org/licenses/by/4.0/).

the alkylammonium chains are interconnected through Van der Waals interactions through the methyl groups, and they are also linked to the $[MX_6]$ layers through N–H···X hydrogen bonds. It is worth noting that the lack of strong chemical bonds between organic–inorganic layers and the larger flexibility of the long-chain organic ligands are key aspects that will allow for phase transitions induced by lower pressures than in 3D hybrid perovskites. Furthermore, the increase in the degrees of freedom of the organic ligands due to their larger length will also favour larger entropy changes through order–disorder transitions than in their 3D 'siblings'.

Using VT-DSC at ambient pressure, these materials reveal thermally-induced phase transitions near room temperature [9, 51, 52]. It should be noted that, as detailed below, polycrystalline and single-crystal samples of 2D hybrid perovskites were studied for comparison purposes.

In the case of polycrystalline $[C_{10}H_{21}NH_3]_2MnCl_4$ samples, this material exhibits a thermally-induced transition with the following thermal parameters (with slight variations depending on the source) [9, 51, 52]: $|\Delta H_t| \sim [71–76]$ kJ kg^{-1}, $|\Delta S_t| \sim [230–260]$ J K^{-1} kg^{-1}, $T_t \sim [310–312]$ K, and $|T_{t(h)} - T_{t(c)}| \sim [1.4–3.0]$ (measured in the onsets) or ~[3.5–9.0] K (measured at the peak's maximum).

Subsequently, when reducing the length of the alkylammonium chain in the case of $[C_9H_{19}NH_3]_2MnCl_4$ polycrystalline samples, the phase transition occurs in two steps (as will be later detailed) showing a main peak with a much smaller peak—or shoulder—very close in temperature. The thermal parameters for the whole transition can be summarized as: $|\Delta H_t| \sim 62$ kJ kg^{-1}, $|\Delta S_t| \sim 212$ J K^{-1} kg^{-1}, $T_t \sim 294$ K, $|T_{t(h)} - T_{t(c)}| = 5.2$ K (obtained from the maximum of the peaks) [52].

In addition, when replacing the inorganic moiety for $[C_9H_{19}NH_3]_2CuBr_4$ powdered samples, there is again a one-step transition, but with a significant decrease of the thermal parameters: $|\Delta H_t| \sim 23$ kJ kg^{-1}, $|\Delta S_t| \sim 76$ J K^{-1} kg^{-1}. However, the transition temperature remains in the same range of $T_t \sim 305$ K with a much smaller hysteresis of $|T_{t(h)} - T_{t(c)}| \sim 0.4$ K (from onsets) or ~2.2 K (from peak's maximum) [51].

Therefore, the thermal properties seem to be very sensitive to small changes in the length of the organic chains, and even more to the change of the inorganic building-blocks.

According to single-crystal XRD, Raman and FT-IR spectroscopy and DFT calculations [9, 51, 52], these phase transitions are mainly related to configurational (or more specifically, conformational) disorder of the alkylammonium chains that change between different rotameric states of the C–C bonds. Additionally, there is also a minor contribution of the chain flipping or reorienting between two different positions in reference with the metal–halide layers. Those processes take place simultaneously for both $[C_{10}H_{21}NH_3]_2MnCl_4$ and $[C_9H_{19}NH_3]_2CuBr_4$ compounds [9, 51]. Nevertheless, the configurational disordering upon the transition is lower in the case of the Cu compound—figures 6.18(a) and (b)—which explains the lower enthalpy and entropy changes for this material [51]. In contrast, for the $[C_9H_{19}NH_3]_2MnCl_4$, the two processes give rise to a two-step phase transition, where the major conformational changes occurs slightly before than the minor orientational changes of the cation [52].

Remarkably, all these 2D hybrid perovskites exhibit very large—and even colossal—barocaloric effects as observed by calorimetry, as will be further explained below.

6.3.1 Barocaloric studies on $[C_{10}H_{21}NH_3]_2MnCl_4$ and $[CH_9H_{19}NH_3]_2CuBr_4$

In particular, $[C_{10}H_{21}NH_3]_2MnCl_4$ and $[CH_9H_{19}NH_3]_2CuBr_4$ polycrystalline samples haven been found to display thermal changes as large as $|\Delta S_{\text{it(q-d)}}| \sim 75$ J kg^{-1} K^{-1} and $|\Delta S_{\text{it(q-d)}}| \sim 68$ J kg^{-1} K^{-1} under the application of pressures of only $p = 150$ bar, see figures 6.19(a) and (b) [51].

Even more, and very interestingly, the authors also demonstrated that these thermal changes can be further enhanced when using single crystals samples. Thereby, these values can increase up to $|\Delta S_{\text{it(q-d)}}| \sim 190$ J kg^{-1} K^{-1} and $|\Delta S_{\text{it(q-d)}}| \sim 78$ J kg^{-1} K^{-1} at the same applied pressure, see figures 6.19(c) and (d). This is attributed to an improved thermal contact between samples and holders, which highlights the impact that extrinsic factors might have when assessing barocaloric effects [51].

This extraordinary barocaloric performance is favored by the very large barocaloric coefficients ($dT_t/dp_{(\text{Mn})} \sim 22.0$ K kbar^{-1}, $dT_t/dp_{(\text{Cu})} \sim 27.0$ K kbar^{-1}) and volume changes across the transition ($\Delta v_{(\text{Mn})} \sim 65.9 \times 10^{-6}$ m^3 kg^{-1}, $\Delta v_{(\text{Cu})} \sim 25.3 \times 10^{-6}$ m^3 kg^{-1}) observed for those materials [51].

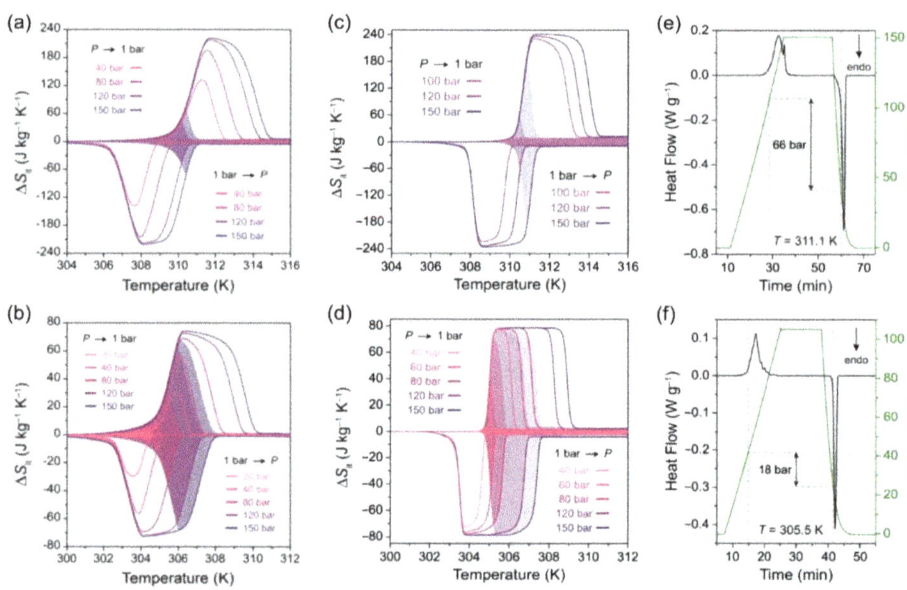

Figure 6.19. Isothermal entropy change obtained by quasi-direct methods ($\Delta S_{\text{it(q-d)}}$) on heating and cooling for $[C_{10}H_{21}NH_3]_2MnCl_4$ and $[CH_9H_{19}NH_3]_2CuBr_4$ polycristalline (a and b) and single-crystal (c and d) samples, respectively. Note: shadowed areas represent the reversible region of the $\Delta S_{\text{it(q-d)}}$. (e and f) VP-DSC curve under isothermal conditions (at 311.1 K and 305.5 K) for the same Mn and Cu single-crystal samples. Figures reproduced from reference [51] (http://creativecommons.org/licenses/by/4.0/).

Additionally, the authors also carried out direct VP-DSC analysis that demonstrated the reversibility of the barocaloric transitions over time and revealed a low mechanical hysteresis (defined as the difference between the transition pressures on pressurization and depressurization) of only 66 and 18 bar for the Mn and Cu materials, see figures 6.19(e) and (f). Accordingly, the pressures required to drive those transitions in a reversible manner can be significantly lower than 150 bar.

It should be noted that those colossal thermal changes of over 100 J K^{-1} kg^{-1} under the remarkably low pressures of 150 bar, pose a great advance towards the future technological implementation, since those operating parameters are coming closer to the market needs. However, the temperature span is still limited to only 3–4 K under such low pressures [51]. Similarly to other hybrid barocalorics, this limitation can be overcome by increasing the applied pressure above 500 bar and up to at least 3600 bar [9, 51, 52].

For instance, polycrystalline samples of the $[C_{10}H_{21}NH_3]_2MnCl_4$ compound have been found to maintain a colossal barocaloric effect of ~100 J K^{-1} kg^{-1} in an operating range of ~13 K when applying 1000 bar, and achieve maximum reversible values of up to $|\Delta S_{it(q\text{-}d)}| \sim 250$ J K^{-1} kg^{-1} [9]. Interestingly, for higher pressures, the material exhibits triple point at $T_{tp} \sim [330–337]$ K and $p_{tp} \sim [1200–1400]$ bar, which highlights a new intermediate phase III between the low-temperature phase II and the high-temperature phase I, see figure 6.20. Nevertheless, the splitting of the VT-DSC peak does not seem to affect the barocaloric performance of the material, see figure 6.20(d).

Finally, regarding the reversible adiabatic temperature change, this value was estimated in a more accurate way than in the previous 3D perovskites, by using VT-DSC and C_p measurements. This parameter was calculated as the maximum temperature distance between the entropy curve at atmospheric pressure on heating and the entropy curve at the higher pressure on cooling: $|\Delta T_{ad}|_{rev} = |T_c(S;p) - T_H(S;p_{atm})|$,

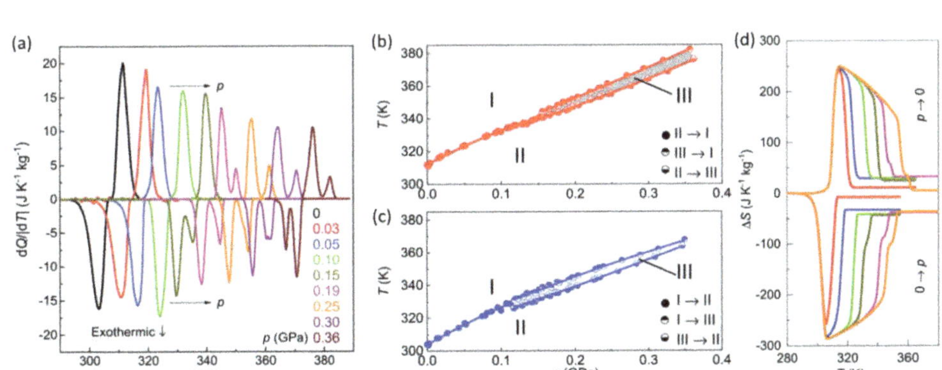

Figure 6.20. (a) VT-DSC curves for the $[C_{10}H_{21}NH_3]_2MnCl_4$ hybrid perovskite at different isobaric conditions from ambient pressure up to 3600 bar (0.36 GPa). (b and c) *T–p* phase diagram obtained on heating (b) and cooling (c) from the maximum of the observed VT-DSC peaks. Note: filled, top-half filled, and bottom-half filled symbols represent the different transitions of II↔I, III↔I and II↔III, respectively. Lines represent data fitting. (d) Barocaloric effect in terms of isothermal entropy change, Δ*S*, obtained from the VT-DSC curves by quasi-direct methods. Figures reproduced form reference [9] (http://creativecommons.org/licenses/by/4.0/).

which already considers the C_p of the materials. Further details of this method can be found in reference [9]. Following this approximation, values of $|\Delta T_{ad}|_{rev} \sim 7$ K and 10 K are obtained at 500 bar for the $[C_{10}H_{21}NH_3]_2MnCl_4$ and $[CH_9H_{19}NH_3]_2CuBr_4$ compounds, respectively [9, 51].

6.3.2 Barocaloric studies on $[C_9H_{19}NH_3]_2MnCl_4$

The $[C_9H_{19}NH_3]_2MnCl_4$ polycrystalline material shows the opposite behaviour than the analogous $[C_{10}H_{21}NH_3]_2MnCl_4$ compound. At ambient pressure, this material exhibits a two-steps phase transition as observed by VT-DSC, see figure 6.21(a). Meanwhile, when increasing the pressure above 200 bar the transition occurs in a single step up to—at least—1000 bar, see figure 6.21(b). This transition exhibits a pressure dependence of $dT_t/dp \sim 17.2$ K kbar^{-1}. Remarkably, the material exhibits a reversible barocaloric effect as large as $|\Delta S_{it(q\text{-}d)}| \sim 212$ J K^{-1} kg^{-1} at 1000 bar as observed in figure 6.21(c), with an estimated $|\Delta T_{ad}| \sim 10$ K [52].

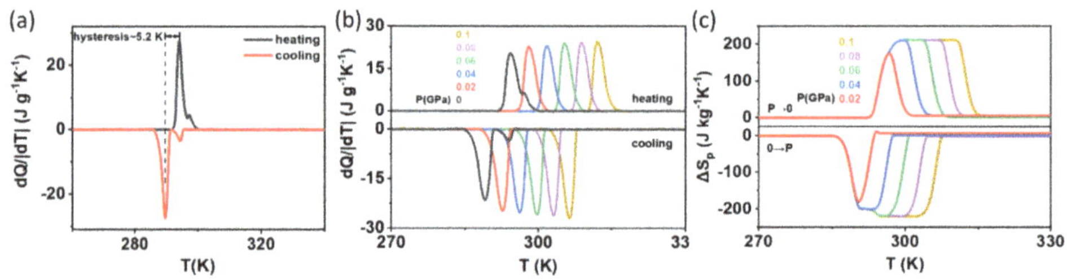

Figure 6.21. VT-DSC curves for the $[C_9H_{19}NH_3]_2MnCl_4$ compound at ambient pressure (a) and at different applied pressures (b) on heating and cooling ramps. (c) Isothermal entropy change calculated by quasi-direct methods using the VT-DSC data. Figures reproduced form reference [52] (http://creativecommons.org/licenses/by/4.0/).

6.3.3 Conclusions

Compiling the main findings of the recent family of barocaloric 2D hybrid perovskites, we can list these key aspects:

- They can exhibit colossal thermal changes of above ~100 J K^{-1} kg^{-1} under remarkably low pressures of 150 bar, which can be attributed to the large configurational degrees of freedom of the long-size organic cations and their large flexibility and compressibility across the transition.
- The enthalpy and entropy change for barocaloric 2D hybrid perovskites is very sensitive not only to the length of the organic cation, but also to the selection of the inorganic $[MX_4]^-$ building-blocks.
- Barocaloric studies in 2D hybrid perovskites have revealed that extrinsic factors such as the morphology of the sample (single-crystal or polycrystalline powder) can largely affect the registered barocaloric parameters (thermal hysteresis, isothermal entropy change, barocaloric coefficient, etc).
- 2D hybrid perovskites can exhibit multiple phase transitions that can merge and/or split in triple points that, in the studied materials, does not seem to affect the barocaloric effects.

6.4 Taking advantage of plastic disorder in 0D hybrid perovskites

Plastic crystals are crystalline dynamic solids formed by rotatable ions or molecules that have been known since the early 20th century [53]. The so-called plastic phase is a singular state of matter where the orientational disorder of crystalline compounds provides a high plasticity to the material, originating the term '*plastic crystal*' to refer to those materials [53, 54]. Such a combination is rare in molecular materials as most crystalline materials tend to be brittle [55]. Moreover, most plastic crystals display phase transitions from an ordered crystalline phase to a disordered plastic phase, which involves very large latent heat that has been very useful for thermal energy storage over the years [56].

More recently, organic molecular plastic crystals—owing to such energetic transitions—have evolved as one of the most promising alternatives for barocaloric cooling. They have been found to present colossal barocaloric effects ($|\Delta S_{it}| > 100$ J K^{-1} kg^{-1}) that lie close to the thermal changes observed in refrigerant gases [57, 58]. Nevertheless, contrary to refrigerant gases that require a few bar of pressure, these organic plastic crystals usually require pressures above 2500 bar to operate [57, 58].

Adopting this same strategy, hybrid organic–inorganic materials with ionic 0D structures and plastic crystal transitions were recently explored trying to find new barocaloric materials. In principle, those 0D hybrids could present barocaloric effects as large as organic plastic crystals, but under the application of pressures as low as in the case of 3D and 2D hybrid materials.

The first family of 0D hybrid materials to be proposed as potential barocaloric materials was that of alkylammonium tetrahalometallates [59], which are a family of compounds already well known for their functional properties, including thermal and multiple energy storage [7, 60, 61].

In this context, very recent preliminary barocaloric studies have been already performed in the specific $[(CH_3)_3(CH_2Cl)N][FeCl_4]$ and $[(CH_3)_3S][FeCl_4]$ compounds [62]. Both materials were identified as potential barocaloric materials given that they present thermally-induced plastic transitions slightly above room temperature, where the organic and inorganic building-blocks can largely increase their orientational disordered, see figure 6.22 as example. This large disorder is favoured by a lack of strong chemical interactions between the organic and inorganic ions.

For instance, the $[(CH_3)_3(CH_2Cl)N][FeCl_4]$ shows a transition at $T_t \sim 326$ K with a relatively low thermal hysteresis ($|T_{t(h)} - T_{t(c)}| \sim 4.0$ K), and very large values of latent heat ($|\Delta H_t| \sim 46$ kJ kg^{-1}) and entropy change ($|\Delta S_t| \sim 140$ J kg^{-1} K^{-1}) [62]. Similarly, the $[(CH_3)_3S][FeCl_4]$ display a transition at $T_t \sim 315$ K with a larger thermal hysteresis of $|T_{t(h)} - T_{t(c)}| \sim 13.0$ K, while keeping comparable thermal values of $|\Delta H_t| \sim 40$ kJ kg^{-1} and $|\Delta S_t| \sim 130$ J kg^{-1} K^{-1} [62].

Subsequently, more detailed isobaric VT-DSC (quasi-direct methods) and isothermal VP-DSC analysis (direct methods) were performed in order to study the barocaloric response of both ionic plastic crystals, figure 6.23. According to quasi-direct methods, both compounds exhibit a barocaloric coefficient of $dT_t/dp \sim 20$ K $kbar^{-1}$, even if the $[(CH_3)_3S][FeCl_4]$ seems to exhibit somehow an initial inertia to the pressure that is overcome above 300 bar [62].

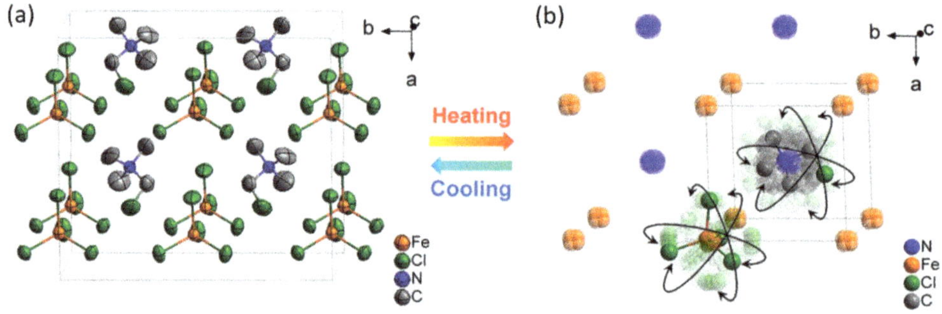

Figure 6.22. Crystal structures of $[(CH_3)_3(CH_2Cl)N][FeCl_4]$ compound (a) in the ordered crystalline phase, and (b) in the disordered plastic crystal phase. Note: all H atoms, as well as Cl and C atoms of the alkylammonium cations in the plastic crystal phase have omitted for clarity. Black lines represent the unit cells of the two polymorphs. Figure reproduced with permission from reference [7], copyright (2018) American Chemical Society.

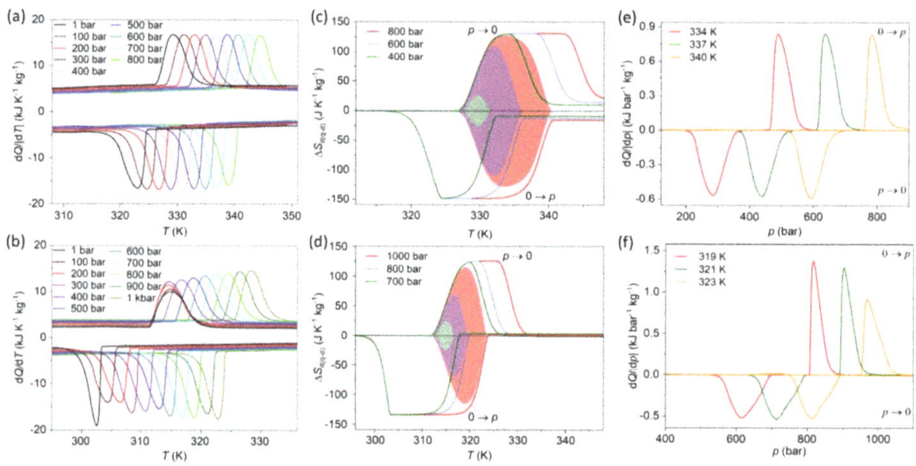

Figure 6.23. (a and b) VT-DSC curves, (c and d) isothermal entropy change curves, and (e and f) VP-DSC curves for the $[(CH_3)_3(CH_2Cl)N][FeCl_4]$ and $[(CH_3)_3S][FeCl_4]$ compounds, respectively. Note: In panels (a and b) heating curves are positive and cooling curves are negative. Panel (b) shows a pressure inertia on heating. Shadowed areas in panels (c and d) represent the reversible isothermal entropy changes. Figures reproduced from reference [62] (https://creativecommons.org/licenses/by-nc-nd/4.0/).

As for the barocaloric effects, depending on the employed method, the $[(CH_3)_3(CH_2Cl)N][FeCl_4]$ material was found to display a reversible effect of between $\Delta S_{it(q\text{-}d)} \sim 123$ J K^{-1} kg^{-1} and $\Delta S_{it(d)} \sim 147$ J K^{-1} kg^{-1} for pressures above 600 bar. Meanwhile, the $[(CH_3)_3S][FeCl_4]$ compound exhibited reversible values of between $|\Delta S_{it(q\text{-}d)}| \sim 125$ J K^{-1} kg^{-1} and $|\Delta S_{it(d)}| \sim 131$ J K^{-1} kg^{-1} under larger pressures of above 900 bar.

It is worth mentioning that these 0D ionic hybrid materials exhibit barocaloric effects of similar magnitude to that of organic plastic crystals under pressures below 1000 bar. Moreover, these effects are also similar to those reported for 2D hybrid perovskites. However, even if the barocaloric coefficients are in the same range, the large thermal hysteresis prevents 0D hybrids from operating at lower pressures than 600 bar. This important inconvenient must be overcome in future studies, where a proper combination of the organic and inorganic units might provide hybrid plastic crystals with enhanced barocaloric performance.

6.4.1 Conclusions

In view of these preliminary results, it can be concluded that:

- 0D ionic and/or molecular hybrid materials are a promising family of barocaloric materials with very large barocaloric effects of over 100 J K^{-1} kg^{-1}, similar to organic plastic crystals and 2D hybrid perovskites, and that can operate under pressures below 1000 bar.
- These 0D hybrids also present a large pressure responsiveness with barocaloric coefficients in the same order of magnitude as their 3D and 2D hybrid 'siblings', although the thermal hysteresis must be reduced in order to decrease the operating pressure.
- Future studies must be performed in order to further understand and enhance the barocaloric performance of 0D hybrid materials.

6.5 Exploiting breathing transitions in 3D porous MOFs

To date, most of the barocaloric studies on hybrid organic–inorganic materials have been performed in dense and non-porous compounds where the pressure was generally applied with oil or an inert gas, such as nitrogen or helium. In these late cases, the gas does not interact with the hybrid materials and their dense structure prevents them from taking any gas adsorption.

On the other side, as indicated in the introduction, there is a vast family of porous hybrid materials, known as metal–organic frameworks (MOFs), that can interact with gases, such as N_2, CO_2, CH_4, CH_3OH or H_2O, among others [13, 14]. These MOFs, which are even more flexible and soft materials than other hybrids (see previous figure 6.6), are also very well known for displaying extraordinary porosity that can serve for adsorbing and storing gases in a reversible manner [14–16].

In view of this exceptional adsorption capacity, MOFs have been proposed as outstanding materials for adsorption—or evaporative—refrigeration [17]. This type of refrigeration makes use of adsorption/desorption enthalpy of a given pair of adsorbate–adsorbent (i.e. gas-MOF in this case) to provide a heating/cooling effect [63–65]. Differently from vapour compression and barocaloric refrigeration, the adsorption cooling technologies do not apply mechanical pressure to induce the adsorption/desorption process. On the contrary, they use an external heat source (often residual heat) to perpetuate the refrigeration cycle. However, the efficiency of such technologies are generally lower than that of vapour compression and they also require relatively more complex designs and high vacuum [66].

In 2021, MOFs with 'breathing transitions' (and more specifically the sub-family of zeolitic imidazole frameworks or ZIFs) were proposed as very promising barocaloric candidates [59], since they were known to present reversible solid–solid phase transitions with very large volume changes [59]. Breathing transitions, firstly described in MOFs, are reversible first-order phase transitions between two polymorphs that present important differences in their pore size [67, 68]. In general, this type of transition occurs between a narrow-pore phase and large-pore phase, which is directly induced by adsorption/desorption processes of a pressurizing gas in those pores, see figure 6.24 [67, 68]. In this transition both processes take place simultaneously and cannot exist one without the other. Curiously, the resemblance of this structural behaviour to the breathing mechanism of our lungs gave rise to the singular name of this transition.

Soon after that, in 2022, pressure-induced caloric effects for heating/cooling applications were experimentally demonstrated in the particular MOF of formula Al(OH)[BDC] (BDC = $[C_6H_4(CO_2)_2]^{2-}$), commonly known as MIL-53(Al) [69]. Interestingly, as further detailed below, this material was found to exhibit a new type of pressure-induced caloric effect related to the breathing transition, which combines thermal changes originated from a solid–solid phase transition and also from gas adsorption/desorption processes when pressurizing it with CO_2. Therefore, this new caloric effect, with some similarities to barocaloric effects but also to adsorption refrigeration, was coined as 'breathing-caloric effect'.

In pursuit of environmental consciousness, it should be mentioned that CO_2 is becoming one of the most eco-friendly and safe refrigerant gases, labelled as R744. This refrigerant presents a much lower global warming potential (GWP) than most of the other refrigerant gases and, in addition it is chemically stable, non-toxic, and non-flammable [71–74]. And what's more, the use of CO_2 as a valuable resource

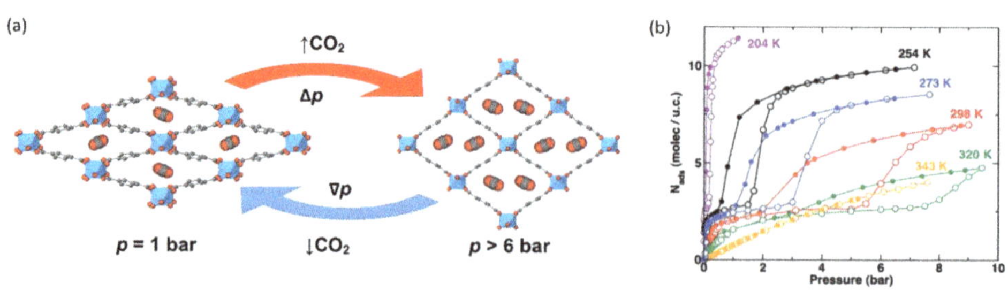

Figure 6.24. (a) Crystal structures of the narrow-pore (left) and large-pore (right) polymorphs of MIL-53(Al) at ambient temperature as a function of CO_2 pressure. Note: CO_2 molecules were randomly placed just for illustration purposes. (b) CO_2 adsorption (open circles) and desorption (full circles) isotherms of MIL-53(Al) at different temperatures. Note: figure (a) reproduced from reference [69] (https://creativecommons.org/licenses/by/4.0/) and figure (b) reproduced with permission from reference [70], copyright (2010) American Chemical Society.

promotes its recovery from residual gases and helps to decrease its emissions [75, 76]. However, CO_2 requires relatively large pressures to operate and present a super-critical transition above 304 K, which limits the operating temperature in subcritical devices or adds extra complexity in transcritical systems [74].

As for the MIL-53(Al), it exhibits a crystal structure formed by unidimensional chains of $[Al(BDC)_4(OH)_2]$ octahedra sharing corners and linked through BDC ligands, giving rise to linear lozenge-shaped channels that can accommodate different guest molecules, see figure 6.24(a). Under CO_2 atmosphere at ambient pressure and temperature, the material exhibits a narrow-pore phase with a few CO_2 molecules embedded in its channels. Meanwhile, when increasing the CO_2 pressure up to 16 bar, the material undergoes a breathing transition (through a 'wine-rack' displacement) towards a large-pore phase with a volume increase of up to 40% [77], which can allocate a larger amount of CO_2 molecules, see figure 6.24 [70].

This transition is reversible and the amount of adsorbed CO_2 molecules depends on the MOF's phase (narrow- or large-pore phase), the working temperature, and the applied CO_2 pressure, as observed by the isothermal gas adsorption studies, see figure 6.24(b) [70]. As can be observed, these adsorption isotherms display a type IV shape, where the adsorption capacity suddenly increases above a given value of pressure, which is the critical pressure than induces the pore opening from the narrow-pore to the large-pore phase. Meanwhile, on depressurization there is a low mechanical hysteresis of a few bar to reverse such a transition, where the CO_2 is desorbed and the MIL-53(Al) goes back to the narrow-pore phase. In that way, MIL-53(Al) presents breathing transitions above atmospheric pressure from 254 K to, at least, 320 K [70].

Remarkably, the CO_2 adsorption/desorption in MIL-53(Al) is particularly energetic due to the creation/breakage of hydrogen bonds between the gas molecules and the framework [78]. However, although the breathing transitions have been very well known for decades, they weren't studied for caloric cooling/heating until very recently.

The pressure-induced caloric effects associated with those breathing transitions were studied by VP-DSC at different temperatures and under CO_2 pressure from 1 to 16 bar, see figure 6.25 [69]. The obtained data revealed two exothermic peaks on pressurization with the corresponding endothermic peaks on depressurization. The first pair of peaks appeared at $p \sim 2.2$ bar ($T = 298$ K) without showing any temperature dependence or hysteresis. They were attributed to CO_2 adsorption/desorption in the channels of the narrow-pore phase, which exhibited enthalpy changes of $\Delta H_1 \sim 10.9$ kJ kg^{-1}. Meanwhile, when increasing the pressure above the breathing transition, the second pair of peaks appeared at $p \sim 8.1$ bar ($T = 298$ K) with a mechanical hysteresis of 4.5 bar^{-1}. These peaks showed an enthalpy change of $\Delta H_2 \sim 81.8$ kJ kg^{-1} that decreased when increasing the temperature [69]. This behaviour would be in agreement with a lower CO_2 adsorption capacity when increasing the temperature, as observed by isothermal adsorption studies [70].

Therefore, this last pair of peaks was undoubtedly attributed to the breathing transition between the narrow-pore and the large-pore phases of MIL-53(Al).

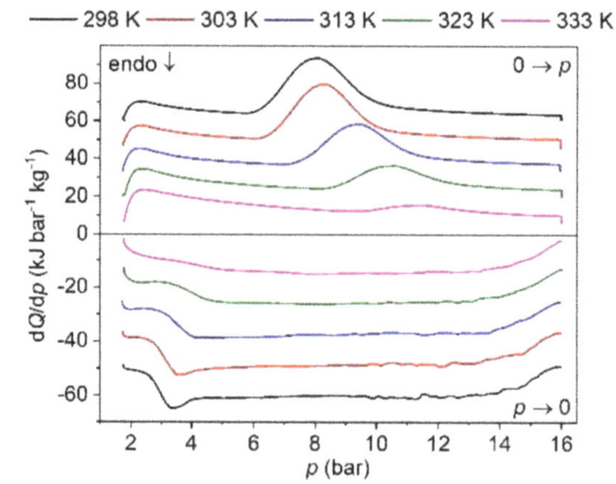

Figure 6.25. VP-DSC curves for the MIL-53(Al) compound under CO_2 pressurization ($0 \rightarrow p$) and depressurization ($p \rightarrow 0$) at different temperatures (298–333 K). Figure reproduced from reference [69] (https://creativecommons.org/licenses/by/4.0/).

As mentioned before, in this breathing transition there are two simultaneous and coexisting mechanisms that provide thermal changes, which upon pressurization are: (i) solid–solid phase transition that opens the pore of MIL-53(Al), and (ii) significant CO_2 adsorption in the large-pore phase [69].

Contrary to barocaloric materials, in MIL-53(Al) the pressure favours a higher-volume phase and, in turn, this process will be endothermic. On the other side, the pressure-induced CO_2 adsorption will be an exothermic process. Therefore, in the specific case of MIL-53(Al) both contributions to the breathing-caloric effect will be opposite in sign, decreasing the total amount of caloric effect that could be attainable if both contributions were additive. Encouragingly, this MOF would not be an isolated example, as many other MOFs present breathing transitions, where in some of them both thermal contributions are expected to be additive [69].

Trying to quantify the magnitude of both contributions in the MIL-53(Al), the solid–solid phase transition was estimated to present an enthalpy change of $|\Delta H_t| \sim 40$ kJ kg^{-1} by using VT-DSC and computational data. Moreover, VT-DSC at different pressures also showed that the transition temperature shifts towards higher temperatures with an extraordinary pressure responsiveness ranging in $dT_t/dp \sim [3.3–7.5]$ K bar^{-1} [69], which is two orders of magnitude larger than in any other barocaloric material [45].

Since this solid–solid transition will be endothermic upon compression and the VP-DSC data shows a large exothermic peak when pressurizing, the CO_2 adsorption (exothermic process) is found to be the major contribution to the breathing-caloric effect in MIL-53(Al). However, this large adsorption could not take place without the structural change from the narrow-pore to the large-pore phase.

In any case, the total pressure-induced thermal changes observed by direct VP-DSC analysis (for the sum of the two pairs of peaks) is as large as $\Delta H_{it(d)} \sim 92.7$ kJ kg^{-1} and $\Delta S_{it(d)} \sim 311$ J K^{-1} kg^{-1} at room temperature and under remarkably low applied

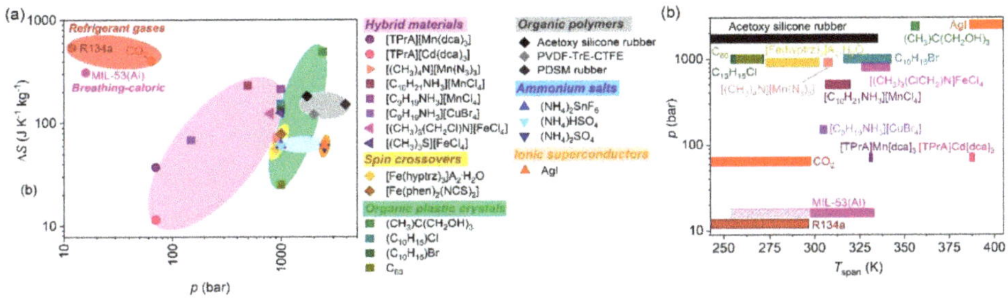

Figure 6.26. Caloric parameters (a) ΔS vs p and (b) T_{span} vs p for MIL-53(Al) in comparison with vapour compression commercial refrigerant gases and selected barocaloric materials. The pink striped area for MIL-53(Al) indicates the T_{span} below room temperature estimated from reported adsorption isotherms [70]. Figures adapted from reference [69] (https://creativecommons.org/licenses/by/4.0/) and updated using data reported in references [45, 51].

pressures of only 16 bar. Moreover, the temperature span can be as large as T_{span} ~ [254–333] K, where the thermal changes are still as large as $\Delta H_{it(d)}$ ~ 49.9 kJ kg^{-1} and $\Delta S_{it(d)}$ ~ 149.8 J K^{-1} kg^{-1} for the upper temperature limit. Figure 6.26 shows a comparison of those thermal parameters for refrigerant gases (specifically the refrigerant R134a and CO_2), most of the reported barocaloric materials and the here described MIL-53(Al) breathing-caloric material.

Outstandingly, these results offer, for the first time, pressure-induced caloric effects as large as those of refrigerant gases under similar operating pressure and temperature conditions. Furthermore, the MIL-53(Al)-CO_2 refrigerant pair improves the thermal parameters of CO_2 stand-alone refrigerant by: (i) reducing the working pressure from 64 bar to only 16 bar, and (ii) expanding the operating temperature limit of 304 K (in the case of subcritical CO_2 systems) up to 333 K.

6.5.1 Conclusions

The here-discussed findings can be summarized in the following key aspects:

- Breathing-caloric effects are a new type of pressure-induced caloric effect that simultaneously combines large thermal changes due to the coexistence of solid–solid phase transitions (barocaloric effects) and adsorption/desorption processes (adsorption cooling).
- The first breathing-caloric effects were found in MIL-53(Al) with values as large as ΔS_{it} ~ [150–311] J K^{-1} kg^{-1} for a wide operating temperature of T_{span} ~ [254–333] K and under extraordinarily low pressures of only 16 bar, which matches the market needs.
- This new type of refrigeration alternative, combines elements from barocaloric cooling, as well as from commercial vapour compression and adsorption refrigeration systems, which could largely facilitate the technological implementation in the near future.

Appendix A: Summary of caloric parameters of hybrid organic–inorganic materials

	Material	$T_{t(h)}$ (K)	$\|T_{t(h)} - T_{t(c)}\|$ (K)	$\|\Delta S_{it(q\text{-}d)}\|_{rev}$ (J K^{-1} kg^{-1})	$\|\Delta S_{it(d)}\|$ (J K^{-1} kg^{-1})	$\|\Delta T_{ad}\|_{rev}$ (K)	p (bar)	dT_t/dp (K $kbar^{-1}$)	Δv (10^{-6} m^3 kg^{-1})	References
Dense 3D hybrids	[TPrA][Mn(dca)$_3$]	331	0.9$^{(onset)}$ 3.0$^{(peak)}$	30.5	35.1	4.1	68.9	23.1	10.0	[32]
	[TPrA][Cd(dca)$_3$]	368	1.4$^{(onset)}$ 3.0$^{(peak)}$	11.5	—	—	68.9	38.2	5.3	[41]
	[$(CH_3)_2NH_2$][Mn(HCOO)$_3$]	264	2.0$^{(peak)}$	41.2	—	8.1	>2000	−4.6	−2.3	[46]
	[$(CH_3)_4N$][Mn(N)$_3$]	305	7.0$^{(peak)}$	70.0	70.0	—	>900	12.0	9.4	[48]
Dense 2D hybrids	[$C_{10}H_{21}NH_3$]$MnCl_4$	[310–312]	[1.4–3.0]$^{(onset)}$ [3.5–9.0]$^{(peak)}$	250.0	—	7.0	500	[22.0–28.0]	65.9	[9, 51, 52]
	[$C_9H_{19}NH_3$]$MnCl_4$	294	5.2$^{(peak)}$	212	—	10.0	1000	17.2	—	[52]
	[$C_9H_{19}NH_3$]$CuBr_4$	305	0.4$^{(onset)}$ 2.2$^{(peak)}$	68.0	—	10.0$^{(a)}$	150	27.0	25.3	[51]
Dense 0D hybrids	[$(CH_3)_3(CH_2Cl)N$]$FeCl_4$	326	4.0$^{(peak)}$	123.0	147.0	—	800	19.7	—	[62]
	[$(CH_3)_3S$]$FeCl_4$	315	13.0$^{(peak)}$	125.0	131.0	—	1000	19.9	—	[62]
Porous MOFs	MIL-53(Al)	298$^{(b)}$	—	—	311.0	—	16.0	[3.3–7.5] $\times 10^3$ $^{(c)}$	40%$^{(d)}$	[69, 77]

Parameters: T_t = transition temperature on heating, $|T_{t(h)} - T_{t(c)}|$ = thermal hysteresis calculated from the peak's maximum (peak) or form the onsets (onset), $|\Delta S_{it(q\text{-}d)}|_{rev}$ = reversible isothermal entropy change obtained by quasi-direct methods, $|\Delta S_{it(d)}|_{rev}$ = reversible isothermal entropy change obtained by direct methods, $|\Delta T_{ad}|_{rev}$ = reversible adiabatic temperature change estimated from quasi-direct methods, p = applied pressure to obtain reversible barocaloric effects, dT_t/dp = barocaloric coefficient, Δv = volume change of the phase transition obtained from powder x-ray diffraction data.

Notes: (a) Value calculated at $p = 500$ bar, (b) temperature at which the MIL-53(Al) presents a pressure-induced transition with a $|\Delta S_{it(d)}| = 311$ J K^{-1} kg^{-1}, (c) value calculated for an equivalent breathing transition observed in MIL-53(Al) at 360 K by VT-DSC (see reference [69]), (d) volume change calculated for the isostructural MIL-53(Cr) compound (see reference [77]).

Acknowledgments

The authors thank María Antonia Señarís-Rodríguez, Manuel Sánchez-Andújar and Socorro Castro-García for useful discussions. The authors also acknowledge financial support from the projects PID2021-122532OB-I00 funded by MCIN/AEI/10.13039/501100011033/and by FEDER 'Una manera de hacer Europa', the project PDC2021-121076-I00 funded by MCIN/AEI/10.13039/501100011033 and by the European Union 'Next GenerationEU/PRTR', and the project ED431C 2022/39 funded by Xunta de Galicia. J M B-G also acknowledges the financial support of the grant RYC2021-033040-I, funded by MCIN/AEI/10.13039/501100011033 and from European Union 'Next GenerationEU/PRTR', the UDC-Inditex InTalent Program, the Postdoctoral Research Project ED481D-2021-012 funded by Xunta de Galicia, and the collaboration agreement 'Development of research strategic actions UDC I+D+i 2021-2022: CICA-Disrupting Projects 2021SEM-A3 (NanoCool). J G-B thanks Xunta de Galicia for a Predoctoral Fellowship, and I D-F acknowledges Ministerio de Universidades for an FPU Predoctoral Fellowship.

References

[1] McNaught A D, Wilkinson A and Chalk S J 2019 *IUPAC. Compendium of Chemical Terminology (the 'Gold Book')* (Oxford: Blackwell Scientific Publications)

[2] Aleman J 2007 Definitions of terms relating to the structure and processing of sols, gels, networks, and inorganic–organic hybrid materials (IUPAC recommendations 2007) *Pure Appl. Chem.* **79** 1801–29

[3] Kickelbick G 2007 *Hybrid Materials: Synthesis, Characterization, and Applications* (New York: Wiley-VCH)

[4] Bermúdez-García J M, Sánchez-Andújar M and Señarís-Rodríguez M A 2017 A new playground for organic–inorganic hybrids: barocaloric materials for pressure-induced solid-state cooling *J. Phys. Chem. Lett.* **8** 4419–23

[5] Rao C N R, Cheetham A K and Thirumurugan A 2008 Hybrid inorganic–organic materials: a new family in condensed matter physics *J. Phys. Condens. Matter* **20** 083202

[6] García-Ben J, McHugh L N, Bennett T D and Bermúdez-García J M 2022 Dicyanamide-perovskites at the edge of dense hybrid organic–inorganic materials *Coord. Chem. Rev.* **455** 214337

[7] Li D, Zhao X M, Zhao H X, Long L S and Zheng L S 2019 Coexistence of magnetic–optic–electric triple switching and thermal energy storage in a multifunctional plastic crystal of trimethylchloromethyl ammonium tetrachloroferrate(III) *Inorg. Chem.* **58** 655–62

[8] Garcia Y, Van Koningsbruggen P J, Bravic G, Guionneau P, Chasseau D, Cascarano G L, Moscovici J, Lambert K, Michalowicz A and Kahn O 1997 Synthesis, crystal structure, EXAFS, and magnetic properties of catena-poly[μ-tris(4-(2-hydroxyethyl)-1,2,4-triazole-N^1, N^2)copper(II)] diperchlorate trihydrate: relevance with the structure of the iron(II) 1,2,4-triazole spin transition molecular material *Inorg. Chem.* **36** 6357–65

[9] Li J, Barrio M, Dunstan D J, Dixey R, Lou X, Tamarit J L, Phillips A E and Lloveras P 2021 Colossal reversible barocaloric effects in layered hybrid perovskite $(C_{10}H_{21}NH_3)_2MnCl_4$ under low pressure near room temperature *Adv. Funct. Mater.* **31** 2105154

[10] Bermúdez-García J M, Sánchez-Andújar M, Yáñez-Vilar S, Castro-García S, Artiaga R, López-Beceiro J, Botana L, Alegría Á and Señarís-Rodríguez M A 2015 Role of temperature and pressure on the multisensitive multiferroic dicyanamide framework [TPrA][Mn(dca)$_3$] with perovskite-like structure *Inorg. Chem.* **54** 11680–7
[11] Gütlich P and Goodwin H A 2004 *Spin Crossover in Transition Metal Compounds I* (New York: Springer)
[12] Tian X, Zhang Y, Zheng R, Wei D and Liu J 2020 Two-dimensional organic–inorganic hybrid Ruddlesden–Popper perovskite materials: preparation, enhanced stability, and applications in photodetection *Sustain. Energy Fuels* **4** 2087–113
[13] Yaghi O M, Li G and Li H 1995 Selective binding and removal of guests in a microporous metal–organic framework *Nature* **378** 703–6
[14] Zhang X, Chen Z, Liu X, Hanna S L, Wang X, Taheri-Ledari R, Maleki A, Li P and Farha O K 2020 A historical overview of the activation and porosity of metal–organic frameworks *Chem. Soc. Rev.* **49** 7406–27
[15] Ding M, Flaig R W, Jiang H L and Yaghi O M 2019 Carbon capture and conversion using metal–organic frameworks and MOF-based materials *Chem. Soc. Rev.* **48** 2783–828
[16] Xu W and Yaghi O M 2020 Metal–organic frameworks for water harvesting from air, anywhere, anytime *Acs Cent. Sci.* **6** 1348–54
[17] Wang S 2018 A robust large-pore zirconium carboxylate metal–organic framework for energy-efficient water-sorption-driven refrigeration *Nat. Energy* **3** 985–93
[18] Cambridge Crystallographic Data Centre, Experimental Crystal Structure Determination: CCDC-938392 (MOF-5), CCDC-1034319 (DUT-8), CCDC-1063411 (VNU-10), CCDC-1441652 (UiO-67), CCDC-755080 (HKUST-1), CCDC-220476 (MIL-53), CCDC-1494752 (MOF-74/CPO-27),CCDC-864309 (ZIF-8)
[19] Li W, Wang Z, Deschler F, Gao S, Friend R H and Cheetham A K 2017 Chemically diverse and multifunctional hybrid organic–inorganic perovskites *Nat. Rev. Mater.* **2** 16099
[20] Zhang W and Xiong R-G 2012 Ferroelectric metal–organic frameworks *Chem. Rev.* **112** 1163–95
[21] Coudert X 2015 Responsive metal–organic frameworks and framework materials: under pressure, taking the heat, in the spotlight, with friends *Chem. Mater.* **27** 1905–16
[22] Mañosa L and Planes A 2017 Materials with giant mechanocaloric effects: cooling by strength *Adv. Mater.* **29** 1603607
[23] Moya X, Kar-Narayan S and Mathur N D 2014 Caloric materials near ferroic phase transitions *Nat. Mater.* **13** 439–50
[24] Hess T, Maier L M, Bachmann N, Corhan P, Schäfer-Welsen O, Wöllenstein J and Bartholomé K 2020 Thermal hysteresis and its impact on the efficiency of first-order caloric materials *J. Appl. Phys.* **127** 075103
[25] Mehling H, Leys J, Glorieux C and Thoen J 2021 Potential new reference materials for caloric measurements on PCM *SN Appl. Sci.* **3** 202
[26] Boström H L B and Goodwin A L 2021 Hybrid perovskites, metal–organic frameworks, and beyond: unconventional degrees of freedom in molecular frameworks *Acc. Chem. Res.* **54** 1288–97
[27] Ji L J, Sun S J, Qin Y, Li K and Li W 2019 Mechanical properties of hybrid organic–inorganic perovskites *Coord. Chem. Rev.* **391** 15–29

[28] Burtch N C, Heinen J, Bennett T D, Dubbeldam D and Allendorf M D 2018 Mechanical properties in metal–organic frameworks: emerging opportunities and challenges for device functionality and technological applications *Adv. Mater.* **30** 1704124

[29] Tan J C and Cheetham A K 2011 Mechanical properties of hybrid inorganic–organic framework materials: establishing fundamental structure–property relationships *Chem. Soc. Rev.* **40** 1059–80

[30] Heinen J, Ready A D, Bennett T D, Dubbeldam D, Friddle R W and Burtch N C 2018 Elucidating the variable-temperature mechanical properties of a negative thermal expansion metal–organic framework *ACS Appl. Mater. Interfaces* **10** 21079–83

[31] Hobday C L 2016 A computational and experimental approach linking disorder, high-pressure behavior, and mechanical properties in UiO frameworks *Angew. Chem. Int. Ed.* **55** 2401–5

[32] Bermúdez-García J M, Sánchez-Andújar M, Castro-García S, López-Beceiro J, Artiaga R and Señarís-Rodríguez M A 2017 Giant barocaloric effect in the ferroic organic–inorganic hybrid [TPrA][Mn(dca)3] perovskite under easily accessible pressures *Nat. Commun.* **8** 15715

[33] Mañosa L, González-Alonso D, Planes A, Bonnot E, Barrio M, Tamarit J-L, Aksoy S and Acet M 2010 Giant solid-state barocaloric effect in the Ni–Mn–In magnetic shape-memory alloy *Nat. Mater.* **9** 478–81

[34] Mañosa L, González-Alonso D, Planes A, Barrio M, Tamarit J-L, Titov I S, Acet M, Bhattacharyya A and Majumdar S 2011 Inverse barocaloric effect in the giant magnetocaloric La–Fe–Si–Co compound *Nat. Commun.* **2** 595

[35] Flerov I N, Gorev M V, Tressaud A and Laptash N M 2011 Perovskite-like fluorides and oxyfluorides: phase transitions and caloric effects *Crystallogr. Rep.* **56** 9–17

[36] Lloveras P 2015 Giant barocaloric effect at low pressure in ferrielectric ammonium sulphate *Nat. Commun.* **6** 1–20

[37] Badding J V 1998 High-pressure synthesis, characterization, and tuning of solid state materials *Annu. Rev. Mater. Sci.* **28** 631–58

[38] Gardner J V, Armstrong A A, Calder B R and Beaudoin J 2014 So, how deep is the Mariana Trench? *Mar. Geod.* **37** 1–13

[39] Henry Royce Institute 2020 *Materials for the Energy Transition Roadmap: Caloric Energy Conversion Materials* (Manchester: Henry Royce Institute)

[40] García-Ben J, Delgado-Ferreiro I, Salgado-Beceiro J and Bermúdez-García J M 2021 Simple and low-cost footstep energy-recover barocaloric heating and cooling device *Materials (Basel)* **14** 5947

[41] Bermúdez-García J M, Yáñez-Vilar S, García-Fernández A, Sánchez-Andújar M, Castro-García S, López-Beceiro J, Artiaga R, Dilshad M, Moya X and Señarís-Rodríguez M A 2018 Giant barocaloric tunability in $[(CH_3CH_2CH_2)_4N]Cd[N(CN)_2]_3$ hybrid perovskite *J. Mater. Chem.* C **6** 9867–74

[42] MącZka M, Ptak M, Gągor A, Sieradzki A, Peksa P, Usevicius G, Simenas M, Leite F F and Paraguassu W 2019 Temperature- and pressure-dependent studies of a highly flexible and compressible perovskite-like cadmium dicyanamide framework templated with protonated tetrapropylamine *J. Mater. Chem.* C **7** 2408–20

[43] Boström H L B, Hill J A and Goodwin A L 2016 Columnar shifts as symmetry-breaking degrees of freedom in molecular perovskites *Phys. Chem. Chem. Phys.* **18** 31881–94

[44] Boström H L B 2020 Tilts and shifts in molecular perovskites *CrystEngComm* **22** 961–8

[45] Lloveras P and Tamarit J-L 2021 Advances and obstacles in pressure-driven solid-state cooling: a review of barocaloric materials *MRS Energy Sustain.* **8** 3–15

[46] Szafrański M, Wei W J, Wang Z M, Li W and Katrusiak A 2018 Research update: tricritical point and large caloric effect in a hybrid organic–inorganic perovskite *APL Mater.* **6** 100701

[47] Pato-Doldán B, Sánchez-Andújar M, Gómez-Aguirre L C, Yáñez-Vilar S, López-Beceiro J, Gracia-Fernández C, Haghighirad A A, Ritter F, Castro-García S and Senaris-Rodriguez M A 2012 Near room temperature dielectric transition in the perovskite formate framework $[(CH_3)_2NH_2][Mg(HCOO)_3]$ *Phys. Chem. Chem. Phys.* **14** 8498–501

[48] Salgado-Beceiro J, Nonato A, Silva R X, García-Fernández A, Sánchez-Andújar M, Castro-Garcia S, Stern-Taulats E, Señarís-Rodríguez M A, Moya X and Bermúdez-García J M 2020 Near-room-temperature reversible giant barocaloric effects in $[(CH_3)_4N]Mn[N_3]_3$ hybrid perovskite *Mater. Adv* **1** 3167–70

[49] Gómez-Aguirre L C 2016 Coexistence of three ferroic orders in the multiferroic compound $[(CH_3)_4N][Mn(N_3)_3]$ with perovskite-like structure *Chem. Eur. J.* **22** 7863–70

[50] Busico V, Carfagna C, Salerno V, Vacatello M and Fittipaldi F 1980 The layer perovskites as thermal energy storage systems *Sol. Energy* **24** 575–9

[51] Seo J, McGillicuddy R D, Slavney A H, Zhang S, Ukani R, Yakovenko A A, Zheng S L and Mason J A 2022 Colossal barocaloric effects with ultralow hysteresis in two-dimensional metal–halide perovskites *Nat. Commun.* **13** 1–15

[52] Gao Y 2022 Reversible colossal barocaloric effect dominated by disordering of organic chains in $(CH_3–(CH_2)_{n-1}–NH_3)_2MnCl_4$ single crystals *NPG Asia Mater.* **14** 34

[53] Timmermans J 1961 Plastic crystals: a historical overview *J. Phys. Chem. Solids* **18** 1–8

[54] Mondal A, Bhattacharya B, Das S, Bhunia S, Chowdhury R, Dey S and Reddy C M 2020 Metal-like ductilityin organic plastic crystals: role of molecular shape and dihydrogen bonding interactions in aminoboranes *Angew. Chem. Int. Ed. Engl.* **59** 10971–80

[55] Das S, Mondal A and Reddy C M 2020 Harnessing molecular rotations in plastic crystals: a holistic view for crystal engineering of adaptive soft materials *Chem. Soc. Rev.* **49** 8878–96

[56] Barrio M, Font J, Muntasell J, Navarro J and Tamarit J L 1988 Applicability for heat storage of binary systems of neopentylglycol, pentaglycerine and pentaerythritol: a comparative analysis *Sol. Energy Mater.* **18** 109–15

[57] Li B 2019 Colossal barocaloric effects in plastic crystals *Nature* **567** 506–10

[58] Lloveras P 2019 Colossal barocaloric effects near room temperature in plastic crystals of neopentylglycol *Nat. Commun.* **10** 1803

[59] Boldrin D 2021 Fantastic barocalorics and where to find them *Appl. Phys. Lett.* **118** 170502

[60] Chen X G, Zhang Y Z, Sun D S, Gao J X, Hua X N and Liao W Q 2019 Above room-temperature dielectric and nonlinear optical switching materials based on $[(CH_3)_3S]_2[MBr_4]$ (M = Cd, Mn and Zn) *Dalt. Trans.* **48** 11292–7

[61] Salgado-Beceiro J, Bermúdez-García J M, Llamas-Saiz A L, Castro-García S, Señarís-Rodríguez M A, Rivadulla F and Sánchez-Andújar M 2020 Multifunctional properties and multi-energy storage in the $[(CH_3)_3S][FeCl_4]$ plastic crystal *J. Mater. Chem.* C **8** 13686–94

[62] Salgado-Beceiro J, Bermúdez-García J M, Stern-Taulats E, García-Ben J, Castro-García , Socorro Sánchez-Andújar M, Moya X and Señarís-Rodríguez M A 2021 Hybrid ionic plastic crystals in the race for enhanced low-pressure barocaloric materials *ChemRxiv (Preprint)*

[63] De Lange M F, Verouden K J F M, Vlugt T J H, Gascon J and Kapteijn F 2015 Adsorption-driven heat pumps: the potential of metal–organic frameworks *Chem. Rev.* **115** 12205–50

[64] Saha B B, Uddin K, Pal A and Thu K 2019 Emerging sorption pairs for heat pump applications: an overview *JMST Adv.* **1** 161–80

[65] Henninger S K, Habib H A and Janiak C 2009 MOFs as adsorbents for low temperature heating and cooling applications *J. Am. Chem. Soc.* **131** 2776–7

[66] Demir H, Mobedi M and Ülkü S 2008 A review on adsorption heat pump: problems and solutions *Renew. Sustain. Energy Rev.* **12** 2381–403

[67] Alhamami M, Doan H and Cheng C H 2014 A review on breathing behaviors of metal–organic-frameworks (MOFs) for gas adsorption *Materials (Basel)* **7** 3198–250

[68] Murdock C R, Hughes B C, Lu Z and Jenkins D M 2014 Approaches for synthesizing breathing MOFs by exploiting dimensional rigidity *Coord. Chem. Rev.* **258–9** 119–36

[69] García-Ben J, López-Beceiro J, Artiaga R, Salgado-Beceiro J, Delgado-Ferreiro I, Kolen'ko Y V, Castro-García S, Señarís-Rodríguez M A, Sánchez-Andújar M and Bermúdez-García J M 2022 Discovery of colossal breathing-caloric effect under low applied pressure in the hybrid organic–inorganic MIL-53(Al) material *Chem. Mater.* **34** 3323–32

[70] Boutin A, Coudert F-X, Springuel-Huet M-A, Neimark A V, Férey G and Fuchs A H 2010 The behavior of flexible MIL-53(Al) upon CH_4 and CO_2 adsorption *J. Phys. Chem.* C **114** 22237–44

[71] Dilshad S, Kalair A R and Khan N 2020 Review of carbon dioxide (CO_2) based heating and cooling technologies: past, present, and future outlook *Int. J. Energy Res.* **44** 1408–63

[72] Bellos E and Tzivanidis C 2019 A comparative study of CO_2 refrigeration systems *Energy Convers. Manag.* X **1** 100002

[73] Bansal P 2012 A review—status of CO_2 as a low temperature refrigerant: fundamentals and R&D opportunities *Appl. Therm. Eng.* **41** 18–29

[74] Emerson 2015 Commercial CO_2 refrigeration systems *Clim. Technol.* **44**

[75] Metz B, Davidson O, Coninck H, Loos M and Meyer L 2005 *Carbon Dioxide Capture and Storage* (Cambridge: Cambridge University Press)

[76] Keith D W 2009 Why capture CO_2 from the atmosphere? *Science* **325** 1654–5

[77] Llewellyn P L, Maurin G, Devic T, Loera-Serna S, Rosenbach N, Serre C, Bourrelly S, Horcajada P, Filinchuk Y and Férey G 2008 Prediction of the conditions for breathing of metal organic framework materials using a combination of X-ray powder diffraction, microcalorimetry, and molecular simulation *J. Am. Chem. Soc.* **130** 12808–14

[78] Stavitski E, Pidko E A, Couck S, Remy T, Hensen E J M, Weckhuysen B M, Denayer J, Gascon J and Kapteijn F 2011 Complexity behind CO_2 capture on NH_2-MIL-53(Al) *Langmuir* **27** 3970–6

IOP Publishing

Barocaloric Effects in the Solid State
Materials and methods
Pol Lloveras

Chapter 7

Colossal barocaloric plastic crystals

Pol Lloveras, Zhe Zhang, Ming Zeng, María Barrio, Yukinobu Kawakita, Dehong Yu, Shangchao Lin, Kuo Li, Xavier Moya, Josep-Lluís Tamarit and Bing Li

7.1 Introduction

Plastic crystals (PCs) are molecular materials that display mesophases between liquid and ordered solid phases, where translational order is kept but orientational or conformational order is broken [1]. These materials are the orientational counterpart of liquid crystals, whose mesophases show orientational order but are devoid of full translational symmetry. Plastic phases, identified for the first time by J Timmermans in 1938 [2], owe their name to their soft mechanical behavior, which is known to rely on molecular diffusion via small densities and lattice defects. The change in orientational degrees of freedom at the ordered-to-plastic phase transition is mediated by globular and/or small, weakly interacting molecules, and typically takes place across strongly first-order phase transitions of reconstructive nature with very large values of latent heat. This typically leads to lower latent heat at melting, which is unusual in polymorphic ordered solids. Eventually, complete orientational ordering takes place sequentially across multiple phase transitions on cooling [3, 4].

Chemically, conventional PCs are organic compounds, which have found many applications in the chemical and pharmaceutical industry, ranging from intermediates for the synthesis of other compounds to antiviral drugs [5–11]. In many conventional PCs, hydrogen bonds are the dominant interaction, and are responsible for stabilizing orientational order in the low-temperature phases. The strength of these hydrogen bonds is believed to determine transition temperatures [12]. Recently, the family of PCs has been extended to include a variety of inorganic [13–16] and organic–inorganic [17, 18] materials (see figure 7.1). The potential applications for these new PCs are many, e.g. inorganic PCs have been proposed as solid electrolytes [19, 20], where dynamic rotors in the plastic phase enhance ionic conductivity [13, 21]. Also, when orientationally ordered, some inorganic PCs display ferroelectricity [22] or magnetism [23].

doi:10.1088/978-0-7503-4690-0ch7

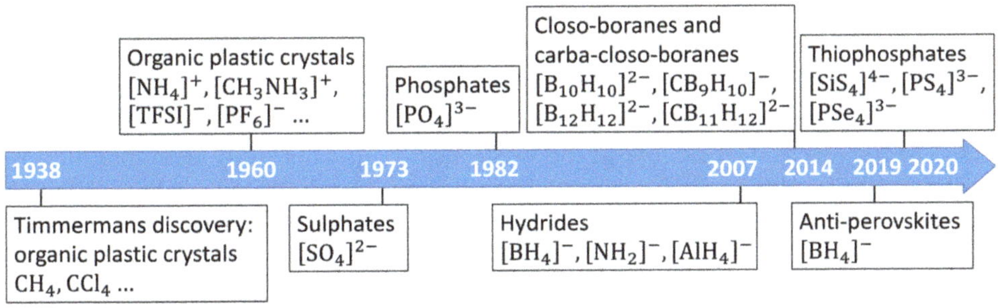

Figure 7.1. Timeline for the discovery of plastic phases, with selected organic and inorganic material examples. Adapted and reproduced from reference [13], copyright (2022) with permission from Springer Nature.

The ordered-to-plastic phase transition in PCs has also sparked interest for technological applications in thermal management, as large values of latent heat can be used for solid-state heat sinks and sources. For this reason, in the 1980s PCs were proposed as solid-state *Phase Change Materials* (PCMs) [24, 25] in passive heat recovery and thermal storage and use of waste heat, an area of research that is regaining interest due to the climate change emergency. The solid-state nature of the ordered-to-plastic phase transitions in these materials is viewed as an advantage because it facilitates the handling of materials compared to melting transitions [26] or sensible heat away from any transition in molten salts [27]. PCs can also be easily incorporated into composite materials, and they have been proposed as active encapsulation agents in solid-liquid PCMs [28]. Composites and binary mixtures of PCs [29, 30] have been explored in a number of solid PCMs (in combination with solid or liquid PCMs) with the aim of tuning transition temperatures and/or enhancing the heat storage capacity.

Interestingly, the very large values of latent heat at ordered-to-plastic phase transitions are accompanied by very large changes in volume, indicating that these transitions can be driven by application of an external hydrostatic pressure (as well as temperature). This allows actively controlling the exchange of large latent heat using pressure, and, thus, exploiting these materials in barocaloric (BC) methods for cooling and heating applications that avoid the harmful gases used in vapor-compression devices. This was shown for the first time in the published patent [31], followed by two recent studies [32, 33], which reported colossal BC effects in the plastic crystal neopentylglycol $[(CH_3)_2C(CH_2OH)_2]$ of about ~400 J K^{-1} kg^{-1} under pressure changes of about ~0.1 GPa. These values outperformed the best BC materials known to date by one order of magnitude under similar pressure changes and brought the performance of solid-state BC materials to the same level as that of fluid refrigerants. Stimulated by these studies, further investigations of the BC response in several different families of organic PCs have confirmed their excellent BC potential, which suggest PCs as a fertile materials playground where abundant and excellent BC properties can be found.

PCs have already become the benchmark BC compounds and the adjective 'colossal' is typically attributed to isothermal entropy changes exceeding 100 J K^{-1} kg^{-1}.

Furthermore, following the principle of BC effects, active thermal management of waste heat based on pressure-driven phase transitions in PCs (in contrast to the aforementioned passive thermally driven phase transitions) has also recently been proposed [34]. Based on an inverse BC effect in NH_4SCN, barocaloric thermal batteries have been realized to harvest and reuse low-grade waste heat [15]. This study demonstrated how pressure can be used for new functionalities other than heating and cooling cycles, thus extending the application of BC materials to rational utilization of thermal energy. On the other hand, many PCs exhibit low toxicity and are made of abundant and cheap elements, which are positive features for applications.

This chapter is aimed at providing an overview of the BC performance of PCs. First, the different material families studied to date will be shortly introduced. Then, the key properties relevant for applications will be discussed and summarized. Finally, insights about the physics underlying BC effects will be described.

7.2 Barocaloric plastic-crystal families

In this section most of the PC materials families studied to date are described. Each family includes composition-related compounds such as different positional isomers, or derivatives from a parent structure obtained by substitution or addition of functional groups. Interestingly, composition-related compounds may show modified properties that fit better certain applications. Snapshots of the molecular structures can be found in figure 7.2, and their crystallographic properties are summarized in table 7.1.

7.2.1 Neopentane derivatives

Neopentane, with chemical formula $C(CH_3)_4$, displays a double carbon branch (methyl radical) in the second carbon position, forming a molecule with achiral tetrahedral symmetry. Neopentane transforms from an ordered tetragonal phase to a plastic cubic phase at 130 K [35]. Neopentane derivatives are obtained by replacing one or more methyl groups by amino and/or hydroxyl radicals [36]. In particular, neopentylglycol (NPG), with chemical formula $(CH_3)_2C(CH_2OH)_2$ (also known as 2,2-dimethylpropane-1,3-diol), is obtained from neopentane by substitution of two methyl groups with two hydroxymethyl groups. Pentaglycerine (PG), with chemical formula $(CH_3)C(CH_2OH)_3$, is synthesized by replacing three methyl groups with three hydroxymethyl groups. In the ordered phase, PG molecules are arranged in layers perpendicular to the tetragonal c axis. Within the layers, the molecules are strongly linked through hydrogen bonds, whereas layers are weakly interacting with each other via van der Waals forces [37, 38]. Upon heating, $(00l)$ layers increase their distance, finally transforming into the cubic plastic phase. Neopentyl alcohol (NPA), with chemical formula $(CH_3)_3C(CH_2OH)$ (2,2-dimethyl-1-propanol) is synthesized by replacing one methyl group of the neopentane molecule with a hydroxyl radical. 2-amino-2-methyl-1,3-propanediol (AMP), with chemical formula $(NH_2)(CH_3)C(CH_2OH)_2$ is obtained by replacing three methyl groups with one amino group and two hydroxyl radicals. Tris(hydroxymethyl)aminomethane (TRIS) with chemical formula $(NH_2)C(CH_2OH)_3$ is a direct derivative of

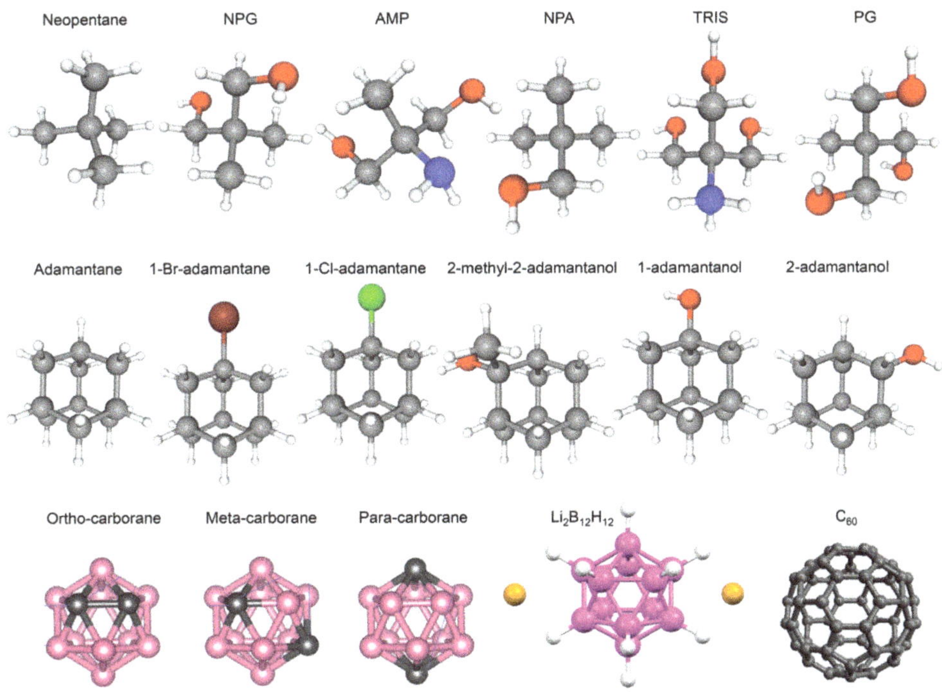

Figure 7.2. Molecules forming barocaloric plastic crystals, and the parent structures neopentane and adamantane. Grey, white, red, blue, dark red, green, pink and orange spheres represent C, H, O, N, Br, Cl, B and Li atoms, respectively. For carboranes, H atoms are not displayed for clarity.

pentaerythrytol $C(CH_2OH)_4$, in which methyl groups of neopentane have been replaced by hydroxymethyl groups. TRIS is then obtained by replacing a hydroxymethyl with an amino group. Similar to PG, TRIS molecules are arranged in layers perpendicular to the *c* axis, with strong intermolecular hydrogen bonds between hydroxyl groups. Contrary to PG, TRIS layers are interconnected by weak hydrogen bonds between amino and hydroxyl groups. At the transition, the distance between layers increases to finally arrange in a cubic lattice [39]. All these derivatives maintain the tetrahedral symmetry and show higher transition temperatures than the parent neopentane [12].

Neopentane and its derivatives are widely used in the chemical and pharmaceutical industry (the reader is referred to PubChem webpage [40] for information about plastic crystal uses). For instance, NPG is used as additive in lubricants, paints and cosmetics, NPA is used for pharmaceutical purposes, PG is used for the production of coats, varnishes and polyester resins [41, 42], AMP is used in cosmetics [43, 44], and TRIS is used in DNA extraction as buffer solution for pH stability [45], for correcting metabolic acidosis [46], for electrophoresis, etc. [47].

7.2.2 Adamantane derivatives

Adamantane, with chemical formula $(CH)_4(CH_2)_6$, is composed of three cyclohexane rings, providing the molecule with a cage-like structure. The PC phase is

Table 7.1. Structural properties of ordered and semiordered (II, III) and plastic (I) phases for PCs reported up to date. Tr, M, O, T and c represent triclinic, monoclinic, orthorhombic, tetragonal and cubic, respectively. Data inside brackets were obtained from numerical simulations.

Family	Compound	Chemical formula	Phase	Structure	Z	$\left(\frac{\partial V}{\partial T}\right)_{p_{atm}} \times 10^4$ cm^3 g^{-1} K^{-1}	Density g cm^3	Reference
Neopentane derivatives	NPG	$C_5H_{12}O_2$	I	fcc, $Fm\bar{3}m$	4	3.8	1.01	[33]
			II	M, $P2_1/c$	4	1.8	1.06	
	PG	$C_5H_{12}O_3$	I	fcc, $Fm\bar{3}m$	4	3.1	1.22	[69]
			II	T, bct, $I\bar{4}$	4	6.0		
	NPA	$C_5H_{12}O$	I	fcc, $Fm\bar{3}m$	4	5.8	0.81	[70]
			II	Tr		3.9		
	TRIS	$C_4H_{11}NO_3$	I	bcc, $Im\bar{3}m$	2	4.5	1.32	[39]
			II	O, $Pna2_1$	4	5.9		
	AMP	$C_4H_{11}NO_2$	I	bcc, $Im\bar{3}m$	2	2.0	1.20	[71]
			II	M		9.0		
Adamantane derivatives	1-Br-ada	$C_{10}H_{15}Br$	I	fcc, $Fm\bar{3}m$	4	5.7	1.39	[50, 72]
			II	O, $Pmcn$	4	2.6	1.51	
			III	M, $P2_1/c$	4		1.55	
	1-Cl-ada	$C_{10}H_{15}Cl$	I	fcc, $Fm\bar{3}m$	4	4.4	1.18	[50, 72]
			II	M, $P2_1/c$	4	1.8	1.25	
	1-ada-ol	$C_{10}H_{16}O$	I	fcc, $Fm\bar{3}m$	4	3.2	1.10	[34]
			II	T, $P4_2/n$	8	1.6	1.15	
	2-ada-ol	$C_{10}H_{16}O$	I	fcc, $Fm\bar{3}m$	4	4.4	1.12	[34]
			II	M, $C2/m$	12	3.1	1.13	
			III	Tr, $P\bar{1}$	4	2.7	1.16	
	2-*m*-2-ada-ol	$C_{11}H_{18}O$	I	fcc, $Fm\bar{3}m$	4	5.2	1.06	[34]
			II	O, $Pbcn$	16	1.9	1.18	
			III					
Closo-dodeca-boranes	M-carborane	$C_2B_{10}H_{12}$	I	T		— —	0.95	[63]
			II	O		— —		
	P-carborane	$C_2B_{10}H_{12}$	I	T		— —	0.95	[63]
			II	O		— —		

(*Continued*)

Table 7.1. (*Continued*)

Family	Compound	Chemical formula	Phase	Structure	Z	$\left(\frac{\partial V}{\partial T}\right)_{p_{atm}} \times 10^4$ cm^3 g^{-1} K^{-1}	Density g cm^3	Reference
	O-carborane	$C_2B_{10}H_{12}$	I	T		— —	0.95	[63]
			II	O		— —		
		$Li_2B_{12}H_{12}$	I	c, $Pa\bar{3}$	4	{2}	1.12	[59, 73]
			II	c, $Pa\bar{3}$	4	{3}		[59, 74]
Fullerite		C_{60}	I	fcc, $Fm\bar{3}m$	4	0.37	1.70	[75]
			II	c, $Pa\bar{3}$	4	0.19		
Inorganic orientationally disordered crystals		NH_4I	I	fcc, $Fm\bar{3}m$	4	0.330	2.455	[14]
			II	c, $Pm\bar{3}m$	1	0.423	3.039	
		KPF_6	I	c, $Fm\bar{3}m$	4	0.683	2.75	[16]
			II	M, $P2/n$	8	2.680		
		NH_4SCN	I	O, $Pbcm$	4	3.600	1.305	[15]
			II	M, $P2_1/c$	4	1.470		

stable within the range 208–543 K, with an fcc crystal structure with $Z = 4$ [48]. Adamantane and some derivatives are used in medical applications, such as in prophylaxis, antiviral, and Parkinson treatment [9–11], and are also used as intermediaries for the synthesis of related compounds. Recently, their large cavities have been reported to have useful physicochemical properties that are raising interest for applications in nano- and biotechnology [49].

1-Br-adamantane (1-Br-ada) and 1-Cl-adamantane (1-Cl-ada) are obtained by replacing a H atom linked to a 3-coordinated carbon by a Br or a Cl atom, respectively [50]. Both at low and high temperatures, 1-Br-ada and 1-Cl-ada are isostructural, exhibiting monoclinic (M) and fcc lattices, respectively. The latter exhibits orientational disorder that has been suggested to consist of a quasi-free rotation [51]. In 1-Cl-ada, this disorder fully develops into a single M → fcc transition, whereas in 1-Br-ada the orientational disorder is achieved after two transitions M → O → fcc, where the intermediate orthorhombic phase (O) displays a sixfold molecular rotation around the threefold molecular axis C_{3v} [51]. Both the M → O and O → fcc transitions take place near room temperature, but only the O → fcc is interesting for BC cooling applications due to its much larger changes in entropy and volume at the transition. 1-Br-ada and 1-Cl-ada enjoy widespread use in the synthesis of other adamantane derivatives exploited as antiviral drugs in the pharmaceutical industry, and for the synthesis of polymers and other materials [52, 53].

1-adamantanol (1-ada-ol, also named 1-hydroxyadamantane) and 2-adamantanol (2-ada-ol, also named as 2-hydroxyadamantane) have both chemical formula $C_{10}H_{16}O$ and are obtained by replacing a H atom linked to the tertiary and secondary carbon, respectively, with a hydroxyl group. In 1-ada-ol, the orientational disorder emerges in a single transition from a tetragonal phase to an fcc phase. In the tetragonal phase the hydrogen bonding between the hydroxyl groups is disordered among two different states [54]. By contrast, 2-ada-ol presents three solid phases from room temperature up to the melting temperature. The low-temperature solid phase is arranged in a triclinic structure which upon heating transforms to an orientationally ordered monoclinic lattice through a phase transition with small latent heat [34]. Upon further heating, 2-ada-ol exhibits another first-order phase transition towards an orientationally disordered cubic lattice. 1-ada-ol and 2-ada-ol can be used to synthesize compounds that show analgesic, antiviral and antibacterial properties [7]. 2-methyl-2-adamantanol (2-m-2-ada-ol) is obtained by replacing the two H atoms attached to the secondary carbon with a methyl and a hydroxyl group. 2-m-2-ada-ol can be used to produce light-resistant polymers for application in photolithography [8]. These compounds are also used in the chemical industry as intermediates for the synthesis of other compounds such as, for instance, 1,3-adamantanediol [5, 55], which itself is used as raw material to make functional materials such as heat-resistant polymers, and resists for the fabrication of semiconductors [6, 56].

7.2.3 Closo-boranes and derivatives

Closo-boranes are species containing $[B_nH_n]^{2-}$ anions ($n \geqslant 5$), and form polyhedral structures with B atoms in their corners [57]. Amongst them, closo-dodecaborane ($n = 12$) is the most stable one and adopts an icosahedral shape, whose nearly globular morphology may facilitate rotor motion and therefore the presence of PC phases. Within this family, the metal-based closo-dodecaborane $Li_2B_{12}H_{12}$ is particularly interesting because it shows a PC phase which, in addition to the orientational disorder of $[B_{12}H_{12}]^{2-}$ anions, exhibits strong diffusion of Li^+ cations, which implies possible applications in solid-state batteries [58]. The icosahedral $[B_{12}H_{12}]^{2-}$ anions maintain an overall fcc structure, which is also displayed in the ordered phase, but rotate around their centers of mass. Recently a molecular dynamics study [59] has predicted colossal BC effects in this compound associated with the enhanced disorder consisting of concomitant rotation and diffusion, which suggests other materials with similar physical mechanisms can be excellent BC materials.

Carboranes are closo-borane derivatives obtained by replacing some B atoms by C atoms, and can be described with the general chemical formula $C_nB_mH_{n+m+p}$ [60]. These materials are of interest in polymer science, catalysis, biomedicine and nanoscience [61, 62]. Amongst them, dicarba-closo-dodecaboranes are obtained by replacing two B atoms with C atoms, and hence have the chemical formula $C_2B_{10}H_{12}$. This permits three positional isomers depending on the specific icosahedral vertices occupied by C atoms: ortho-carborane (1,2-dicarba-closo-dodecaborane), where C atoms are placed in neighboring sites, meta-carborane

(1,7-dicarba-closo-dodecaborane), where C atoms are separated by one B atom, and para-carborane (1,12-dicarba-closo-dodecaborane), where the two C atoms are placed on opposite vertices. Recently, colossal BC effects have been reported across the transition between ordered and plastic phases in these three closely related compounds [63]. Interestingly, the better performance shown by para-carborane compared to the others suggests that higher symmetry could enhance entropy changes.

7.2.4 Fullerite C_{60}

This compound is in fact a paradigmatic orientationally disordered crystal because x-ray [64] and NMR [65] measurements, alongside the Pauling–Fowler model [66] indicate that in the orientationally disordered phase the quasi-spherical molecules rotate nearly freely. However, compared to typical plastic crystals, C_{60} do not show soft mechanical behavior due to stress-induced diffusion but is included here for completeness. Below the transition at 257 K, the number of possible orientations decreases, leading to a symmetry reduction from an fcc structure to a primitive cubic structure where the different molecules become distinguishable [67].

7.2.5 Inorganic orientationally disordered crystals

Similar to organic molecules, inorganic molecules can also form orientation-disordered crystals but, as in the case of C_{60}, they do not show soft mechanical behavior typical from plastic crystals. The most common case is probably ammonium salts. An ammonium cation has a tetrahedral configuration. Under certain conditions, it is subjected to some reorientations over the lattice sites, which can be described by various models like twofold or threefold rotation, tetrahedral tumbling, cubic tumbling, and isotropic rotation [68]. Reorientation modes of NH_4^+ are usually coupled with the lattice symmetry and in turn lead to a series of structural phase transitions including the plastic crystal transition. For example, NH_4SCN undergoes successive phase transitions from monoclinic to orthorhombic, and to tetragonal phases at about 360 and 390 K, respectively [15]. The low-temperature phase transition is accompanied by the disorder of orientations of NH_4^+ with large entropy changes of about 128 J kg^{-1} K^{-1}. Such a phase transition can be shifted to lower temperatures under external pressures, leading to an inverse colossal barocaloric effect. The high-temperature phase transition is mostly related to the reorientation disorder of SCN^-, with small entropy changes. The details will be discussed in section 7.5. In NH_4I, a phase transition takes place from cubic β phase ($Pm\bar{3}m$) to cubic α phase ($Fm\bar{3}m$) at about 260 K with a large volumetric discontinuity of about 16% [14]. Even if the crystal symmetry is the same, the coordination environments of NH_4^+ and the hydrogen networks are very different. In the β phase, the NH_4^+ tetrahedron resides in a cube cage formed by eight I ions and leads to a T_d configuration with two orientational degrees of freedom, which match the cubic tumbling model. In the α phase, each NH_4^+ ion is surrounded by six I ions, which form an octahedral cage with six orientational degrees of freedom. The entropy change at this phase transition is 70 J kg^{-1} K^{-1}, while the driving pressure is as low as

20 MPa. Thus, the corresponding pressure-normalized entropy changes are the highest among known inorganic barocaloric compounds.

In addition to inorganic molecules, some inorganic structural blocks like octahedra have active reorientational degrees of freedom. P and F atoms are arranged in PF_6 octahedra in KPF_6 [16]. At room temperature, the octahedra are orientationally disordered over the cubic lattice, accommodating with high symmetry, $Fm\bar{3}m$. When cooled down to about 260 K, the cubic phase transforms into a monoclinic phase $P2/n$. Under pressures above 80 MPa, these two phases both transform to a rhombohedral phase $R\bar{3}$. Due to the emergence of this high-pressure phase, the entropy changes are tripled to about 140 J kg^{-1} K^{-1}.

7.3 Determination of barocaloric effects in plastic crystals

So far, the most used method to determine the BC response in PCs is the quasi-direct method. Here we give a short description of this method (for a more extensive explanation the reader can consult chapter 1). The quasi-direct method consists of subtracting isobaric entropy curves at different pressures, following adiabatic or isothermal paths. For this purpose, the entropy function $S(T, p)$ must be determined under isobaric conditions. The ultimate purpose is to subtract entropy curves at different pressures, so $S(T, p)$ is normally established with respect to a reference state, e.g. at atmospheric pressure and at temperature T_0:

$$S(T, p) = S(T_0, p_{\mathrm{atm}}) + \int_{T_0}^{T} \frac{1}{T}\left(C_p + \frac{dQ}{|dT|}\right) dT - \int_{p_{\mathrm{atm}}}^{p} \left(\frac{\partial V}{\partial T}\right)_{p;T_0} dp \tag{7.1}$$

where C_p is the heat capacity, evaluated for the relevant phase, depending on temperature and pressure. To determine the temperature- and pressure-dependent heat capacity C_p in each phase, one strategy consists of combining modulated calorimetry measurements at atmospheric pressure with the thermodynamic relation:

$$\left(\frac{\partial C_p}{\partial p}\right)_T = -T\left(\frac{\partial^2 V}{\partial T^2}\right)_p \tag{7.2}$$

Equation (7.2) indicates that the pressure-dependence of C_p is proportional to the convexity of $V(T)$. In many cases, in the relatively narrow temperature range of BC interest, the temperature dependence of V is approximately linear at atmospheric pressure, which according to the previous equation and assuming that the linearity is maintained at high pressure, means that C_p can be considered roughly independent of pressure. It is worth noticing here that the inclusion of C_p in the calculation of $S(T, p)$ leads to a nonzero slope of the $S(T, p)$ curve outside the transition region. This may have small impact on the estimation of isothermal entropy changes ΔS, and sometimes this term can be left out [63]. Instead, its effect on adiabatic temperature changes may become very significant [76].

The temperature- and pressure-dependent entropy change associated with the transition $dQ/|dT|$ will be nonzero only within the transition region and it is typically measured using isobaric calorimetry, for which commercial [32] or custom-built [33] variable-pressure calorimeters are used. Finally, the last term in equation (7.1) accounts for the pressure-dependence of the entropy at T_0 and, again, requires $V(T)$ data. In fact, this term is nothing but the *indirect method* [77] for caloric effects applied to a single phase. It is thus clear that, in addition to variable-pressure calorimetry, pressure-dependent volume is highly desirable for the study of BC effects. However, $V(T,p)$ data require dilatometry or diffraction experiments at high pressures, and the latter are typically available at synchrotron and neutron large-scale facilities only [33, 78].

Once the entropy function is determined, isothermal entropy changes ΔS and adiabatic entropy changes ΔT are calculated as subtractions. Considering that for most PCs reported so far the emergence of rotational disorder is associated with a large volume expansion, we focus the present analysis on conventional BC effects[1]. Therefore, for PCs exhibiting $dT/dp > 0$, transitions on decompression are endothermic, whereas transitions on compression are exothermic. Since endothermic and exothermic phase transitions will take place at different temperature–pressure conditions due to the transition hysteresis, then BC effects on first decompression and on first compression (which are typically irreversible) must be computed from entropy curves calculated on heating (S_H) and on cooling (S_C), respectively:

$$\Delta S(T_s, p_{atm} \to p) = S_C(T, p) - S_C(T, p_{atm}) \tag{7.3}$$

$$\Delta S(T, p \to p_{atm}) = S_H(T, p_{atm}) - S_H(T, p) \tag{7.4}$$

$$\Delta T(T_s, p_{atm} \to p) = T(S_C, p) - T_s(S_C, p_{atm}) \tag{7.5}$$

$$\Delta T(T_s, p \to p_{atm}) = T(S_H, p_{atm}) - T_s(S_H, p) \tag{7.6}$$

where ΔT is conveniently expressed as a function of the starting temperature T_s because temperature changes along an adiabatic process.

To determine the BC effects that will be obtained reversibly along sustained cycles, it has been shown [79] that ΔT should be computed by subtracting the entropy curve on heating at atmospheric pressure from the entropy curve on cooling at high pressure:

$$\Delta T_{rev}(T_s, p_{atm} \to p) = T(S_C, p) - T_s(S_H, p_{atm}) \tag{7.7}$$

$$\Delta T_{rev}(T_s, p \to p_{atm}) = T(S_H, p_{atm}) - T_s(S_C, p) \tag{7.8}$$

whereas reversible ΔS can be computed from overlapping between irreversible ΔS on compression and on decompression. An example of a set of experiments and derived quantities used for the quasi-direct method is shown in figure 7.3 for the particular case of the PC NPG [33].

[1] A discussion concerning inverse effects can be found in chapter 1 of the present book.

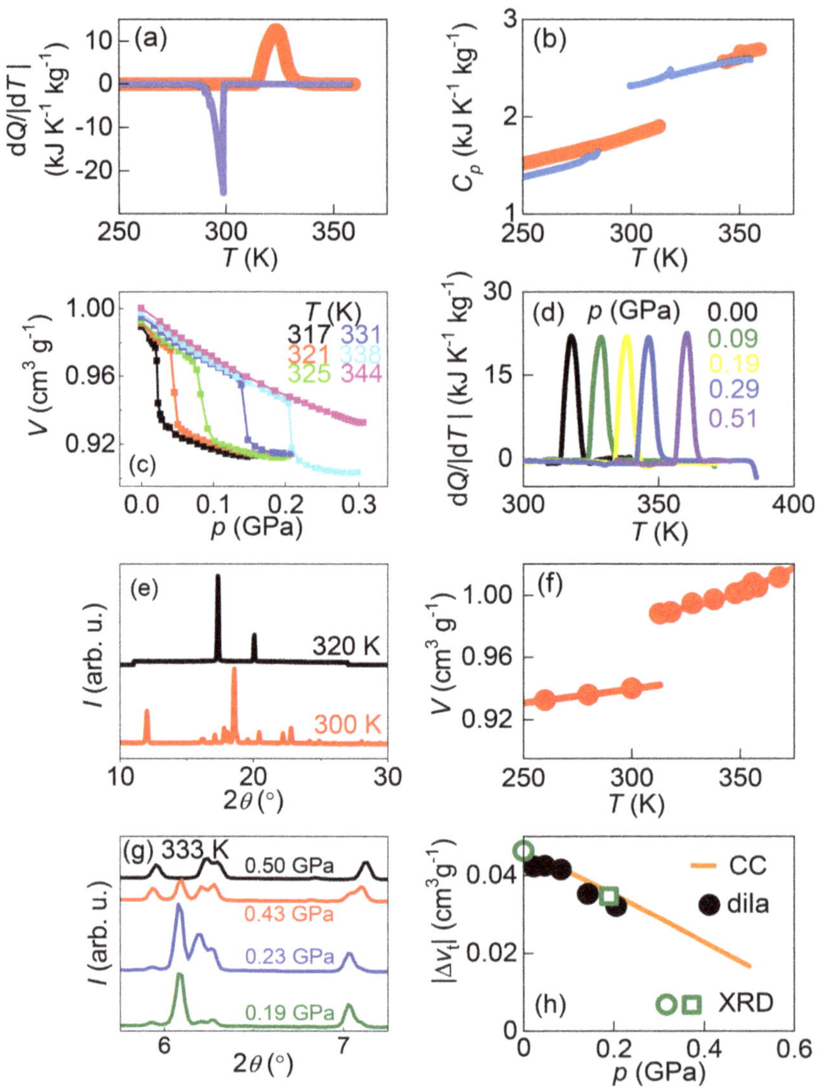

Figure 7.3. Representative measurements for the comprehensive BC characterization of the PC NPG. (a) $dQ/|dT|$ and (b) specific heat at p_{atm}. (c) Dilatometry: isothermal $V(p)$. (d) Pressure-dependent calorimetry. (e) X-ray powder diffraction at normal pressure. (f) Temperature-dependent volume obtained from x-ray powder diffraction. (g) Synchrotron x-ray powder diffraction at different pressures. (h) p-dependent transition volume change obtained from (c) (black symbols), from (e-g) (empty symbols) and Clausius–Clapeyron equation (orange line). Adapted and reproduced from reference [33] CC BY 4.0.

Recently, direct measurements of ΔT have been reported in PCs [63]. These measurements are difficult because generating a fast pressure change rate is challenging, which hinders adiabatic conditions. Consequently, measured values for ΔT may be smaller than expected.

7.4 Barocaloric effects in plastic crystals: state of the art and thermodynamic origin

In this section we provide a summary of the BC response in the PCs reported up to date (see section 7.2). Values for the transition properties and BC effects have been listed in table 7.2, whereas a visual comparison of BC performance is shown in figure 7.4. As can be seen, BC effects larger than 100 J K^{-1} kg^{-1} have been reported so far primarily only for PCs. The exception are some polymers [80], a spin crossover complex [81] and a layered hybrid organic–inorganic pseudoperovskite exhibiting a first-order phase transition with a strong disordering of the organic chain based on, primarily, its melting-like process and, secondarily, a small axial rotation [82, 83]. Among PCs, NH_4I is special because its BC strength is quite large due to the ultrasensitivity to pressure even if its entropy change is smaller than 100 J K^{-1} kg^{-1}. In addition, the estimated ΔT for this compound is amongst the highest [14].

The reasons for the good BC response in PCs are twofold: the contribution of the phase transition and the contribution of the single phases. The importance of the contributions from individual phases was first recognised in the study of giant BC

Table 7.2. BC properties in PCs. Transition temperature T (defined using peaks in calorimetry) and dependence of T on pressure, dT/dp, determined using endothermic transitions. Thermal hysteresis is at normal pressure. Data for ΔV_t in brackets are calculated using the Clausius–Clapeyron equation. Data for BC effects ΔS and ΔT in brackets are irreversible data. Pressure change Δp refers to the value for which ΔS and ΔT were obtained. Data for ΔT_{rev} in square brackets are calculated through the approximation $\Delta T_{rev} = |\Delta S_{rev}| T / C_p$. Data in curly brackets are obtained from modelling. Dashed lines refer to unavailable data.

Material	T K	dT/dp K GPa^{-1}	ΔV_t 10^{-6} m^3 kg^{-1}	hyst. K	ΔT_{rev} K	ΔS_{rev} J K^{-1} kg^{-1}	Δp GPa	Reference
NPG	315	103	46	24	7.5	421	0.25	[33]
PG	354	86	38	12	10	490	0.24	[79]
NPA	232	170	45.5	40	16	293	0.26	[79]
TRIS	406.8	26	41.9	80	(8)	(600)	0.25	[79]
AMP	352.7	75	48.1	55	(15)	(690)	0.25	[79]
1-Br-ada	316	355	40	8	16	138	0.1	[72]
1-Cl-ada	254	274	47	9	20	160	0.1	[72]
1-ada-ol	361	179	45	15	11	175	0.16	[34]
2-ada-ol (II-I)	390	48	5.4	1	8	100	0.185	[34]
2-ada-ol (III-II)	326	210	2.7	2				[34]
2-*m*-2-ada-ol	375	241	88	11	7	300	0.08	[34]
o-carborane	277	330	(28.4)	8	[14]	79	0.05	[63]
m-carborane	286	340	(26.9)	9	[13]	71	0.05	[63]
p-carborane	308	380	(39.1)	11	[19]	97	0.05	[63]
$Li_2B_{12}H_{12}$	630 {470}	{104}	77.7 {39}	—	{43}	{376}	0.1	[63]
C_{60}	257	170	4.6	3	9.7	30	0.1	[75]
NH_4I	275	810	58.7	~26–20	~34	71	0.04	[14]

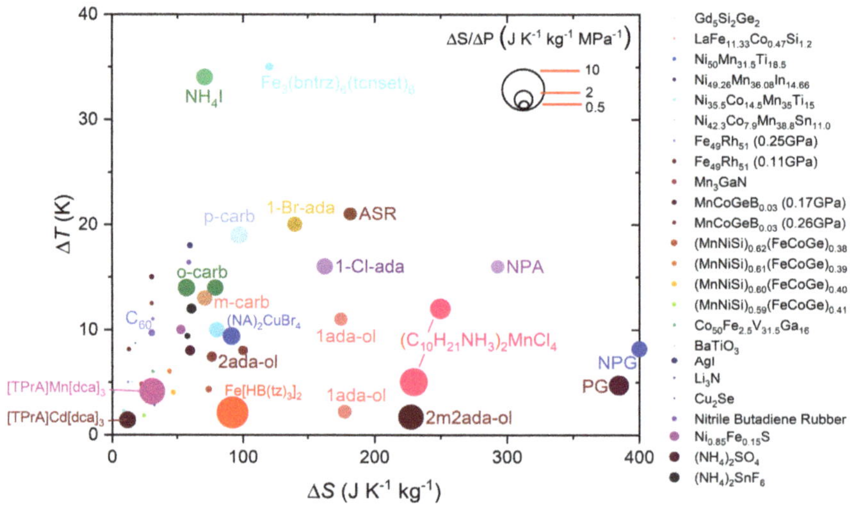

Figure 7.4. Comparison between different BC materials: for each compound, values for ΔT_{rev} (y axis), ΔS_{rev} (x axis) and BC strength $\frac{\Delta S_{\text{rev}}}{\Delta p}$ (circle area) for a given pressure change are displayed. The best BC compounds are labelled inside the figure and near the corresponding symbol, the others appear in the outer legend. ASR stands for Acetoxy Silicone Rubber. For some compounds there are two symbols because two values of pressure have been considered. Data have been taken from references [14, 34, 63, 72, 81, 84, 85].

effects in ammonium sulphate [78]. These two contributions are discussed in the following sections.

7.4.1 Barocaloric effects associated with the transition

Thermodynamic origin of the entropy change

Typically, phase transitions from ordered to PC phases involve two main sources of entropy change: (i) the change in the number of accessible configurations by the entities forming the crystal and (ii) the change in the lattice which, for many PCs, can be approximated to the volume change. These two contributions are discussed below:

- *Configurational entropy change.* The entropy change due to a change in the number of accessible configurations can be calculated as:

$$\Delta S_c = \mathrm{R}M^{-1}\ln\left(\frac{N_{\mathrm{I}}}{N_{\mathrm{II}}}\right) \tag{7.9}$$

  where R is the universal gas constant, M is the molar mass, and N_{I} and N_{II} are the number of accessible configurations in the plastic phase (I) and the ordered phase (II), respectively, which can be determined from single crystal x-ray diffraction. For completely ordered phases, $N_{\mathrm{II}} = 1$ but the difficulty of finding out the structure of these materials may hinder the determination of N_{I} for the plastic phases. Unfortunately, refined structures from powder patterns cannot account for the actual disorder in full. Alternatively, the maximum number of possible configurations can be deduced from compatibility constraints between the molecular and lattice symmetries, but in this case the actual active rotations remain elusive. Conversely, if the configurational entropy change can be determined using other methods, then the number of configurations can be estimated by inverting equation (7.9).

- *Volumetric entropy change.* In a (p, V, T) physical system, any change in volume entails a change in entropy. In the solid state, this change ultimately depends on the lattice vibrations. This can be seen by taking into account the entropy S as a state function of, for instance, T and V. Then, the differential form of $S(T, V)$ can be written as

$$dS = \left(\frac{\partial S}{\partial T}\right)_V dT + \left(\frac{\partial S}{\partial V}\right)_T dV \tag{7.10}$$

where

$$\left(\frac{\partial S}{\partial V}\right)_T = \left(\frac{\partial p}{\partial T}\right)_V \tag{7.11}$$

Independently, using the cyclic chain rule:

$$\left(\frac{\partial p}{\partial T}\right)_V \left(\frac{\partial T}{\partial V}\right)_p \left(\frac{\partial V}{\partial p}\right)_T = -1 \rightarrow \left(\frac{\partial p}{\partial T}\right)_V = -\frac{\frac{1}{V}\left(\frac{\partial V}{\partial T}\right)_p}{\frac{1}{V}\left(\frac{\partial V}{\partial p}\right)_T} = \frac{\alpha_V}{\beta_T} \tag{7.12}$$

where α_V and β_T are the volumetric thermal expansion and isothermal compressibility, respectively. If a change in volume with no configurational change is considered at constant temperature

$$dS = \left(\frac{\partial p}{\partial T}\right)_V = \frac{\alpha_V}{\beta_T} dV \tag{7.13}$$

Then, considering a change from an initial state with $(\alpha_{V1}, \beta_{T1})$ to a final state with $(\alpha_{V2}, \beta_{T2})$ and assuming $\frac{\alpha_{V1}}{\beta_{T1}} \sim \frac{\alpha_{V2}}{\beta_{T2}}$ we obtain:

$$\Delta S_V = \int_1^2 \left(\frac{\alpha_V}{\beta_T}\right) dV \sim \left\langle \frac{\alpha_V}{\beta_T} \right\rangle \Delta V_{\mathrm{t}} \tag{7.14}$$

where averages are taken over the two states. This last equation has been used to estimate the entropy change due to a volume change at a solid-state first-order phase transition [86–88]. Although it may entail large uncertainties, as it involves numerical derivation of several experimental datasets, this expression is useful to obtain an insight about the order of magnitude of ΔS_V. For example, it was found that for adamantane and some cyclohexane derivatives this contribution follows an average trend of $\Delta S_V \sim 0.4\Delta S_{\mathrm{t}}$ [51, 89–91], leading to $\Delta S_c \sim 0.6\Delta S_t$. For the BC PCs presented here, this trend has been approximately confirmed only for a few compounds, all of them adamantane derivatives: 1-Br-ada, 1-Cl-ada and 2-m-2-ada-ol.

Alternatively, in an elastic (harmonic) system this change can be more accurately calculated from the volumetric contribution to the Helmholtz free energy [75, 79]:

$$F_V = \frac{K}{2\rho}\left(\frac{\Delta V}{V}\right)^2 \tag{7.15}$$

where $K = \frac{C_{11}+2C_{12}}{3}$ is the bulk modulus and C_{11} and C_{12} are (harmonic) elastic constants. Using the thermodynamic relation $S = -(\partial F/\partial T)_V$ then we can write:

$$\Delta S_V = -\left(\frac{\partial F_V}{\partial T}\right)_V = -\frac{1}{2\rho}\left(\frac{\partial K}{\partial T}\right)\left(\frac{\Delta V}{V}\right)^2 \tag{7.16}$$

where the high-temperature phase has been considered as the reference unstrained state. However, the experimental determination of this term is difficult to perform or involves a significant uncertainty.

A summary of the estimated contributions ΔS_c and ΔS_V for several PCs is shown in table 7.3, which reveals that in most cases the major contribution to ΔS_t is given by the configurational entropy change ΔS_c. Nevertheless, recent investigations based on quasi-elastic neutron scattering and/or theoretical calculations (based on density functional theory, molecular dynamics or *ab*

Table 7.3. Estimated contributions to the total transition entropy change, ΔS_t, estimated configurational (ΔS_c) and volumetric (ΔS_V) entropy changes. Overall (irreversible) isothermal entropy changes ΔS upon pressure changes Δp and isothermal entropy changes ΔS_+ in single phases II (ordered or semiordered) and I (plastic).

Material	ΔS_t	ΔS_c	ΔS_V	ΔS	Δp	ΔS_+^{II} (0.1 GPa)	ΔS_+^{I} (0.1 GPa)	Reference
	J K^{-1} kg^{-1}				MPa	J K^{-1} kg^{-1}	J K^{-1} kg^{-1}	
NPG	380	330	60	370	50	18	38	[33]
PG	485	325	80	300	40	60	31	[79]
NPA	204	321	52	250	100	39	58	[79]
TRIS	682	287	90	330	50	59	45	[79]
1 AMP	632	315	100	600	100	90	20	[79]
1-Br-ada	136	62	76	120	50	26.3	57	[72]
1-Cl-ada	104	112	51	140	20	18.3	44	[72]
1-ada-ol	210	174	40	220	160	16	32	[34]
2-ada-ol	66.5	98–136	7	100	185	31	44	[34]
2-*m*-2-ada-ol	371	194	110	360	80	19	52	[34]
C_{60}	26	—	1	30	100	1.9	3.7	[75]

initio molecular dynamics) on order–disorder transitions in PCs [92] and in other materials [93] suggest that changes in the vibrational entropy could contribute more significantly than expected. Interestingly, theoretical methods may allow one to identify particular phonon modes or rotational disorder, with the *exact* motion of the molecules, which cannot be derived from neutron scattering experiments.

Implications of the Clausius–Clapeyron equation

It is typically observed that the transition entropy change decreases with increasing pressure, which may be detrimental for the BC response. This could be explained because the larger compressibility [32] of the plastic phase as compared to the ordered phase may lead to a reduction of the volume change at the transition, ΔV_t. Hence, given that it is very often observed that dT/dp is constant within the pressure range of interest, according to the Clausius–Clapeyron expressed as $\Delta S_t = \Delta V_t(dT/dp)^{-1}$, the reduction in ΔV_t results in a reduction of the ΔS_t. Therefore, the reduction in ΔS_t occurs mainly due to a reduction of the volumetric entropy change at the transition. This is consistent with the fact that as ΔS_c is calculated only from structural information; changes in this contribution with increasing pressure are not expected if the two phases involved in the transition are unchanged.

For first-order phase transitions, the adiabatic temperature changes under pressure changes Δp can be approximately estimated as $\Delta T \sim (dT/dp)\Delta p$, where dT/dp is the slope of the two-phases coexistence line. The isothermal entropy changes ΔS are of the order of ΔS_t. According to the Clausius–Clapeyron equation $dT/dp = \Delta V_t/\Delta S_t$ these two terms are inversely proportional to each other and their simultaneous optimization requires maximizing as much as possible the volume change. Such a dilemma is also seen between ΔS and Relative Cooling Power (RCP [94]). As mentioned above, ΔS is proportional to dp/dT, but RCP is inversely proportional to dp/dT under a given pressure. In NH_4SCN, the large ΔV at the transition leads to simultaneously large ΔS and RCP [15].

7.4.2 Barocaloric effects in single phases

Plastic crystals are often characterized by a large thermal expansion in the high-temperature disordered phase (see table 7.1 and figure 7.3(d)). These large values permit large BC effects away from the transition in single phases, as seen via the last term in equation (7.1): $\Delta S_+(T, p_0 \rightarrow p_1) = -\int_{p_0}^{p_1} (\partial V/\partial t)_p dp$. In most materials, including PCs, BC contributions associated with the transition and single phases are of the same sign (conventional in this case) and therefore add up in such a way that the overall BC response of the material is enhanced. Moreover, BC effects in single phases are always reversible because there is no (significant) hysteresis associated with the single-phase dynamics. Also, BC effects in single phases are not expected to saturate within the pressure range of interest (well below 1 GPa), cf. BC contribution from the transition.

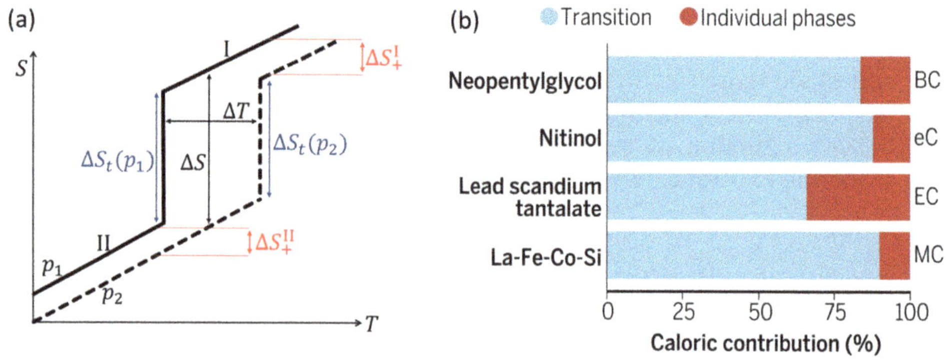

Figure 7.5. (a) Qualitative isobaric entropy curves in the entropy-temperature diagram at two pressures p_1 and p_2. Transition entropy changes ΔS_t at p_1, and isothermal entropy changes ΔS upon pressure change $p_1 \rightarrow p_2$, that involve the transition entropy change. The isothermal entropy changes at single phases, ΔS_+^{I} and ΔS_+^{II} are indicated. (b) Percentage contributions to caloric effects originated due to first-order phase transitions (blue) and in each phase outside the transition (red), for one representative material of each caloric type. Reproduced from reference [95] with permission from AAAS.

In the case that the C_p of the ordered and plastic phases are similar (and so are the slopes of $S(T)$ in a relatively narrow temperature interval), according to figure 7.5(a), the following relation should hold as a first approximation:

$$\left|\Delta S\right| \sim \left|\Delta S_t(p_0)\right| + \left|\Delta S_+^{II}(p_0 \rightarrow p_1)\right| \sim \left|\Delta S_t(p_1)\right| + \Delta S_+^{I}(p_0 \rightarrow p_1)\right| \tag{7.17}$$

This relation indicates that in order to estimate the total BC effects, the single-phase contribution in only one phase should be added to the transition entropy change. Table 7.3 shows a summary of the quantities discussed in this section.

This phenomenon occurs in all caloric types and the relative weight of the single-phase contribution to the overall caloric response is similar, as shown in figure 7.5(b). However, the absolute values of single-phase contributions are significantly larger for BC PCs, providing another advantageous feature.

7.5 Spectroscopic insights into colossal barocaloric effects

7.5.1 Introduction to quasi-elastic neutron scattering (QENS)

While the phenomenological thermodynamic description is well established, and the outstanding performance of many BC materials has been demonstrated, what happens in PCs at the atomic and molecular level when external pressures are applied remains largely unexplored [32]. This microscopic knowledge is vital for explaining the behaviour of BC materials, and can underpin the discovery of new materials. QENS is the ideal tool to investigate this, as incoherent QENS can deliver information on self-dynamics or single-particle dynamics of hydrogen atoms in PCs, due to its large incoherent scattering cross-sections [68].

At first, a very brief introduction to QENS is given [96–98]. Neutron scattering describes an interaction process between the incident neutrons and nuclei of target samples, where the scattering intensity is proportional to a sum of coherent and incoherent double-differential scattering cross-sections, expressed as

$$\left(\frac{d^2\sigma}{d\Omega d\omega}\right)_{\text{tot}} = \left(\frac{d^2\sigma}{d\Omega d\omega}\right)_{\text{coh}} + \left(\frac{d^2\sigma}{d\Omega d\omega}\right)_{\text{inc}} \tag{7.18}$$

Here, σ is the scattering cross section, Ω is the solid angle and ω is the energy transfer during a scattering event between neutrons and sample. The subscripts 'tot', 'coh' and 'inc' represent total, coherent and incoherent, respectively. The coherent and incoherent scattering cross-sections are proportional to individual dynamical structure factors $S(\mathbf{Q}, \omega)$ with respect to momentum transfer $\mathbf{Q}$ and energy transfer ω,

$$\left(\frac{d^2\sigma}{d\Omega d\omega}\right)_{\text{coh}} = \frac{k'}{k}\frac{\sigma_{\text{coh}}}{4\pi}S_{\text{coh}}(\mathbf{Q}, \omega) \tag{7.19}$$

$$\left(\frac{d^2\sigma}{d\Omega d\omega}\right)_{\text{inc}} = \frac{k'}{k}\frac{\sigma_{\text{inc}}}{4\pi}S_{\text{inc}}(\mathbf{Q}, \omega) \tag{7.20}$$

Here, k' and k are the wave vectors of scattered and incident neutrons. Following the van Hove formalism, the dynamical structure factors are directly related to atomic correlation functions through spatial and temporal Fourier transforms,

$$S_{\text{coh}}(\mathbf{Q}, \omega) = \frac{1}{2\pi}\int G(\mathbf{r}, t)e^{i(\mathbf{Qr}-\omega t)}d\mathbf{r}dt \tag{7.21}$$

$$S_{\text{inc}}(\mathbf{Q}, \omega) = \frac{1}{2\pi}\int G_s(\mathbf{r}, t)e^{i(\mathbf{Qr}-\omega t)}d\mathbf{r}dt \tag{7.22}$$

$$G(\mathbf{r}, t) = G_d(\mathbf{r}, t) + G_s(\mathbf{r}, t) \tag{7.23}$$

The distinct correlation function $G_d(\mathbf{r}, t)d\mathbf{r}$ is the conditional probability that under a given condition that a particle was at time $t = 0$ at the origin $\mathbf{r} = 0$, another particle is found at time t at the position $\mathbf{r}$ in a volume element $d\mathbf{r}$. The self-correlation function $G_s(\mathbf{r}, t)d\mathbf{r}$ is the conditional probability that the same particle is found at time t at the position $\mathbf{r}$ in a volume element $d\mathbf{r}$, under a given condition that the particle was at time $t = 0$ at the origin $\mathbf{r} = 0$. The coherent intensity is partially related to the $G_d(\mathbf{r}, t)$ term such that accounts for collective motions. The incoherent intensity is only connected to $G_s(\mathbf{r}, t)$ and it represents the single-particle dynamics.

Incoherent scattering can be separated into elastic, quasi-elastic, and inelastic scattering, as shown in figure 7.6. The elastic component appears as a delta function convoluted by the instrumental resolution centered at $E = \hbar\omega = 0$, which is related to the static distribution of the nuclei. The quasi-elastic component is usually described by a Lorentzian function centered at $\hbar\omega = 0$ with a finite linewidth. It reflects the diffusional or rotational motions of the system. The inelastic component is usually a pair of peaks centered at $\pm\hbar\omega \neq 0$, corresponding to whether scattered

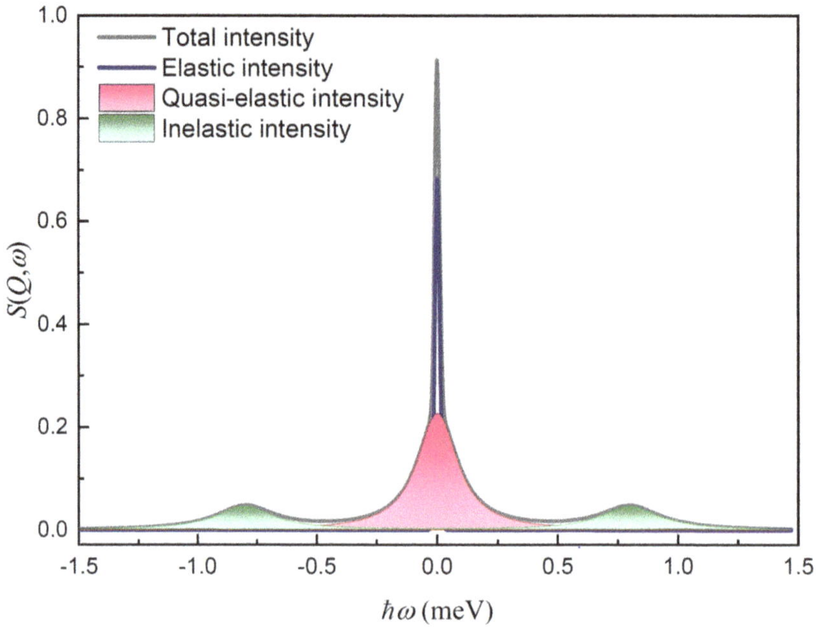

Figure 7.6. Schematic diagram of a neutron scattering spectrum including elastic, quasi-elastic, and inelastic components.

neutrons gain or lose energy during the scattering process. They are usually attributed to molecular vibrational and tunneling modes. Typically, the ratio between the intensity of the elastic component and the sum of elastic and quasi-elastic components leads to the elastic incoherent scattering factor (EISF), whose **Q** dependence is quite informative in reflecting the atomic and molecular motions. The combination of EISF and the linewidth of the quasi-elastic intensity allows understanding the details of the atomic motions [68].

QENS measurements are usually performed at high-resolution triple-axis spectrometers [99], time-of-flight spectrometers [100, 101], and backscattering spectrometers [102] since these instruments have a wide range of energy resolutions ranging from μeV to meV, which match the typical timescales of these atomic motions [98]. Such a technique is well demonstrated in the atomic dynamics study of the famous inorganic–organic hybrid perovskite $CH_3NH_3PbI_3$ [103]. This compound exhibits outstanding photovoltaic performances, but very different electronic and thermal transport properties compared to conventional inorganic semiconductors. High-resolution QENS suggests that the ultrafast orientational disorder of molecular dipoles of $[CH_3NH_3]^+$, activated at ~165 K, acts as an additional scattering source for optical phonons as well as for charge carriers. It is revealed that acoustic phonons dominate the thermal transport, rather than optical phonons due to sub-picosecond lifetimes. Such a fourfold reorientation mode of $[CH_3NH_3]^+$ and its antiferromagnetic-like freezing at the tetragonal-to-orthorhombic phase transition make the molecular orientation a pseudo-spin [104]. Thus, this disordered state is also called plastic crystal [105]. Typically, the orthorhombic-to-tetragonal phase transition can be induced by external pressures and the driving pressure can be as small as 100 MPa, in terms of its well-defined temperature–pressure phase diagram [106]. These results inspired Bing *et al* to explore colossal BC effects in NPG and related PCs [32]. Recent developments in low-background pressure cells have extended the capability of neutron spectroscopy in studying BC materials. [107, 108].

7.5.2 QENS of NPG

Temperature- and pressure-dependent QENS measurements were conducted on polycrystalline NPG, as the prototype compound of colossal barocaloric effect at the cold neutron disk chopper direct-geometry spectrometer AMATERAS of J-PARC in Japan, as shown in figure 7.7 [32, 100]. With the incident neutron energy E_i of 2.64 meV, a superior energy resolution of 25 μeV was achieved. It is known that NPG undergoes a first-order phase transition from the low-temperature monoclinic phase to the high-temperature face-centered-cubic (fcc) phase at about 313 K. This phase transition shows a strong asymmetry, as the phase transition in the cooling process is much faster [109]. In the fcc phase, intense QENS signals are

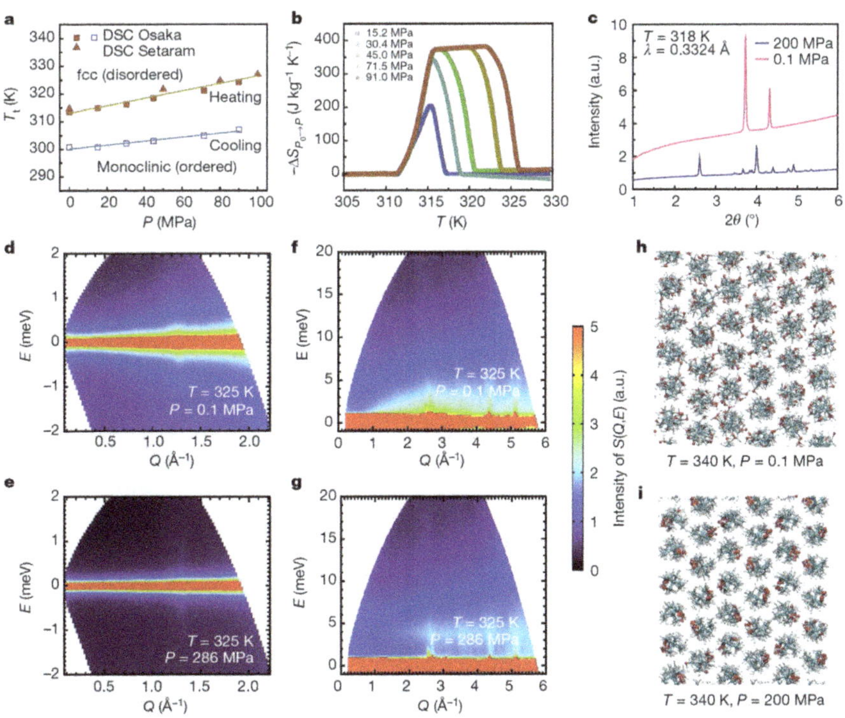

Figure 7.7. Responses to external pressure of NPG. (a) The temperature–pressure phase diagram. (b) Entropy changes as a function of temperature under a constant pressure. (c) Pressure-dependent x-ray diffraction patterns. (d–g) $S(Q, E)$ of pressure-dependent neutron scattering. (h and i) Pressure-dependent atomic structures in molecular dynamics (MD) simulations. Reproduced from reference [32], copyright (2019), with permission from Springer.

observed in the vicinity of the elastic lines. A careful analysis of EISF suggests that the NPG molecules are subjected to nearly isotropic rotation over the fcc lattice with typical energy scales of about 0.1 meV. Compared to the fcc phase, there is only an intramolecular motion in the monoclinic phase. Under application of external pressures of 200 MPa, the QENS signals related to the isotropic rotation mode is effectively suppressed. These experimental observations are well reproduced by several molecular dynamics simulations [32, 110–112]. The spatial distribution of NPG molecular orientations in the monoclinic phase is characteristic of four poles, corresponding to the tetrahedral configuration of molecules, whereas the distribution becomes completely random in the fcc phase, which can be well restored to the ordered state under external pressure. Separately, in density functional theory simulations, the activation barrier of the isotropic rotation of NPG molecules in the monoclinic phase is three times higher than that in the fcc phases [113]. The estimated molecular rotation-rate well matches the experimental results.

The direct observations of extensive molecular orientational disorder and its sensitive response to external pressures in the PC state not only underlie colossal BC effects, but also have profound physical implications in the fundamental understanding of solids. A plastic crystal state may define the upper bound for the degree of disorder in solids. Most atomic degrees of freedom are disordered while only the mass centers of molecules are ordered to maintain the rigid lattice. When further disorder of the center of mass is developed, the rigidity is lost and the system becomes a liquid. The high sensitivity of PCs to external pressures is another advantage for colossal BC PCs.

PCs usually have much greater compressibility (smaller bulk moduli) than regular materials. The extensive disorder indicates weaker intermolecular interactions (usually hydrogen bonds) as well as lower energy barriers. Hence, the phonons in these systems are usually low-lying and the top of acoustic branches is below 1 meV, such as in NPG. Such phonon band structures with relatively flat acoustic phonons branches result in small bulk moduli that are proportional to the slopes of the acoustic dispersions. Consequently, the pressure sensitivity is essentially rooted in weak interactions. Similar situations are also found in superionic conductors $AgCrSe_2$ and Ag_2Se [114, 115]. In these compounds, Ag^+ ions are subjected to extensive dynamic occupational disorder and exhibit diffusion hopping above the superionic transition temperature. They are ductile/plastic unlike conventional inorganic semiconductors and may have promising applications in flexible electronics [116].

In addition, the near-picosecond timescale of the molecular orientational disorder is comparable to the typical timescales of phonons (frequency as well as lifetime), which enhances an intrinsically intimate coupling between the orientational and vibrational degrees of freedom, similar to the spin-lattice (magnon–phonon) coupling in magnetic materials. As shown for NH_4I, the reorientation–phonon coupling is so strong that the driving pressure can be reduced to 20 MPa [14]. Understanding this issue in more depth may help design new materials with even lower driving pressures.

The *in situ* QENS study of NPG PC not only microscopically confirms the mechanistics of colossal BC effects, but also extends the applications of QENS techniques. In a series of renowned conferences such as QENS/WINS 2022 [117] and the Gordon Research Conference [118], extensive attention has been given to this study.

7.5.3 Inverse colossal BCE

Unlike conventional BC materials whose temperature increases upon pressurization, a material with an inverse BC effect displays a temperature drop upon pressurization. Similar behaviors have been observed in other caloric counterparts, e.g. inverse magnetocaloric effects in Ni–Mn–Sn alloys [119]. NH_4SCN, as the first orientationally disordered crystal, shows inverse colossal BC effects, with the entropy changes higher than 100 J kg^{-1} K^{-1} [15]. NH_4SCN undergoes a first-order phase transition from monoclinic to orthorhombic at T_t of about 360 K with a large negative volumetric thermal expansion of 5%. This value of negative thermal expansion is ranked as one of the largest among current state-of-the-art materials. The phase transition temperature is shifted to the lower temperatures at larger pressures, which leads to a negative slope of the phase boundaries on the T–p phase diagram. When pressure changes from 0.1 to 100 MPa, the transition temperature changes from 364 down to 334 K.

This compound has unique barocaloric properties not only because it is the first inverse colossal BC system, but also because large entropy changes and large RCP are simultaneously obtained. Its maximum entropy change is about 128.7 J kg^{-1} K^{-1}, twice larger than those of other current inverse BC materials. In addition, its saturation pressure is about 80 MPa, much smaller than those of other systems, usually higher than 500 MPa. Compared to the previous reports where a smaller $|dT_t/dp|$ leads to a larger entropy change and to a smaller RCP, large entropy changes and RCP are simultaneously achieved in NH_4SCN, due to the huge lattice volume changes. These properties make this material not only promising for refrigeration applications but also for thermal storage applications.

In situ QENS was employed to understand this inverse BC effect. Shown in figure 7.8 are the contour plots of $S(\mathbf{Q}, \omega)$ at 300, 380, and 400 K, where NH_4SCN

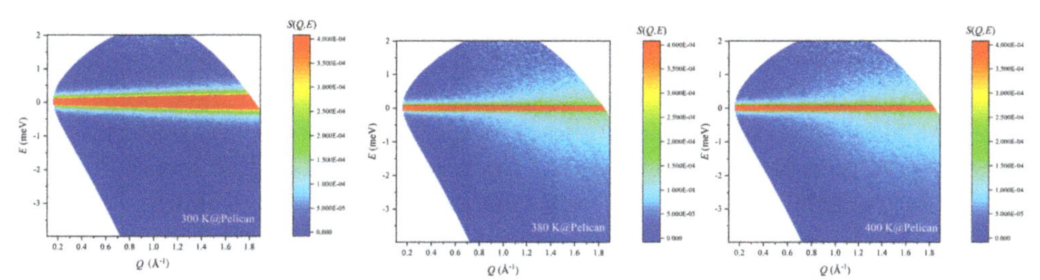

Figure 7.8. QENS spectra of NH_4SCN at 300, 380, and 400 K, where it crystallizes in monoclinic, orthorhombic and tetragonal phases. The QENS signals of the latter two phases are much stronger. Data taken at Pelican, ANSTO.

lies in the monoclinic, orthorhombic, and tetragonal phases, respectively. It can be seen that the intensity of QENS at 380 and 400 K are almost identical but they are much stronger than the intensity at 300 K. Therefore, the monoclinic-to-orthorhombic phase transition is accompanied by modifications of the reorientation dynamics, whereas the orthorhombic-to-tetragonal one is not. The analysis of EISF indicates that the 300 K data are close to the tetrahedral tumbling model, where the NH_4^+ tetrahedron is as a whole randomly oriented over four equivalent configurations. By contrast, the 380 K data points are most likely described by the cubic tumbling, where the orientation of the tetrahedron of NH_4^+ is random over a cubic lattice. Molecular dynamics simulations suggest that the distribution is not a simple cubic lattice, but with preference. More careful analysis of the crystal structure indicates that the hydrogen-bonding environment of NH_4^+ is not exactly symmetric. The average timescales show that the preferred cubic tumbling mode is much faster than the tetrahedral tumbling. The weak **Q** dependence indicates the local nature of the modes. MD predicted that the de-coherence rates of NH_4^+ rotations are in accordance with the experimentally estimated orientational relaxation times for both the monoclinic and orthorhombic phases. Pressure-dependent measurements suggest that the QENS profile becomes wider, indicating enhanced disorder upon applying pressure. As it is known that atomic disorder is usually suppressed by pressure, the pressure-enhanced atomic disorder in NH_4SCN is considerably unique. This counter-intuitive unique pressure-induced atomic disorder in NH_4SCN can be well understood through MD simulations. Pressure promotes transverse vibrations of SCN^- and weakens the hydrogen-bonding network of the monoclinic phase, and eventually drives the compound to the more stable orthorhombic phase through tilting of SCN^- and disordering of NH_4^+.

Finally, it should be noted that negative thermal expansion usually coexists with an inverse barocaloric effect. In terms of $(\partial T/\partial p)_S = T(\partial V/\partial T)_p/C_p$, negative $(\partial V/\partial T)_p$ means negative $(\partial T/\partial p)_S$, which corresponds to the inverse barocaloric effect [120].

7.6 Shortcomings of plastic crystals for applications

- *Transition hysteresis*. Although not always, the strong first-order character of the phase transitions from the plastic phase is often associated with a highly hysteretic exothermic transition (see section 7.2). In some cases, thermal hysteresis can reach dozens of K and, eventually, orientational disorder can even freeze, leading to orientational glasses [121, 122]. In fact, some plastic crystals show easy glass-formation ability. This large thermal hysteresis leads to large pressure hysteresis, and as discussed previously (see section 7.3), this is detrimental for material and device performance. Figure 7.9 shows irreversible and reversible effects in plastic crystals derived from neopentane determined from quasi-direct measurements.
- *Thermal conductivity*. One of the most important drawbacks that commonly affect PCMs is the fact that a large transition latent heat is usually associated with the emergence of structural disorder, which cuts the lifetimes and mean

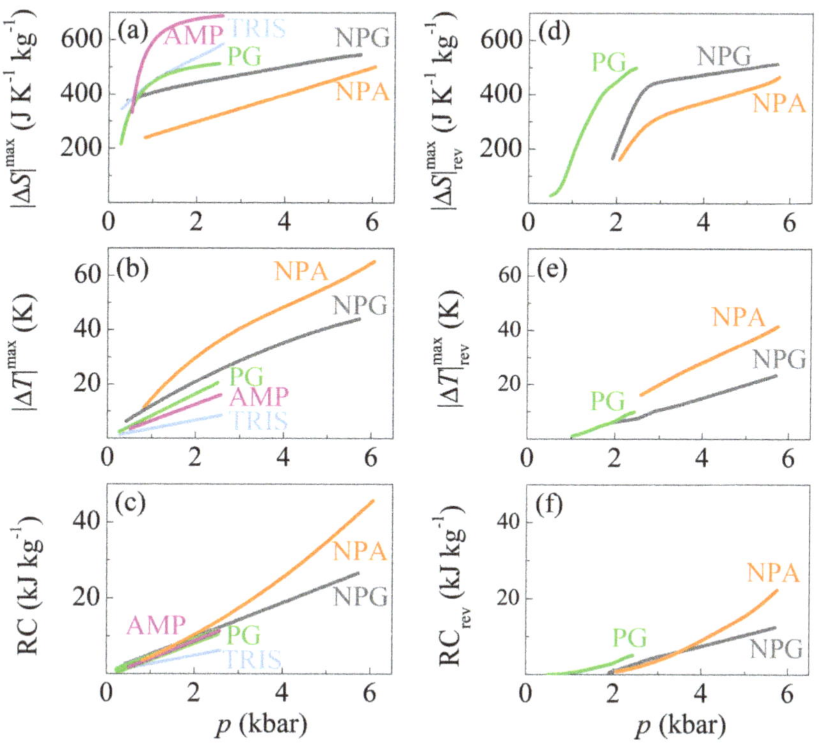

Figure 7.9. Irreversible and reversible (a and d) maximum isothermal entropy changes, (b and e) maximum adiabatic temperature changes and (c and f) maximum refrigerant capacity for different plastic crystals derived from neopentane. Reproduced from reference [79] with permission from the Royal Society of Chemistry.

free paths of heat-carrying phonons [103]. In addition, plastic crystals usually have large compressibility or small bulk modulus [32], which reflect their small phonon velocities and weak intramolecular interactions [123]. These two factors lead to significant reduction of thermal conductivity [124] around 0.2–0.3 W K^{-1} m^{-1} [125]. Thermal conductivities of plastic crystals are usually anisotropic dependent on the propogation direction of hydrogen bonding networks [110]. Consequently, they finally result in a poor heat exchange, which is detrimental to the device performance. To minimize this problem, inclusion of nanoparticles [126, 127] or embedding the PCM in porous highly conductive matrices [128, 129] have been performed, leading to a substantial increase in thermal conductivity [130], of course at the expense of reducing the energy density.

In neopentylglycol-50%-wt-graphene nanosheets composites, an improvement from 0.42 to 18.4 W m^{-1} K^{-1} was obtained while still exhibiting colossal $\Delta S = 122$ J K^{-1} kg^{-1} upon compression of 100 MPa from normal pressure [127].

- *Segregation of binary systems.* Many studies have addressed the determination of binary phase diagrams of PCs, with the purpose of tuning the transition properties. However, these binary systems often appear to be unstable and segregate immediately or after a few or a number of cycles, which clearly restricts their use for BC applications [131–134].

- *High vapor pressure*. Organic PCs are often readily subliming materials, which may complicate their handling for technological applications [135]. For instance, it might lead to the release of gases with the corresponding loss of mass or possible inhalation. In the latter case, despite the overall low toxicity of these compounds, inhalation might provoke irritation of the breathing system and/or to bad smell.
- *Low density*. Organic PCs exhibit densities of the order of that of water, typically ranging from 0.8 to 1.3 g cm^{-3}. Therefore, this results in BC effects per unit volume that, although still larger, are closer to those exhibited by other compounds such as magnetostructural alloys enjoying densities that can be 5 to 10 times larger.

7.7 Concluding remarks

Recent investigations have placed PCs in the spotlight of research on BC effects. Under moderate pressure changes of ~100 MPa these materials display colossal isothermal entropy changes well above 100 J K^{-1} kg^{-1} that are unrealizable in other materials families. Alongside large adiabatic temperature changes, the BC response in PCs offers an outstanding platform with which to develop novel solid-state methods for thermal management driven by pressure, mainly cooling and heating methods. Despite several shortcomings, it is hoped that these materials will not suffer from any unsurmountable barrier that will frustrate their applicability. Separately, barocaloric effects in plastic crystals are also attracting great interest from condensed matter physicists, as these materials display strong molecular orientational disorder and large volume changes, which lead to very rich molecular dynamics. These phenomena have stimulated plenty of experiments such as neutron scattering or Raman spectroscopy, and advanced theoretical calculations and modelling, which provide relevant insights not only in such mesophases at the limit between solids and liquids, but also for a better overall understanding of the physical behavior of solids. We therefore envisage exciting years ahead in the progress of this field.

Acknowledgements

This work was supported by MINECO Projects No. PID2020-112975GB-I00 and TED2021-129952B-C31 (Spain), DGU Project No. 2017SGR-42 (Catalonia) and the Key Research Program of Frontier Sciences of Chinese Academy of Sciences (Grant No. ZDBS-LY-JSC002). X M acknowledges funding from ERC Starting Grant (no. 680032) and the Royal Society.

References

[1] Tamarit J L, Barrio M, Pardo L C and Negrier P 2021 *Plastic Crystals* (Cham: Springer International Publishing) pp 163–90

[2] Timmermans J 1938 *J. Chim. Phys.* **35** 331–44

[3] Pardo L, Barrio M, Tamarit J L, López D, Salud J and Oonk H 2005 *Chem. Mater.* **17** 6146–53

[4] Reuter J, Büsing D, Tamarit J L and Würflinger A 1997 *J. Mater. Chem.* **7** 41–6

[5] Mitsukura K, Sakamoto H, Kubo H, Yoshida T and Nagasawa T 2010 *J. Biosci. Bioeng.* **109** 550–3

[6] Tanaka N and Yamaguchi M 2003 Method for producing 1,3-adamantanediol *Japan Patent* JP2005082563A

[7] Orzeszko A, Kamińska B and Starościak B J 2002 *Il Farmaco* **57** 619–24

[8] Nozaki K, Watanabe K, Yano E, Kotachi A, Takechi S and Hanyu I 1996 *J. Photopolym. Sci. Technol.* **9** 509–22

[9] Aigami K, Inamoto Y, Takaishi N, Hattori K, Takatsuki A and Tamura G 1975 *J. Med. Chem.* **18** 713–21

[10] Rabey J, Nissipeanu P and Korczyn A 1992 *J. Neural Transm.-Parkinson's Dis. Dementia Sect.* **4** 277–82

[11] Wishnok J S 1973 *J. Chem. Educ.* **50** 780

[12] López D, Salud J, Tamarit J L, Barrio M and Oonk H 2000 *Chem. Mater.* **12** 1108–14

[13] Zhang Z and Nazar L F 2022 *Nat. Rev. Mater.* **7** 389–405

[14] Ren Q *et al* 2022 *Nat. Commun.* **13** 2293

[15] Zhang Z *et al* 2023 *Sci. Adv.* **9** eadd0

[16] Zhang Z *et al* 2023 *Mater. Horiz.* **10** 977

[17] Salgado-Beceiro J, Bermúdez-García J M, Llamas-Saiz A L, Castro-García S, Señarís-Rodríguez M A, Rivadulla F and Sánchez-Andújar M 2020 *J. Mater. Chem.* C **8** 13686–94

[18] Walker J, Marshall K P, Salgado-Beceiro J, Williamson B A, Løndal N S, Castro-Garcia S, Andújar M S, Selbach S M, Chernyshov D and Einarsrud M-A 2022 *Chem. Mater.* **34** 2585–98

[19] Pringle J M, Howlett P C, MacFarlane D R and Forsyth M 2010 *J. Mater. Chem.* **20** 2056–62

[20] Zhu H, MacFarlane D R, Pringle J M and Forsyth M 2019 *Trends Chem.* **1** 126–40

[21] Zachariah M, Romanini M, Tripathi P, Barrio M, Tamarit J L and Macovez R 2015 *J. Phys. Chem.* C **119** 27298–306

[22] Harada J, Shimojo T, Oyamaguchi H, Hasegawa H, Takahashi Y, Satomi K, Suzuki Y, Kawamata J and Inabe T 2016 *Nat. Chem.* **8** 946–52

[23] Mochida T, Ishida M, Tominaga T, Takahashi K, Sakurai T and Ohta H 2018 *Phys. Chem. Chem. Phys.* **20** 3019–28

[24] Benson D, Burrows R and Webb J 1986 *Sol. Energy Mater.* **13** 133–52

[25] Usman A, Xiong F, Aftab W, Qin M and Zou R 2022 *Adv. Mater.* **34** 2202457

[26] Jamekhorshid A, Sadrameli S and Farid M 2014 *Renew. Sustain. Energy Rev.* **31** 531–42

[27] Suresh C and Saini R P 2020 *Int. J. Energy Res.* **44** 4743–58

[28] Serrano A, Dauvergne J-L, Doppiu S and Palomo Del Barrio E 2019 *Materials* **12** 3169

[29] Barrio M, López D O, Tamarit J L, Negrier P and Haget Y 1995 *J. Mater. Chem.* **5** 431–9

[30] Serrano A, Duran M, Dauvergne J-L, Doppiu S and Del Barrio E P 2021 *Sol. Energy Mater Sol. Cells* **220** 110848

[31] Moya X, Avramenko A, Mañosa L, Tamarit J and Lloveras P 2018 Use of barocaloric materials and barocaloric devices *UK Patent* WO2018/069506

[32] Li B *et al* 2019 *Nature* **567** 506–10

[33] Lloveras P *et al* 2019 *Nat. Commun.* **10** 1803

[34] Salvatori A, Negrier P, Aznar A, Barrio M, Tamarit J L and Lloveras P 2022 *APL Mater.* **10** 111117
[35] Mones A H and Post B 1952 *J. Chem. Phys.* **20** 755–6
[36] Tamarit J L, Legendre B and Buisine J 1994 *Mol. Cryst. Liq. Cryst. Sci. Technol., Sect.* A **250** 347–58
[37] Eilerman D, Lippman R and Rudman R 1983 *Acta Crystallogr.* B **39** 263–6
[38] Barrio M, Font J, López D, Muntasell J, Tamarit J L, Negrier P, Chanh N and Haget Y 1993 *J. Phys. Chem. Solids* **54** 171–81
[39] Eilerman D and Rudman R 1980 *J. Chem. Phys.* **72** 5656–66
[40] Kim S *et al* 2021 *Nucleic Acids Res.* **49** D1388–95
[41] Lewis R A 2016 *Hawley's Condensed Chemical Dictionary* (New York: Wiley)
[42] Sheftel V O 2000 *Indirect Food Additives and Polymers: Migration and Toxicology* (Boca Raton, FL: CRC Press)
[43] Sugimoto N, Patel K and Abe H 2013 Cosmetic composition for eyelashes *US patent* US20130295035A1
[44] Yamaguchi K and Ishitobi S 2016 Water-in-oil emulsion sunscreen cosmetic composition *US patent* US20120288458A11
[45] Miller D, Bryant J, Madsen E and Ghiorse W 1999 *Appl. Environ. Microbiol.* **65** 4715–24
[46] Kallet R H, Jasmer R M, Luce J M, Lin L H and Marks J D 2000 *Am. J. Respir. Crit. Care Med.* **161** 1149–53
[47] Sevigny P 2016 Electrophoresis buffer for extending the useful electrophoresis life of an electrophoresis gel *US Patent* US20160356742A11
[48] Nordman C and Schmitkons D 1965 *Acta Crystallogr.* **18** 764–7
[49] McIntosh G C, Yoon M, Berber S and Tománek D 2004 *Phys. Rev.* B **70** 045401
[50] Barrio M, Levit R, Lloveras P, Aznar A, Negrier P, Mondieig D and Tamarit J-L 2018 *Fluid Phase Equilibr.* **459** 219–29
[51] Bazyleva A B, Blokhin A V, Kabo G J, Kabo A G and Paulechka Y U 2005 *J. Chem. Thermodyn.* **37** 643–57
[52] Wanka L, Iqbal K and Schreiner P R 2013 *Chem. Rev.* **113** 3516–604
[53] Marchenko R and Potapov A 2017 *Molbank* **2017** M968
[54] Harvey P D, Gilson D F and Butler I S 1987 *Can. J. Chem.* **65** 1757–60
[55] Négrier P, Hassine B B, Barrio M, Romanini M, Mondieig D and Tamarit J-L 2020 *CrystEngComm* **22** 1230–8
[56] Whittaker A K, Blakey I, Blinco J, Jack K S, Lawrie K, Liu H, Yu A, Leeson M, Yeuh W and Younkin T 2009 *Advances in Resist Materials and Processing Technology XXVI* (Bellingham, WA: SPIE) **7273** 585–94
[57] Porterfield W W, Jones M E, Gill W R and Wade K 1990 *Inorg. Chem.* **29** 2914–9
[58] Zhu M, Pang Y, Lu F, Shi X, Yang J and Zheng S 2019 *ACS Appl. Mater. Interfaces* **11** 14136–41
[59] Sau K, Ikeshoji T, Takagi S, Orimo S i, Errandonea D, Chu D and Cazorla C 2021 *Sci. Rep.* **11** 11915
[60] Grimes R N 2016 *Carboranes* (New York: Academic)
[61] Juan Z, Pengcheng H, Gong C and Maosheng Z 2012 *Prog. Chem.* **24** 556
[62] Chen Y *et al* 2022 *Mol. Ther. Oncolytics* **24** 400–16
[63] Zhang K, Song R, Qi J, Zhang Z, Zhang Z, Yu C and Li K 2022 *Adv. Funct. Mater.* **32** 2112622

[64] Heiney P A, Fischer J E, McGhie A R, Romanow W J, Denenstein A M, McCauley J P Jr, Smith A B and Cox D E 1991 *Phys. Rev. Lett.* **66** 2911
[65] Tycko R, Dabbagh G, Fleming R, Haddon R, Makhija A and Zahurak S 1991 *Phys. Rev. Lett.* **67** 1886
[66] André D, Dworkin A, Szwarc H, Céolin R, Agafonov V, Fabre C, Rassat A, Straver L, Bernier P and Zahab A 1992 *Mol. Phys.* **76** 1311–7
[67] David W I, Ibberson R M, Matthewman J C, Prassides K, Dennis T J S, Hare J P, Kroto H W, Taylor R and Walton D R 1991 *Nature* **353** 147–9
[68] Bee M 1988 *Quasielastic Neutron Scattering: Principles and Applications in Solid State Chemistry, Uiology and Materials Science* (Boca Raton, FL: CRC Press)
[69] Barrio M 1993 Miscibilidad en fase plástica: la sincristalización entre los derivados del neopentano *PhD thesis* Universitat Politècnica de Catalunya, Barcelona, Spain
[70] Salud J, Barrio M, López D, Tamarit J L and Alcobé X 1998 *J. Appl. Crystallogr.* **31** 748–57
[71] Mekala P and Chen T T 2015 *TT Chen Honorary Symosium on Hydrometallurgy, Electrometallurgy and Materials Characterization* (New York: Wiley) p 461
[72] Aznar A, Negrier P, Planes A, Manosa L, Stern-Taulats E, Moya X, Barrio M, Tamarit J-L and Lloveras P 2021 *Appl. Mater. Today* **23** 101023
[73] Paskevicius M, Pitt M P, Brown D H, Sheppard D A, Chumphongphan S and Buckley C E 2013 *Phys. Chem. Chem. Phys.* **15** 15825–8
[74] Her J-H, Yousufuddin M, Zhou W, Jalisatgi S S, Kulleck J G, Zan J A, Hwang S-J, Bowman R C J and Udovic T J 2008 *Inorg. Chem.* **47** 9757–9
[75] Li J, Dunstan D, Lou X, Planes A, Mañosa L, Barrio M, Tamarit J-L and Lloveras P 2020 *J. Mater. Chem.* A **8** 20354–62
[76] Stern-Taulats E, Gràcia-Condal A, Planes A, Lloveras P, Barrio M, Tamarit J-L, Pramanick S, Majumdar S and Mañosa L 2015 *Appl. Phys. Lett.* **107** 152409
[77] Moya X, Kar-Narayan S and Mathur N D 2014 *Nat. Mater.* **13** 439–50
[78] Lloveras P *et al* 2015 *Nat. Commun.* **6** 8801
[79] Aznar A, Lloveras P, Barrio M, Negrier P, Planes A, Mañosa L, Mathur N D, Moya X and Tamarit J-L 2020 *J. Mater. Chem.* A **8** 639–47
[80] Imamura W, Usuda É O, Paixão L S, Bom N M, Gomes A M and Carvalho A M G 2020 *Chin. J. Polym. Sci.* **38** 999–1005
[81] Romanini M *et al* 2021 *Adv. Mater.* **33** 2008076
[82] Li J, Barrio M, Dunstan D J, Dixey R, Lou X, Tamarit J-L, Phillips A E and Lloveras P 2021 *Adv. Funct. Mater.* **31** 2105154
[83] Seo J, McGillicuddy R D, Slavney A H, Zhang S, Ukani R, Yakovenko A A, Zheng S-L and Mason J A 2022 *Nat. Commun.* **13** 2536
[84] Lloveras P and Tamarit J-L 2021 *MRS Energy Sustain.* **8** 3–15
[85] Seo J, Braun J D, Dev V M and Mason J A 2022 *J. Am. Chem. Soc.* **144** 6493–503
[86] Wisotzki K and Würflinger A 1982 *J. Phys. Chem. Solids* **43** 13–20
[87] Sandrock R and Schneider G 1983 *Ber. Bunsen. Phys. Chem.* **87** 197–201
[88] Jenau M, Reuter J, Tamarit J L and Würflinger A 1996 *J. Chem. Soc., Faraday trans.* **92** 1899–904
[89] Jenau M and Würflinger A 1997 *Z. Phys. Chem.* **199** 255–65
[90] Kabo G, Kozyro A, Frenkel M and Blokhin A 1999 *Mol. Cryst. Liq. Cryst. Sci. Technol., Sect.* A **326** 333–55

[91] Kabo G, Blokhin A, Charapennikau M, Kabo A and Sevruk V 2000 *Thermochim. Acta* **345** 125–33

[92] Meijer B, Dixey R, Demmel F, Perry R, Walker H and Phillips A 2022 *Phys. Chem. Chem. Phys.* **25** 9282–93

[93] Yuan S, Meijer B E, Cai G, Dixey R J C, Demmel F, Dove M T, Liu J, Playford H Y, Walker H C and Phillips A E 2022 *Adv. Funct. Mater.* **32** 2207717

[94] Stern-Taulats E, Planes A, Lloveras P, Barrio M, Tamarit J-L, Pramanick S, Majumdar S, Frontera C and Mañosa L 2014 *Phys. Rev.* B **89** 214105

[95] Moya X and Mathur N D 2020 *Science* **370** 797–803

[96] Embs J P, Juranyi F and Hempelmann R 2010 *Z. Phys. Chem.* **224** 5–32

[97] Berrod Q, Lagrené K, Ollivier J and Zanotti J-M 2018 *EPJ Web of Conferences* 188 (Les Ulis: EDP Sciences) p 05001

[98] Gardner J S, Ehlers G, Faraone A and García Sakai V 2020 *Nat. Rev. Phys.* **2** 103–16

[99] Wu C-M, Deng G, Gardner J, Vorderwisch P, Li W-H, Yano S, Peng J-C and Imamovic E 2016 *J. Instrum.* **11** P10009

[100] Nakajima K *et al* 2011 *J. Phys. Soc. Jpn.* **80** SB028

[101] Yu D, Mole R, Noakes T, Kennedy S and Robinson R 2013 *J. Phys. Soc. Jpn.* **82** SA027

[102] Shibata K *et al* 2015 *JPS Conf. Proc.* **8** 036022

[103] Li B *et al* 2017 *Nat. Commun.* **8** 16086

[104] Onoda-Yamamuro N, Matsuo T and Suga H 1990 *J. Phys. Chem. Solids* **51** 1383–95

[105] Carignano M A, Kachmar A and Hutter J 2015 *J. Phys. Chem.* C **119** 8991–7

[106] Onoda-Yamamuro N, Yamamuro O, Matsuo T and Suga H 1992 *J. Phys. Chem. Solids* **53** 277–81

[107] Yuan B, Mole R, Wang C-W, Shumack A, White R, Li B, Tong X and Yu D 2022 *EPJ Web of Conferences* 272 (Les Ulis: EDP Sciences) p 02009

[108] Hattori T, Ohira-Kawamura S and Kawasaki T 2022 *High Press Res.* **42** 226–35

[109] Zhang Z, Chen Y-N, Qi J, Zhang Z, Ohara K, Sakata O, Zhang Z-D and Li B 2022 *Chin. Phys.* B **31** 036802

[110] Wang S, Sun L, Li B and Dai L 2021 *J. Phys. Chem.* C **125** 15853–62

[111] Li F, Li M, Niu C and Wang H 2022 *Appl. Phys. Lett.* **120** 073902

[112] Li F, Niu C, Xu X, Li M and Wang H 2022 *Appl. Phys. Lett.* **121** 223902

[113] Li F, Li M, Xu X, Yang Z, Xu H, Jia C, Li K, He J, Li B and Wang H 2000 *Nat. Commun.* **11** 4190

[114] Kajimoto R, Yokoo T, Nakamura M, Kawakita Y, Matsuura M, Endo H, Seto H, Itoh S, Nakajima K and Ohira-Kawamura S 2019 *Phys. B: Condens. Matter* **562** 148–54

[115] Shi X, Chen H, Hao F, Liu R, Wang T, Qiu P, Burkhardt U, Grin Y and Chen L 2018 *Nat. Mater.* **17** 421–6

[116] Yang Q, Yang S, Qiu P, Peng L, Wei T-R, Zhang Z, Shi X and Chen L 2022 *Science* **377** 854–8

[117] http://qens-wins2022.dipc.org/ (accessed January 2023)

[118] https://www.grc.org/neutron-scattering-conference/2023/. (accessed January 2023)

[119] Krenke T, Duman E, Acet M, Wassermann E F, Moya X, Mañosa L and Planes A 2005 *Nat. Mater.* **4** 450–4

[120] Rajput R 2009 *Engineering Thermodynamics: A Computer Approach (SI Units Version)* (Burlington, MA: Jones & Bartlett Publishers)

[121] Brand R, Lunkenheimer P and Loidl A 2002 *J. Chem. Phys.* **116** 10386–401

[122] Drozd-Rzoska A, Rzoska S J, Pawlus S and Tamarit J L 2006 *Phys. Rev.* B **73** 224205
[123] Qi J *et al* 2020 *Nat. Commun.* **11** 5197
[124] Krivchikov A I, Vdovichenko G A, Korolyuk O, Bermejo F J, Pardo L, Tamarit J L, Jezowski A and Szewczyk D 2015 *J. Non-Cryst. Solids* **407** 141–8
[125] Akbulut S, Ocak Y, Keşlioğlu K and Maraşlı N 2009 *J. Phys. Chem. Solids* **70** 72–8
[126] Praveen B and Suresh S 2018 *Eng. Sci. Technol. an Int. J.* **21** 1086–94
[127] Liu Y, Zhou H, Xu Z, Liu D, Li J, Hu F and Ma T 2022 *Mater. Res. Lett.* **10** 675–81
[128] Wang X, Guo Q, Zhong Y, Wei X and Liu L 2013 *Renew. Energ.* **51** 241–6
[129] Zhang N, Song Y, Du Y, Yuan Y, Xiao G and Gui Y 2018 *Adv. Eng. Mater.* **20** 1800237
[130] Son C and Morehouse J 1991 *J. Thermophys. Heat Trans.* **5** 122–4
[131] Font J, Muntasell J, Navarro J, Tamarit J L and Lloveras J 1987 *Sol. Energy Mater.* **15** 299–310
[132] Font J, Muntasell J, Navarro J and Tamarit J 1987 *Sol. Energy Mater.* **15** 403–12
[133] Font J, Muntasell J, Navarro J and Tamarit J L 1988 *Thermochim. Acta* **136** 55–71
[134] Muntasell J, Barrio M, Font J, López D O, Tamarit J L, Diarte M A C, Guion J, Teisseire M, Chanh N B and Haget Y 1991 *J. Therm. Anal.* **37** 2395–8
[135] Font J and Muntasell J 1994 *Mater. Res. Bull.* **29** 1091–100

IOP Publishing

Barocaloric Effects in the Solid State
Materials and methods
Pol Lloveras

Chapter 8

Atomistic simulation of barocaloric effects

Claudio Cazorla

There is a need to abandon current refrigeration technologies employing compression of greenhouse gases because these are environmentally unfriendly. Solid-state cooling exploits the thermal response of caloric materials to external stimuli and represents a promising refrigeration alternative to standard cooling methods. The barocaloric (BC) effect induced by isotropic compression turns out to be particularly auspicious from an applied point since it renders larger caloric figures of merit than other solid-state cooling variants exploiting the application of electric and magnetic fields. Through the modeling of BC effects it is possible to obtain useful physical insights able to guide, and in some cases even lead, solid-state cooling experiments. Atomistic modeling of BC effects comprises methodologies extending from computationally highly efficient, but not always physically reliable, interatomic potentials to quantitatively exact, but computationally highly demanding, first-principles approaches. In this chapter, we overview atomistic methods and strategies that are well suited for modeling of BC effects, including for instance classical molecular dynamics and the quasi-harmonic approximation. Several case studies are surveyed that illustrate the great potential of atomistic simulation techniques in predicting and simulating BC phenomena.

8.1 Introduction

Large isothermal entropy changes in materials, $|\Delta S_T| \sim 10$–100 kJ K^{-1} kg^{-1}, associated with variations in their structural, ferroelectric and magnetic features may drive substantial caloric effects (e.g., adiabatic temperature changes, $|\Delta T_S|$, of the order of 1–10 K). Solid-state refrigeration is an energy efficient and environmentally friendly technology that exploits caloric effects in crystals for cooling purposes [1–8]. Upon the application of external fields, it is possible to achieve reversible temperature shifts in caloric materials that can be implemented in refrigeration cycles not involving the use of greenhouse gases. Materials undergoing abrupt structural phase transitions under small or moderate hydrostatic pressure

doi:10.1088/978-0-7503-4690-0ch8

shifts are appropriate for developing solid-state cooling applications relying on the BC effect [1–3].

The physical origins and size of BC effects can be accurately determined with atomistic simulation methods and theory. Atomistic simulation techniques can characterize pressure-induced phase transitions in materials and provide valuable insights into the underlying physical mechanisms at the nano scale. In addition, atomistic simulation methods may be inexpensive from a resources point of view. For instance, with the use of ordinary computational resources and open-source software sometimes it is possible to reproduce complex BC phenomena and quantitatively assess the resulting adiabatic temperature and isothermal entropy shifts. Modeling of BC effects, therefore, serves as an invaluable pilot and accompaniment to experimentation.

The degree of reliability of atomistic simulations, however, is proportional to the simplifications made on the necessary interatomic interactions and structural models. In general, improving the reliability of structural models implies reducing the accuracy on the interatomic forces owing to practical computational limitations. For instance, when the adopted simulation approach is highly accurate first-principles methods, most estimations are likely to be obtained for perfectly ordered atomic (or molecular) systems at zero temperature. These simulation conditions obviously differ from the real conditions under which experiments are carried out. Meanwhile, to mimic BC effects under realistic $T \neq 0$ conditions in crystals that contain defects (e.g., vacancies and dislocations) and inhomogeneities (e.g., interfaces and grain boundaries) one should employ computationally affordable interatomic modeling methods (e.g., classical force fields), which typically entail transferability issues and have limited predictive capability. Nonetheless, well-established simulation strategies exist that allow for a reasonable balance between computational affordability and structural model reliability and which can render physically meaningful outcomes.

In this chapter, we review computational techniques based on atomistic simulation methods that may be used to theoretically estimate and foresee BC effects. We start by briefly describing genuine first-principles methods, like density functional theory (DFT), and other practical approaches that provide a fair description of the interactions among atoms in materials, namely, classical force fields and machine learning potentials. Next, several computational techniques designed to describe phase transitions and estimate thermodynamic properties of materials (e.g., entropies and free energies) are summarized (e.g., the quasi-harmonic approximation and thermodynamic integration methods). Subsequently, several simulation strategies that may be used for direct, quasi-direct and indirect evaluation of BC effects are succinctly covered. The chapter finalizes with some representative examples of studies in which atomistic simulation approaches have been employed to predict and rationalize original BC effects in families of materials that are technologically important, namely, fast-ion conductors [4, 9–11], complex hydrides [12] and multiferroic perovskite oxides [13, 14]. The present chapter is addressed to experienced computational materials researchers with interests on caloric materials and solid-state refrigeration as well as to novice scientists wishing to learn free-energy methods and atomistic modeling of phase transitions.

8.2 Modeling of atomic interactions

The basics of DFT, an *ab initio* (also called first-principles) approach broadly used in materials and condensed matter science, and analytical interatomic potentials, also known as classical force fields, are briefly reviewed next. The latter modeling approaches approximate and typically rely on the results of accurate first-principles methods like DFT. On the other hand, the computational cost of *ab initio* techniques is several orders of magnitude higher than that of analytical interatomic potentials, thus in many cases DFT methods may not be straightforwardly employed for the analysis of BC phenomena. Machine learning techniques are also succinctly described since they constitute a modern atomistic modeling method in the middle way between DFT and classical potential approaches as far as computational expense and physical accuracy are concerned.

8.2.1 *Ab initio* methods

The dynamics of nuclei and electrons in condensed matter systems can be decoupled because their respective masses are different by several orders of magnitude. Consequently, the wave function of a many-electron system, $\Psi(\mathbf{r}_1, \mathbf{r}_2, \ldots, \mathbf{r}_N)$, can be determined by solving the Schrödinger equation corresponding to the following non-relativistic Born–Oppenheimer Hamiltonian:

$$
\begin{aligned}
H = &-\frac{1}{2}\sum_i \nabla_i^2 - \sum_{\mathrm{I}}\sum_i \frac{Z_{\mathrm{I}}}{|\mathbf{R}_{\mathrm{I}} - \mathbf{r}_i|} \\
&+ \frac{1}{2}\sum_i\sum_{j\neq i}\frac{1}{|\mathbf{r}_i - \mathbf{r}_j|},
\end{aligned}
\tag{8.1}
$$

where $\mathbf{r}_i$ are the positions of the electrons, Z_{I} the nuclear charges and $\mathbf{R}_{\mathrm{I}}$ the nuclear positions (which are considered to be fixed). Ψ is a complex mathematical function that for crystals and most realistic systems is unknown. The main purpose of any *ab initio* approach is to identify a good approximation for Ψ, or a derived quantity like the electronic density, that can be used to perform calculations. Examples of first-principles methods include DFT, quantum Monte Carlo (QMC), Møller–Plesset perturbation theory (MP2) and the coupled-cluster method with single, double and perturbative triple excitations [CCSD(T)], to cite a few. From all these approaches, DFT is the one most frequently applied to the study of materials and pressure-induced phase transitions, owing to a remarkable balance between computational affordability and physical accuracy, thus we summarize its basic features in what follows.

In 1965, Kohn and Sham conceived a revolutionary theory that was able to effectively calculate the energy and related properties of many-electron systems without relying on the explicit knowledge of Ψ [15, 16]. The theory was named density functional theory (DFT) and the main idea underlying it is that the exact many-electron ground-state energy, E, and density, $n(\mathbf{r})$, can be estimated by solving the following the one-electron Schrödinger equation involving the Kohn–Sham (KS) Hamiltonian:

$$H_{\mathrm{KS}}\Psi_{i\sigma} = \epsilon_{i\sigma}\Psi_{i\sigma}, \tag{8.2}$$

in which the index i labels different one-electron orbitals and σ different spin states. The KS Hamiltonian adopts the form:

$$H_{\mathrm{KS}} = -\frac{1}{2}\nabla^2 + V_{\mathrm{ext}}(\mathbf{r}) + \int \frac{n(\mathbf{r}')}{|\mathbf{r} - \mathbf{r}'|}d\mathbf{r}' + V_{\mathrm{xc}}(\mathbf{r}), \tag{8.3}$$

in which

$$n(\mathbf{r}) = \sum_{i\sigma}|\Psi_{i\sigma}(\mathbf{r})|^2, \tag{8.4}$$

$V_{\mathrm{xc}}(\mathbf{r}) = \delta E_{\mathrm{xc}}/\delta n(\mathbf{r})$ being the exchange–correlation potential and V_{ext} an external field.

The exchange–correlation energy may be defined like the interaction energy difference between a classical many-electron system and its quantum counterpart, thus it has a quantum mechanical origin. Even though E_{xc} amounts to a relatively small portion of the total energy, it turns out to be crucial for all molecules and materials since it acts directly on the bonding between atoms. Nevertheless, $E_{\mathrm{xc}}[n]$ is generally unknown and consequently needs to be approximated. In principle, this is the main source of uncontrollable and fundamental errors in DFT approaches. In standard DFT methods, $E_{\mathrm{xc}}[n]$ usually is approximated with the expression:

$$E_{\mathrm{xc}}^{\mathrm{approx}}[n] = \int \epsilon_{\mathrm{xc}}^{\mathrm{approx}}(\mathbf{r})n(\mathbf{r})d\mathbf{r}, \tag{8.5}$$

in which $\epsilon_{\mathrm{xc}}^{\mathrm{approx}}$ depends on $n(\mathbf{r})$, $\nabla n(\mathbf{r})$ and/or the electronic kinetic energy $\tau(\mathbf{r}) = \frac{1}{2}\sum_{i\sigma}|\nabla\Psi_{i\sigma}(\mathbf{r})|^2$. The basic features of the most popular $E_{\mathrm{xc}}[n]$ functionals used for modeling of materials and phase transitions are summarized next.

In local DFT approaches (e.g., local density approximation—LDA–), $E_{\mathrm{xc}}^{\mathrm{approx}}$ in equation (8.5) is evaluated by adopting the exchange–correlation energy of a uniform gas of electrons with density $n(\mathbf{r})$, $\epsilon_{\mathrm{xc}}^{\mathrm{unif}}$, which is exactly known from quantum Monte Carlo calculations [17, 18]. The non-uniformity of real electronic systems is tackled by partitioning the space into infinitesimal volume elements that are regarded as locally uniform. In semi-local approaches (e.g., generalized gradient approximation—GGA), E_{xc} is approximated in a similar fashion to that in local approaches but with the difference that $\epsilon_{\mathrm{xc}}^{\mathrm{approx}}$ is also made to depend on the $n(\mathbf{r})$ gradient [19, 20]. Local and semi-local approximations satisfy few exact E_{xc} constraints and may work reasonably well for systems in which the real-space variation of the electronic density is low and moderate (e.g., inorganic crystals). Meta-GGA functionals conform to an extension of the GGA approach in which the kinetic energy density is considered as an additional input (e.g., the recently proposed meta-GGA SCAN [21, 22]).

Hybrid DFT functionals consist of a mixture of semi-local correlation energies and semi-local and non-local exact Hartree–Fock (HF) exchange energies. The fraction in which local and non-local exchange densities are combined generally respond to empirical rules. For instance, the well-known B3LYP approximation [23] takes 80% of the GGA and LDA functionals and the rest of the exact HF exchange energy. Other examples of popular hybrid functionals are the range-separated HSE06 [25] and PBE0 [24]. Hybrid DFT functionals may describe to some extent the delocalisation of the exchange–correlation hole around an electron, in contrast to local and semi-local functionals, hence may partially correct for the electronic self-interaction errors that are ubiquitous in standard DFT approaches [26]. Such a characteristic turns out to be particularly useful when simulating strongly correlated systems, like transition-metal oxides, that contain highly localised d and f electronic orbitals [27–29].

8.2.2 Classical interatomic potentials

Ab initio approaches can describe the interactions between electrons and ions with high accuracy but they also require dedicated computational resources. The interactions between atoms in materials sometimes can be satisfactorily approximated with analytical functions known as force fields or classical interatomic potentials, which are based on physical knowledge and intuition. In such cases, the simulations can be dramatically accelerated with respect to first-principles calculations. Force fields typically contain a number of parameters, the value of which are adjusted to reproduce *ab initio* or experimental data. An example of a classical potential fitting technique is the force matching approach due to Ercolessi and Adams [30], which is widely employed in the fields of materials science and condensed matter physics [31–34]. The ways in which force fields are constructed, however, are neither unique nor straightforward, and the thermodynamic ranges in which they provide correct results are limited.

An example of a popular pairwise interaction model is the Coulomb–Buckingham (CB) potential, which adopts the simple form [35–38]:

$$V_{\alpha\beta}(r_{ij}) = A_{\alpha\beta} e^{-\frac{r_{ij}}{\rho_{\alpha\beta}}} - \frac{C_{\alpha\beta}}{r_{ij}^6} + \frac{Z_\alpha Z_\beta}{r_{ij}}, \tag{8.6}$$

in which r_{ij} represents the radial distance between a pair of α and β atoms labelled i and j, subscripts α and β different atomic species, Z nuclear charges, and A, C and ρ potential parameters. The exponential term in equation (8.6) accounts for the short-range repulsive forces resulting from electron–electron interactions. The second term in the right-hand-side represents the long-range attractive interactions deriving from dispersion van der Waals forces. Finally, the third term is the Coulomb interaction between point charges. It is worth noting that 'core–shell' modeling can be performed on top of the CB potential to describe atomic polarizability effects [36–39]. In typical core–shell approaches, each atom is decomposed into a charged core, which interacts with other cores via V_{ij} terms analogous to equation (8.6), and a charged shell that is harmonically bound to the core.

Some technologically relevant materials (e.g., oxide and hybrid organic–inorganic perovskites) are characterized by a subtle balance between long-range and short-range forces that can be respectively originated by the Coulomb interactions between permanent electric dipoles and higher-order moments [40] and orbital hybridizations. The simplicity of the CB pairwise interaction model may be not adequate to cast out the complexity of such a diversity of interatomic forces. Bond-valence (BV) potentials, for example, provide an improvement with respect to the CB potential since can reproduce chemical bonding in complex materials with augmented reliability [41]. (Many other interatomic interaction models going beyond CB exist, like the embedded atom potential [42, 43] and multibody force fields [31], but for the sake of focus those are not covered here.)

A general BV potential is [44–47]:

$$V_{\rm BV}(r,\,\theta) = V_{\rm bind}(r) + V_{\rm charge}(r) + V_{\rm rep}(r) + V_{\rm nl}(\theta), \tag{8.7}$$

in which the first term in the right-hand side accounts for the bond-valence potential energy, the second for the Coulomb potential energy, the third for repulsive interactions, and the fourth for an angle potential energy that prevents unphysically large distortions of covalently connected structural units (e.g., the oxygen octahedra in perovskite oxides and molecular cations in hybrid organic–inorganic compounds).

The bond-valence energy term generally is expressed as:

$$V_{\rm bind}(r) = \sum_{\alpha=1}^{N_s} S_\alpha \sum_{i=1}^{N_\alpha} |V_{i\alpha}(r_i) - V_\alpha|^{\gamma_\alpha}, \tag{8.8}$$

with

$$V_{i\alpha}(r_i) = \sum_{\beta=1}^{N_s} \sum_{j}^{NN} \left(\frac{r_0^{\alpha\beta}}{r_{ij}^{\alpha\beta}} \right)^{C_{\alpha\beta}}, \tag{8.9}$$

where N_s represents the number of atomic species in the system, S_α are fitting parameters, N_α the number of α atoms, V_α the desired atomic valence for ion α, γ_α fitting parameters typically set to 1, j an atomic index that runs over nearest-neighbour (NN) ions, $r_0^{\alpha\beta}$ and $C_{\alpha\beta}$ parameters determined by empirical rules, and $r_{ij}^{\alpha\beta}$ the radial distance between ions i and j.

For the repulsive energy term, $V_{\rm rep}$, the following expression normally is employed:

$$V_{\rm rep}(r) = \epsilon \sum_{i=1}^{N_\alpha} \sum_{j=1}^{N_\beta} \sum_{\alpha=1}^{N_s} \sum_{\beta=1}^{N_s} \left(\frac{B_{\alpha\beta}}{r_{ij}^{\alpha\beta}} \right)^{12}, \tag{8.10}$$

where ϵ and $B_{\alpha\beta}$ are fitting parameters. Meanwhile, a harmonic function is used for the angle potential energy that reads:

$$V_{\rm nl}(\theta) = k \sum_{i=1}^{N_{\rm bun}} \left(\theta_{i,\,x}^2 + \theta_{i,\,y}^2 + \theta_{i,\,z}^2 \right), \tag{8.11}$$

where k is a fitting parameter, N_{bun} the number of covalently bonded units in the simulated system (e.g., oxygen octahedral in oxide perovskites and molecular cations in hybrid organic–inorganic compounds), and $\{\theta_{i,\gamma}\}$ the angular degrees of freedom of such entities.

Reliable BV, or formally analogous, interatomic potentials have been developed for archetypal ferroelectric materials (e.g., $BaTiO_3$, $PbTiO_3$, and $PbZr_{0.2}Ti_{0.8}O_3$) [44–48], hybrid organic–inorganic perovskites (e.g., $MAPbI_3$) [32, 33] and complex hydrides (e.g., $Li_2B_{12}H_{12}$ and $LiCB_{11}H_{12}$) [49], among others. Recently, the outcomes of molecular dynamics simulations performed with some of those classical force fields have predicted the existence of novel and giant caloric effects [12, 50, 51].

8.2.3 Machine learning potentials

When the use of *ab initio* methods may be prohibitive and the versatility of the accessible force fields may be too limited to reproduce the phase transitions and materials behaviours of relevance, machine learning (ML) techniques may be very useful. Machine learning (ML) is an artificial intelligence subclass that systematically identifies correlations within sets of data to perform analysis and make predictions [52, 53].

Simulations of similar systems, or of the same system under different conditions, contain redundant information. For example, during a molecular dynamics run the atomic forces and total internal energy are successively computed at each iteration; for a long enough period of time, atomic configurations that are located closely on the same potential-energy surface will be sampled. Such a redundancy on the computation of energies and atomic forces may be handled in such a way that the computationally intensive calculations (e.g., based on DFT) are performed only for a reduced set of configurations while ML techniques are used to interpolate between them. This kind of approach will be successful only if a reasonably good balance between invested computational effort and incurred errors stemming from interpolation is attained.

ML approaches can be used to obtain atomic forces and energies directly from the system atomic positions, thus they can be considered as a particular type of interatomic potential [54]. Nonetheless, in contrast to classical force fields, ML potentials do not rely on physically motivated analytic functions but on flexible mathematical objects. Consequently, the result of ML modeling can be highly versatile regardless of the prior amount of knowledge of the simulated system. Several promising approaches have been employed to design ML potentials including Gaussian processes, permutation invariant polynomials, modified Shepard method using Taylor expansions, artificial neural networks and graph neural networks. The popular artificial neural networks, for example, have been shown to conform to a class of 'universal approximators' [55] since for a given set of layer values they can approximate unknown multidimensional functions with practically unlimited precision.

ML potentials are being employed with increasing frequency to solve problems in materials science via atomistic simulations due to their ability to reach high accuracy levels at moderate computational expense (see work [56] for a general review on this topic). In the context of caloric effects, only few works relying on ML techniques can be found in the literature (see, for instance, works [57, 58]); however, based on the recent progression realized in other similar research fields, we foresee a surge in the application of ML methods to the prediction and analysis of caloric effects and materials in the near future.

8.3 Free-energy methods

BC effects associated with first-order phase transitions that entail abrupt volume and structural changes typically are large. The two main physical descriptors of BC effects are the isothermal entropy change, ΔS_T, and adiabatic temperature change, ΔT_S, associated with the phase transition. These two quantities are thermodynamically related, as we explain next. Consider that the entropy of a condensed matter system depends on temperature and pressure, $S(T, P)$; an infinitesimal entropy variation then can be expressed as:

$$dS = \left(\frac{\partial S}{\partial T}\right)_P dT + \left(\frac{\partial S}{\partial P}\right)_T dP. \tag{8.12}$$

In an adiabatic process the entropy is conserved, $dS = 0$, thus from the equation above it follows that:

$$\frac{C_P}{T}\, dT = -\left(\frac{\partial S}{\partial P}\right)_T dP, \tag{8.13}$$

in which the constant-pressure heat capacity $C_P \equiv T(\partial S/\partial T)_P$ has been introduced.

By integrating both sides of the latter equality, one obtains:

$$\Delta T_S = -\int \frac{T}{C_P}\, dS_T. \tag{8.14}$$

which typically is approximated like [1, 2]:

$$\Delta T_S \approx -\frac{T_t}{C_P} \cdot \Delta S_T, \tag{8.15}$$

where T_t corresponds to the phase transition temperature and C_P to the averaged heat capacity of the involved phases close to the transition point.

Atomistic simulations generally are performed in the (N, V, T) and (N, P, T) ensembles hence the entropy and other related thermodynamic quantities can be theoretically determined as a function of the state variables V, P, T and ρ $(\equiv N/V)$. Concerning BC effects, once the thermodynamic function $S(P, T)$ of the involved phases are known, both the ΔS_T and ΔT_S descriptors can be straightforwardly calculated (see equations (8.12)–(8.15) above). Meanwhile, the entropy of a given system is related to its Gibbs free energy, $G(P, T)$, through the expression:

$$S = -\left(\frac{\partial G}{\partial T}\right)_P, \tag{8.16}$$

thus it follows, without too much surprise, that the fundamental quantity allowing for the theoretical determination of BC effects is $G(P, T)$. For this reason, in this section we explain several computational approaches that allow for direct estimation of free energies in crystals by using atomic interaction modeling techniques like those introduced in the previous section (i.e., *ab initio*, classical potential and machine learning methods).

8.3.1 The quasi-harmonic approximation

Within the quasi-harmonic approximation (QHA) [59–61] it is assumed that the potential energy of a solid can be expressed with a quadratic expansion around the equilibrium atomic configuration of the form:

$$E_{\mathrm{qh}} = E_{\mathrm{eq}} + \frac{1}{2}\sum_{l\kappa\alpha, l'\kappa'\alpha'} \Phi_{l\kappa\alpha, l'\kappa'\alpha'} u_{l\kappa\alpha} u_{l'\kappa'\alpha'}, \tag{8.17}$$

in which E_{eq} is the energy of the system at equilibrium, $\boldsymbol{\Phi}$ the crystal force-constant matrix, and $u_{l\kappa\alpha}$ the displacement of atom κ in the lattice site l along the Cartesian direction α. The atomic displacements normally are expressed as:

$$u_{l\kappa\alpha}(t) = \sum_{q} u_{q\kappa\alpha} \exp\left[i(\omega t - \boldsymbol{q}\cdot(\boldsymbol{l} + \boldsymbol{\tau}_\kappa)\right], \tag{8.18}$$

in which $\boldsymbol{q}$ is a wave vector within the first Brillouin zone corresponding to the equilibrium unit cell and $\boldsymbol{l} + \boldsymbol{\tau}_\kappa$ the position vector of atom κ in cell l. The normal modes of the crystal then are found by diagonalizing the dynamical matrix:

$$D_{\boldsymbol{q};\kappa\alpha,\kappa'\alpha'} = \frac{1}{\sqrt{m_\kappa m_{\kappa'}}}\sum_{l'} \Phi_{0\kappa\alpha, l'\kappa'\alpha'} \exp\left[i\boldsymbol{q}\cdot(\boldsymbol{\tau}_\kappa - \boldsymbol{l}' - \boldsymbol{\tau}_{\kappa'})\right], \tag{8.19}$$

which allows one to describe the crystal as a collection of non-interacting harmonic oscillators with positively defined frequencies $\omega_{\boldsymbol{q}s}$ and energy levels:

$$E_{\boldsymbol{q}s} = \hbar\omega_{\boldsymbol{q}s}\left(\frac{1}{2} + n\right), \tag{8.20}$$

n being the Bose–Einstein occupation number. Accordingly, the potential energy of a crystal can be written within the QHA like:

$$E_{\mathrm{qh}} = \sum_{\boldsymbol{q}s} \hbar\omega_{\boldsymbol{q}s}\left[\frac{1}{2} + \frac{1}{\exp\left(\hbar\omega_{\boldsymbol{q}s}/k_{\mathrm{B}}T\right) - 1}\right] \tag{8.21}$$

and the constant-volume heat capacity like:

$$
\begin{aligned}
C_V(T) &= \left(\frac{\partial E_{\mathrm{qh}}}{\partial T}\right)_V \\
&= \sum_{qs} \frac{\left(\hbar\omega_{qs}\right)^2}{k_{\mathrm{B}}T^2} \frac{\exp\left(\hbar\omega_{qs}/k_{\mathrm{B}}T\right)}{\left[\exp\left(\hbar\omega_{qs}/k_{\mathrm{B}}T\right) - 1\right]^2}.
\end{aligned}
\tag{8.22}
$$

Likewise, the Helmholtz free energy, $F(V, T)$, is analytically expressed as:

$$
F(V, T) = E_0(V) + k_{\mathrm{B}}T\sum_{qs} \ln\left[2\sinh\left(\frac{\hbar\omega_{qs}(V)}{2k_{\mathrm{B}}T}\right)\right], \tag{8.23}
$$

where E_0 corresponds to the energy of the static lattice and the dependence of the vibrational frequencies on volume has been explicitly noted. The Helmholtz and Gibbs free energies are related through the Legendre transformation $G = F + PV$, in which the pressure is equal to $P = -(\partial F/\partial V)_T$. Thus, by knowing $F(V, T)$, $C_V(T)$ and the equation of state $P(V, T)$, one can deduce the value of G and C_P as a function of P and T.

Crystals in which anharmonic effects are important (e.g., materials in which imaginary phonon frequencies appear in the vibrational phonon spectra calculated at zero temperature), however, are not adequately described by the QHA (e.g., the Helmholtz free energy expression in equation (8.23) is ill-defined and cannot be evaluated). In such a case, one should resort instead to genuine anharmonic free-energy approaches for the calculation of $F(V, T)$ like molecular dynamics, thermodynamic integration from reference models [42, 43, 62] and self-consistent phonon approaches [63–66], to cite some examples. The computational cost of these anharmonic free-energy methods, however, typically is orders of magnitude higher than that of quasi-harmonic approaches. Next, we briefly describe some of them.

8.3.2 Molecular dynamics

Molecular dynamics (MD) is a computer simulation method in which the trajectories of atoms in a many-body interacting system are determined numerically by solving the Newton's equations of motion [67, 68]. For this purpose, one needs to know the forces acting between the particles and their potential energies, which can be calculated by any of the methods described in section 8.2. Thermostat and barostat techniques are employed in MD simulations to render constant temperature and pressure conditions [69]. Temperature and anharmonic effects in materials are naturally accounted for in MD simulations (quantum nuclear effects [59], on the other hand, are systematically disregarded in this approach). Estimation of quantities like T-renormalized phonon frequencies, $\omega(V, T)$, and vibrational density of states, $g(\omega)$, can be obtained from MD simulations by using phonon-mode decomposition [70], force-constant optimization [71] and effective Hamiltonian fitting [72] techniques, to cite some examples.

When the interatomic interactions in MD simulations are described with first-principles methods, typically DFT potentials, they are referred to as *ab initio* molecular dynamics (AIMD). AIMD simulations certainly are several orders of magnitude computationally more intensive than MD simulations performed with classical potentials. Nevertheless, thanks to the current sustained growth in computational capability and improved algorithms design (e.g., linear scaling DFT methods [73]), reliable AIMD simulation of complex materials currently is within reach (see, for instance, works [74–78]). ML based molecular dynamics, on the other hand, represents an excellent alternative to AIMD simulations (provided that a reliable interaction model can be generated) due to its great balance between numerical accuracy and computational expense [79].

A standard approach to estimate T-renormalized $g(\omega)$'s from MD simulations consists in calculating the Fourier transform of the velocity–velocity autocorrelation function as follows [74–76]:

$$g(\omega) = \frac{1}{N_{\rm ion}} \sum_{i}^{N_{\rm ion}} \int_0^{\infty} \langle \mathbf{v}_i(t) \cdot \mathbf{v}_i(0) \rangle e^{i\omega t} dt, \tag{8.24}$$

where $\mathbf{v}_i(t)$ stands for the velocity of the atom i at time t, and $\langle \cdots \rangle$ denotes statistical average in the (N, V, T) ensemble. By knowing $g(\omega)$, one can easily estimate $F(V, T)$ from the integral expression that is equivalent to equation (8.23), namely:

$$F(V, T) = E_0(V) + \int_0^{\infty} k_{\rm B}T \ln\left[2\sinh\left(\frac{\hbar\omega}{2k_{\rm B}T}\right)\right] g(\omega) d\omega. \tag{8.25}$$

Likewise, the constant-volume heat capacity can be computed by replacing the summation in equation (8.22) with the corresponding integral expression, and the entropy with the formula [80]:

$$\begin{aligned} S(V, T) = -\left(\frac{\partial F}{\partial T}\right)_V = \int_0^{\infty} \frac{\hbar\omega}{2T} \coth\left(\frac{\hbar\omega}{2k_{\rm B}T}\right) - \\ k_{\rm B} \ln\left[2\sinh\left(\frac{\hbar\omega}{2k_{\rm B}T}\right)\right] g(\omega) d\omega. \end{aligned} \tag{8.26}$$

Thus, just like in the QHA method, one can deduce the value of the thermodynamic functions $G(P, T)$ and $C_P(T)$, and in turn of the BC descriptors ΔS_T and ΔT_S, from the knowledge of $F(V, T)$, $C_V(T)$ and $P(V, T)$.

8.3.3 Thermodynamic integration

More often than not, accurate estimation of the Helmholtz free energy within an ample thermodynamic interval based on equations (8.24)–(8.25) may turn out to be computationally too intensive. In such a situation, thermodynamic integration techniques, in their different variants, may turn out to be very useful since they rely on the fact that thermodynamic quantities other than $F(V, T)$ (e.g., the internal energy and pressure) can be numerically converged more efficiently in practice. Two popular modalities of thermodynamic integration (TI) are (1) standard TI and (2) TI from reference models [42, 43, 62, 81].

Standard TI exploits two well-known thermodynamic relations. First, by considering the hydrostatic pressure expression:

$$\left(\frac{\partial F}{\partial V}\right)_T = -P, \tag{8.27}$$

it is possible to estimate Helmholtz free energy shifts as a function of volume at a fixed temperature like:

$$F(V_1, T) - F(V_0, T) = \int_{V_1}^{V_0} P \; dV. \tag{8.28}$$

Likewise, based on the following expression deduced from statistical mechanics:

$$\left(\frac{\partial[\beta F]}{\partial \beta}\right)_V = E, \tag{8.29}$$

where $\beta \equiv 1/k_B T$, it is possible to estimate Helmholtz free energy shifts as a function of temperature at a fixed volume like:

$$\beta_1 F(V, T_1) - \beta_0 F(V, T_0) = \int_{\beta_0}^{\beta_1} E \; d\beta. \tag{8.30}$$

In view of equations (8.28)–(8.30), it is possible then to extend the value of the Helmholtz free energy by performing MD simulations over a dense grid of (V, T) points, from which P and E can be determined, and subsequently compute the value of the involved integrals numerically.

In the standard TI approach, however, only Helmholtz free energy differences can be estimated hence for the computation of total free energies of different phases, and therefore of BC effects, the value of $F(V, T)$ should be known at least at one thermodynamic state. TI from a reference model can be very useful for this end. The general principle in this approach is that the change in Helmholtz free energy is computed as the total energy function $E_\lambda(\mathbf{r}_1, \ldots \mathbf{r}_N)$ varies adiabatically from E_0 to E_1, the free energies associated with these energy functions being F_0 and F_1. Specifically, the formula for TI from a reference model is [62, 81]:

$$F_1 - F_0 = \int_0^1 \langle E_1 - E_0 \rangle_\lambda, \tag{8.31}$$

in which $\langle \cdots \rangle_\lambda$ represents the thermal average estimated for the system governed by the energy function $E_\lambda = (1 - \lambda)E_0 + \lambda E_1$. In practice, one takes E_1 to be the total energy function whose free energy wish to calculate (e.g., a specific DFT functional), and E_0 the total energy function of a 'reference' system whose free energy is exactly known. For solids, one typically chooses the reference system to be a perfectly harmonic system, the corresponding Helmholtz free energy expression being that in equations (8.23)–(8.25). Meanwhile, the integral in equation (8.31) should be computed numerically and for this end one can perform a series of MD simulations in the (N, V, T) ensemble governed by the energy function E_λ at different λ values.

8.4 Estimation of BC effects

In analogy to the experiments, BC effects can be estimated with atomistic simulations in an indirect, quasi-direct and direct fashion. In the indirect approach, one exploits the Maxwell relation:

$$-\left(\frac{\partial S}{\partial P}\right)_T = \frac{\partial^2 G}{\partial T \partial P} = \left(\frac{\partial V}{\partial T}\right)_P, \tag{8.32}$$

to estimate the adiabatic temperature and isothermal entropy shifts like (see equation (8.13)):

$$\begin{aligned} \Delta T_S &= \int \frac{T}{C_P} \left(\frac{\partial V}{\partial T}\right)_P dP \\ \Delta S_T &= - \int \left(\frac{\partial V}{\partial T}\right)_P dP. \end{aligned} \tag{8.33}$$

Nevertheless, the derivatives with respect to temperature in equation (8.33) are ill-defined for the case of first-order phase transitions since the system volume presents a discontinuity at the phase transition point. In the case of second-order like phase transitions, in which the system volume varies continuously rather than abruptly [4, 9–11], indirect approaches based on the Maxwell relation turn out to be valid.

As mentioned previously, large BC effects typically are associated with first-order phase transitions. In such a case, one can approximately estimate the size of ΔS_T and ΔT_S with the Clausius–Clapeyron method or exactly with quasi-direct and direct techniques. We briefly describe each of these approaches in the following sections.

8.4.1 The Clausius–Clapeyron method

To a first approximation, the BC isothermal entropy change associated with a first-order P-induced phase transition can be estimated with the Clausius–Clapeyron relation like [2]:

$$\Delta S_T \approx \Delta S = \Delta V \cdot \frac{dP_t}{dT_t}, \tag{8.34}$$

where ΔV is the volume change that the system experiences during the phase transformation occurring at conditions $P_t(T_t)$ and the derivative is estimated on the corresponding phase boundary line. Meanwhile, the corresponding adiabatic temperature change can be roughly determined with equation (8.15).

The Clausius–Clapeyron (CC) method is not exact and entails a few technical problems in practice. First, in experiments the entropy change of the phase transformation, ΔS, is not necessarily equal to the BC descriptor ΔS_T due to the presence of pervasive hysteresis and phase coexistence effects [1], which in general are not reproducible with atomistic simulation techniques (see next section 8.4.2). And second, the value of the derivative and volume change entering equation (8.34) typically are accompanied by large numerical uncertainties that are propagated into the evaluation of ΔS. Having said this, the CC method represents a fair approach to estimate BC effects and it has been already employed in numerous experimental and theoretical works [12, 82, 83]. In order to estimate BC effects with the highest accuracy, however, it is recommended to employ quasi-direct and direct methods.

8.4.2 Quasi-direct estimation of BC effects

The theoretical quasi-direct method emulates the analogous experimental quasi-direct approach (see, for instance, works [3, 84, 85]) in which essentially the pressure and temperature dependences of the system entropy, $S(P, T)$, are determined (figure 8.1). Upon derivation of the full entropy curves (i.e., exhibiting a discontinuity, ΔS, at the first-order phase transition points) obtained at constant pressures P_0 and P_1 and expressed as a function of temperature, it is straightforward to graphically deduce the size of the BC descriptors ΔS_T and ΔT_S on the corresponding S–T diagram, as is illustrated in figure 8.1. From a practical point of view, the $S(P, T)$ curves can be calculated numerically with any of the free-energy methods described in section 8.3, and the phase transition points be determined under the

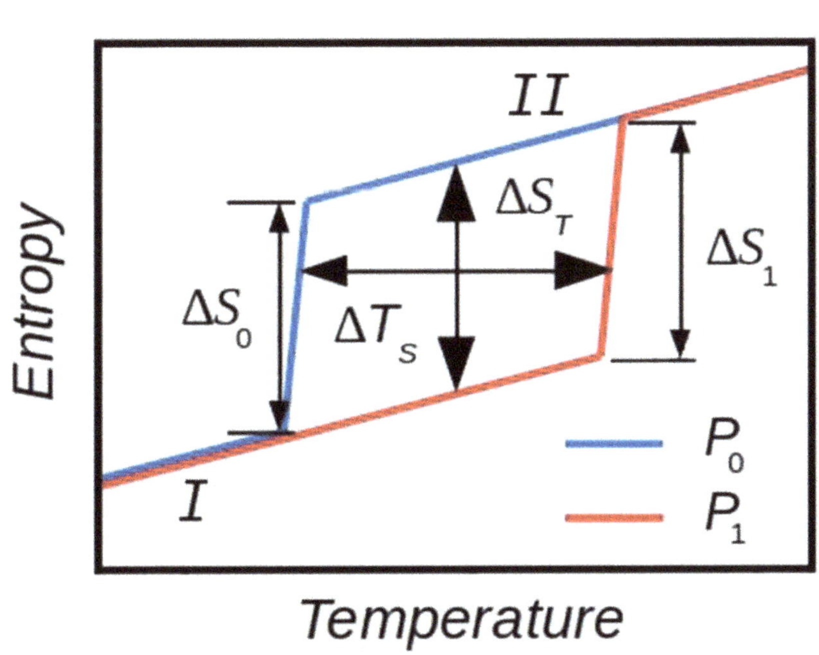

Figure 8.1. Sketch of the entropy curves $S(P, T)$ that can be theoretically estimated with computational free-energy methods for a system undergoing a pressure-induced first-order phase transition, I → II. Hysteresis effects have been disregarded. The BC descriptors ΔS_T and ΔT_S associated with the phase transition can be straightforwardly deduced from these curves. The entropy changes associated to the phase transition, ΔS's, are not necessarily equal to the BC isothermal entropy change ΔS_T.

condition $G_{\rm I}(P, T_t) = G_{\rm II}(P, T_t)$ (I and II being the two phases involved in the first-order transformation).

It is worth mentioning, however, that there are important differences between the theoretical and experimental quasi-direct BC approaches. In experiments, the full entropy curves, typically determined with calorimetric techniques, inevitably include hysteresis and phase coexistence effects. Hysteresis, for instance, originates from the fact that in practice transition paths occur under nonequilibrium conditions; consequently, irreversible processes take place within the material. Irreversibility entails severe limitations to the overall performance of the refrigeration cycle upon successive iterations, hence this is an important aspect to take into consideration when envisaging possible solid-state cooling applications [86]. Unfortunately, hysteresis, and in general any nonequilibrium process, cannot be reproduced with the simulation techniques reviewed in section 8.3 since perfect equilibrium and reversibility conditions are always assumed in them. Thus, in most cases the value of the theoretically estimated BC descriptors should be regarded as upper bounds of the real ΔS_T and ΔT_S shifts (i.e., those measured in the experiments).

8.4.3 Direct estimation of BC effects

In this simulation approach, the system is first thermalized at the desired initial temperature and pressure in the (N, P, T), ensemble. Upon thermalization of the system, the simulated ensemble is changed to the isobaric-isoenthalpic, (N, P, H) (where H represents the enthalpy of the system). At this point the hydrostatic pressure is ramped up to the targeted final value and the resulting temperature change, $\Delta T_{\rm on}$, is monitored. This step has to be done sufficiently slowly in order to guarantee adiabaticity. Subsequently, P is ramped down to its initial value, also slowly enough to guarantee adiabaticity, and the accompanying temperature change, $\Delta T_{\rm off}$, and final temperature, T_f, are monitored. If the system has remained in thermal equilibrium during the entire simulation cycle, then it follows that $T_i = T_f$ and $\Delta T_{\rm on} = \Delta T_{\rm off}$ within the corresponding numerical uncertainties. In that case, either $\Delta T_{\rm on}$ or $\Delta T_{\rm off}$ can be identified with the adiabatic temperature change ascribed to the simulated BC process.

The main advantages of using the direct BC simulation method are the little supervision required and the possibility to combine pressure with additional external fields (e.g., electric bias) in order to explore original multicaloric effects [1]. The disadvantages, on the other hand, are large computational expense, no direct access to the isothermal entropy change, and the fact that implementation of the (N, P, H) ensemble is rare in most molecular dynamics software packages. To the best of our knowledge, direct estimation of BC effects has been only performed for the hybrid organic–inorganic perovskite $CH_3NH_3PbI_3$ [51], also known as MAPI.

8.5 Representative examples

In this section, several illustrative studies in which original BC effects were reproduced and quantified with atomistic simulation methods are discussed. The selected BC effects were predicted for technologically relevant materials like fast-ion conductors (FICs) (LiN_3), orientationally disordered complex hydrides ($Li_2B_{12}H_{12}$) and multiferroic perovskite oxides ($BiFe_{1-x}Co_xO_3$). In these cases, the involved materials modeling techniques comprised molecular dynamics relying on classical force fields (LiN_3 and $Li_2B_{12}H_{12}$) and the quasi-harmonic approximation performed with DFT methods ($BiFe_{1-x}Co_xO_3$). Analogous BC simulation success can be achieved also for other families of pressure responsive materials.

8.5.1 Fast-ion conductor LiN_3

FICs, also known as *superionic* conductors and solid-state electrolytes, are materials that despite being solid display high ionic mobilities [87]. Examples of typical FICs include alkali-rich compounds ($Li_{10}GeP_2S_{12}$ and $Na_3Zr_2Si_2PO_{12}$), alkali-earth metal fluorides (CaF_2) and oxides (doped CeO_2). Superionicity in general manifests with a significant increase in the concentration of point defects (e.g., Frenkel pair defects) for a particular type of atoms (e.g., alkali ions). In view of their high ionic conductivity, FICs are suitable materials for the realization of all-solid-state electrochemical batteries [88], electrocatalysts and fuel cells [89]. Due to the large entropy change associated with the order–disorder superionic phase transition (i.e., $\sim$100 J K^{-1}) [34, 90, 91], FICs have been also established as promising caloric materials for utilization in solid-state refrigeration applications [3, 4, 10, 11, 92]. Figure 8.2 shows the two common polymorphs of bulk Li_3N, a structurally simple and well-known lithium-based FIC. The α phase (hexagonal, space group $P6/mmm$) presents a layered structure that alternates planes of hexagonal Li_2N with planes of lithium caions. The β phase (hexagonal, space group $P6_3/mmmc$) possesses an additional layer of Li^+ ions that is intercalated between consecutive Li_2N planes and which leads to the doubling of the unit cell as compared to the α polymorph. Very high ionic conductivities amounting to $\sim 10^{-4}$–10^{-3} S cm^{-1} have been experimentally measured in Li_3N at room temperature [93–95]. In molecular dynamics simulations employing first-principles methods and classical force fields [10], it was found that perfectly stoichiometric α–Li_3N (that is, with no defects) displays fast-ionic conductivity already at $T = 300$ K; in contrast, room temperature superionicity in β–Li_3N was only observed upon the addition of a small concentration of Li^+ vacancies ($\sim$1%). Figure 8.2 also encloses the theoretical estimation of BC effects in Li_3N under pressure shifts of up to 1 GPa at temperatures of 300 and 400 K [10]. Those calculations relied on molecular dynamics simulations performed with classical interatomic potentials (section 8.2.2) and the set of equations (8.33). For the β polymorph, a peak room temperature entropy shift of -24 J K^{-1} kg^{-1} was estimated which slightly increases in absolute value at higher temperatures (figure 8.2(a)). The size of this BC descriptor can be regarded as 'giant' since it fulfills the condition $|\Delta S| > 10$ J K^{-1} kg^{-1}. The concomitant adiabatic temperature shifts, however, are large but not giant since $|\Delta T| < 10$ K, figure 8.2(b). The main cause of such relatively modest $|\Delta T|$ values is the huge heat capacity of Li_3N, which amounts to $\sim 4 \times 10^3$ J K^{-1} kg^{-1} at room temperature. Analogous BC results were found for α-Li_3N (figures 8.2(c) and (d)) although in this case the computed adabatic

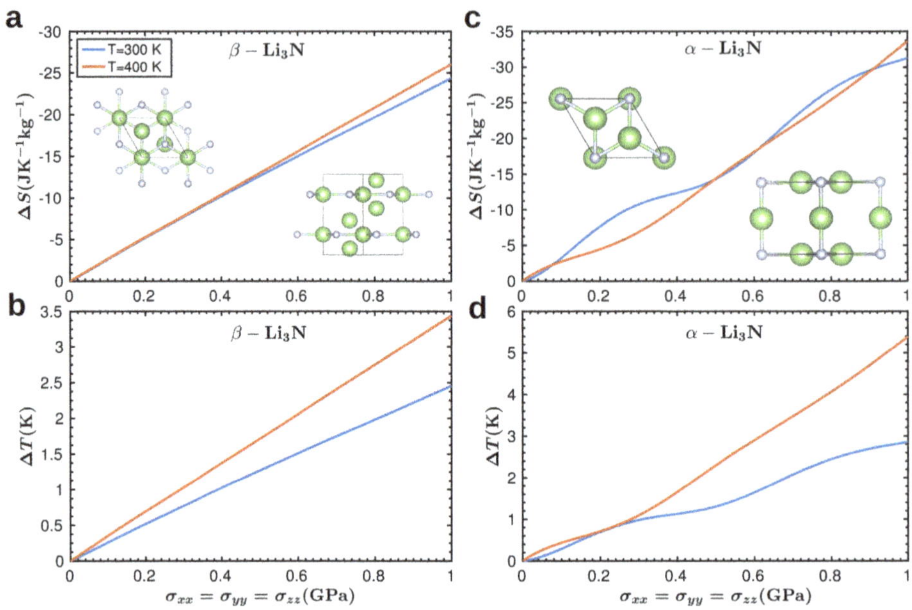

Figure 8.2. BC effects in bulk Li_3N estimated with classical molecular dynamics simulations. (a) Isothermal entropy and (b) adiabatic temperature shifts estimated for β-Li_3N. (c) Isothermal entropy and (d) adiabatic temperature shifts estimated for α-Li_3N. Ball-stick representation of the two analyzed Li_3N polymorphs; Li and N ions are represented with green and blue coloured spheres. Adapted from [10] CC BY 4.0.

temperature and isothermal entropy were slightly superior (e.g., $\Delta S = -32$ J K^{-1} kg^{-1} and $\Delta T = +2.8$ K at $T = 300$ K for a pressure shift of 1 GPa).

It is worth noting that the large BC effects disclosed in Li_3N were not driven by any structural phase transformation. This claim was confirmed, for instance, by the radial distribution functions calculated for all pairs of ionic species in Li_3N at $T = 300$ K since they rendered almost identical structural traits independently of the applied pressure [10]. The absence of an underlying phase transformation is in stark contrast with the usual mechanisms observed for other BC materials in which most of the caloric response is concentrated around the phase transition points [2]. Which are then causes of the giant ΔS values disclosed for Li_3N? It is now well-established that hydrostatic compression may significantly deplete the ionic mobility and thus the entropy of the high-T disordered phase [34, 87, 90, 91]. Fundamentally, the volume that is available to the mobile ions is effectively reduced at high pressures and consequently the diffusion energy barriers and defects formation energies governing ion migration are drastically raised. As a result, giant entropy changes can be attained in superionic crystals under increasing compression without the need of a structural phase transition happening (which in practice may be beneficial for avoiding materials fatigue issues).

Thus, work [10] introduces an original and rational strategy for achieving large (and possibly also reversible) BC effects at ambient temperature based on FICs, which was theoretically demonstrated for bulk Li_3N. Instead of focusing on the triggering of a structural phase transition, which only serendipitously will take place around room temperature, the starting point in this new approach is a material that is superionic at normal conditions. Large BC effects then can be attained through the application of hydrostatic compression based on the fact that it significantly and sustainedly influences the ionic conductivity, and in turn the entropy and volume, of solid-state electrolytes.

8.5.2 Complex hydride $Li_2B_{12}H_{12}$

In recent years, colossal BC effects (i.e., $|\Delta S| \sim 100$ J K^{-1} kg^{-1}) have been experimentally measured in two different families of materials displaying order–disorder phase transitions [3, 82, 84]. First, giant BC effects were theoretically predicted [9] and experimentally observed in the archetypal FIC AgI [3]. This compound displays a first-order normal (low-temperature and low-entropy) to superionic (high-temperature and high-entropy) phase transition that can be induced by either temperature or pressure and which entails the appearance of highly mobile Ag cations in the high-temperature superionic state [87]. In the same way, the entropy changes reported for other normal to superionic phase transitions are in general also sizeable [4, 10, 11]. And second, colossal BC effects have been found in the molecular solid neopentylglycol [82, 84], $(CH_3)_2C(CH_2OH)_2$, and other plastic crystals [85]. In such molecular solids, the molecules rotate around their center of mass, which remain fluctuating around well-defined equilibrium lattice positions. These molecular rotations lead to orientational disorder, thus rendering high conformational entropy. By applying hydrostatic compression, it is possible to alter (i.e., partially or fully frustrate) the molecular reorientational motion thus inducing a partial or fully ordered state with a much lower entropy than the high-temperature phase [96]. The described order–disordered phase transition typically is of first-order type (i.e., is accompanied by a large relative change in volume) and the BC effects resulting from it are huge in size and practically comparable to those attained in conventional coolers that employ greenhouse gases [82, 84, 85].

The recent theoretical work [12], reports a prediction for the occurrence of colossal BC effects (that is, $|\Delta S| \sim 100$ J K^{-1} kg^{-1}) in the energy compound $Li_2B_{12}H_{12}$ (LBH), which is a complex hydride already known from the fields of solid-state batteries [100, 101] and hydrogen storage [97–99]. By relying on classical molecular dynamics simulations [102], a pressure-induced isothermal entropy shift of $|\Delta S| = 367$ J K^{-1} kg^{-1} and an adiabatic temperature change of $|\Delta T| = 43$ K were identified at a temperature of 480 K. Such remarkable entropy and temperature span values were obtained under relatively small hydrostatic pressure shifts of ~0.1 GPa (figure 8.3), thus rendering gigantic BC strengths of $|\Delta S|/P \sim 10^3$ J K^{-1} kg^{-1} GPa^{-1} and $|\Delta T|/P \sim 10^2$ K GPa^{-1}. The colossal BC effects disclosed in bulk LBH were shown to be originated by simultaneous P-induced frustration of Li^+ diffusion and $(BH)_{12}^{-2}$ icosahedra reorientational motion.

Under ambient conditions, $Li_2B_{12}H_{12}$ (LBH), namely, lithium dodecahydrododecaborate, displays an ordered phase with cubic symmetry ($Pa\bar{3}$, $Z = 4$), referred to as α, that is characterized by lithium cations residing on near-trigonal-planar sites

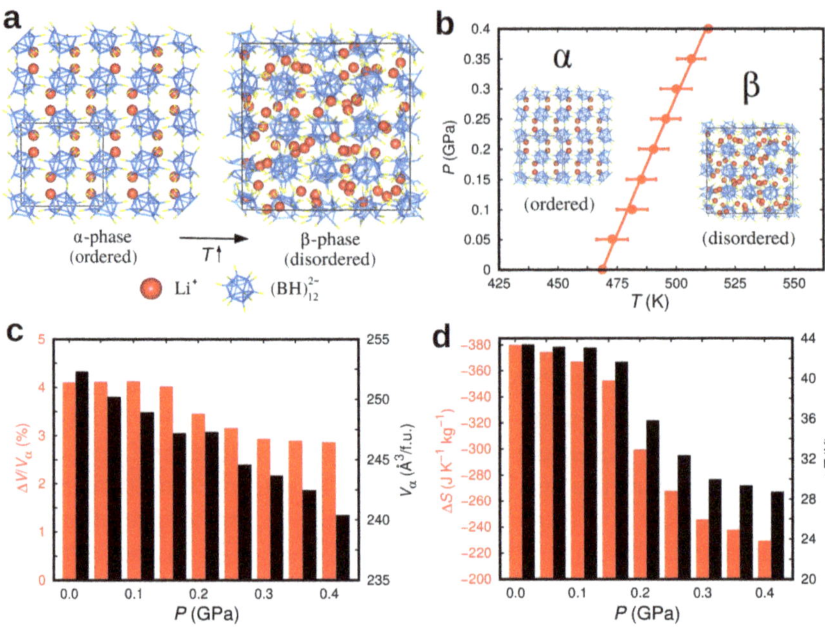

Figure 8.3. (a) Ball-stick representation of the ordered low-T and disordered high-T phases of $Li_2B_{12}H_{12}$. Li, B, and H ions are represented with red, blue, and yellow spheres, respectively. (b) Estimated P–T phase diagram of $Li_2B_{12}H_{12}$. (c) Pressure-dependence of the relative volume change for the T-induced $\alpha \rightarrow \beta$ phase transition; volume per formula unit (f.u.) of the α phase, V_α, at phase-transition points (black). (d) BC isothermal entropy (red) and adiabatic temperature (black) shifts of bulk $Li_2B_{12}H_{12}$ as a function of pressure. Adapted from [12] CC BY 4.0.

surrounded by several $(BH)_{12}^{-2}$ icosahedron anions. At the same time, each $(BH)_{12}^{-2}$ anion is located in an octahedral cage surrounded by six Li^+ cations (figure 8.3(a)). At temperatures of approximately 600 K, a symmetry preserving order–disorder phase transition takes place that gives rise to a disordered state, referred to as β, in which the lithium ions are highly mobile and the $(BH)_{12}^{-2}$ anions reorient around their centers of mass (figure 8.3(a)). The experimental volume expansion ascribed to such a first-order order–disorder phase transition amounts to $\Delta V^{\mathrm{expt}}/V_\alpha^{\mathrm{expt}} \approx 8\%$ [100] and the accompanying phase transition enthalpy is equal to $\Delta H^{\mathrm{expt}} \approx 130$ kJ kg^{-1} [103]. These large thermodynamic descriptor variations associated with the $\alpha \leftrightarrow \beta$ transformation could be suitable for BC applications if the underlying phase transition could be driven by relatively small hydrostatic pressure shifts of the order of 0.1 GPa. Figure 8.3(b) shows the P–T phase diagram of bulk LBH obtained with classical MD simulation techniques [102]. The phase boundary between the α and β phases was determined by performing numerous MD simulations at small P–T shifts of 0.05 GPa and 12.5 K, and monitoring the volume, lithium cation diffusion, and $(BH)_{12}^{-2}$ reorientational properties of the system. In analogy to the experiments, a point in the α–β coexistence line of LBH shown in figure 8.3(b) was identified with sharp and coincident variations in the volume (figure 8.3(c)), Li^+ diffusion coefficient (D_{Li}), and $(BH)_{12}^{2-}$ reorientational frequency ($\lambda_{B_{12}H_{12}}$). Overall, it was found that the critical temperature of the $\alpha \leftrightarrow \beta$ phase transition could certainly be displaced by hydrostatic pressure (i.e., approximately by 13 K under a pressure shift of 0.1 GPa).

In view of these promising results, the BC ΔS and ΔT shifts induced by compressions of $0 \leqslant P \leqslant 0.4$ GPa were theoretically determined for bulk LBH in work [12]. Owing to the fact that the nature of the $\alpha \leftrightarrow \beta$ phase transition is of first-order type [100], the Clausius–Clapeyron method was used for this end (section 8.4.1). The estimated ΔS and ΔT shown in figure 8.3(d) certainly render colossal BC effects. At $T = 480$ K and $P = 0.1$ GPa, for example, an isothermal entropy change of -367 J K^{-1} kg^{-1} and an adiabatic temperature change of +43 K were estimated. The size of ΔS and ΔT shifts were found to steadily decrease under increasing pressure (e.g., at $T = 515$ K and $P = 0.4$ GPa, the same quantities were found to amount to -229 J K^{-1} kg^{-1} and +28 K). It is worth noting that the BC effects predicted for LBH were direct in all the cases, that is, $\Delta T > 0$ under the application of pressure, since the low-entropy ordered phase (α) was stabilized over the high-entropy disordered phase (β) by compression ($\Delta S < 0$).

The phase transition underlying the colossal BC effects enclosed in figure 8.3 is quite remarkable since it combines defining features of FICs (ionic diffusion) and molecular crystals (reorientational motion), families of materials that individually have individually been shown to hold very promising BC potential. Therefore, alkali-metal complex borohydrides with general chemical formula $A_2B_{12}H_{12}$ (A = Li, Na, K, Cs) [104, 105]) conform to a new family of BC materials in which the salient phase-transition characteristics of solid-state electrolytes and molecular plastic crystals coexist, widening in this way the range of BC compounds with potential for realizing solid-state refrigeration applications.

8.5.3 Supertetragonal oxide perovskite $BiFe_{1-x}Co_xO_3$

Super-tetragonal ($\mathcal{T}$) perovskite oxides comprise a family of materials that are very interesting from a fundamental point of view and simultaneously exhibit great potential for ferroelectric, sensor, and energy conversion applications [106–108]. Super-tetragonal phases are characterized by giant electric polarizations of the order of 100 μC cm^{-2}, which in some cases are also accompanied by magnetism. The coexistence of ferroelectricity and magnetism in solids, known as multiferroics, opens prospects for controlling the magnetization via electric fields by exploiting their cross-order magnetoelectric coupling. Magnetoelectric couplings may be used, for example, to engineer ultra efficient logic and memory devices and also miniaturized antennas and sensors. Moreover, phase transitions involving $\mathcal{T}$ phases in general exhibit giant relative changes in volume of the order of 10% (e.g., $PbVO_3$ and related solid solutions), which could be utilized in mechanical degradation applications [109, 110]. Examples of $\mathcal{T}$ multiferroic materials include bulk $BiCoO_3$ (BCO) and $BiFeO_3$ (BFO) thin films [13, 111, 112].

Recently, it has been demonstrated with first-principles simulations based on DFT methods that $BiFe_{1-x}Co_xO_3$ solid solutions (BFCO) with $0.25 \leqslant x \leqslant 0.50$ are ideal magnetoelectric systems in which the full potential of multiferroic $\mathcal{T}$ phases could be realized [113]. In particular, it has been found that for moderate hydrostatic compressions of $0.1 \lesssim P \lesssim 1$ GPa (depending on the relative Fe/Co content) it is actually possible to drive a phase transition from a low-T rhombohedral ($\mathcal{R}$) phase to a high-T $\mathcal{T}$ phase (figures 8.4(a)–(c)) near room temperature. The discovered pressure-induced $\mathcal{R} \rightarrow \mathcal{T}$ phase transition entails (i) a huge electric polarization increase of ~150%, (ii) the existence of robust magnetization with a net magnetic moment of ≈ 0.13 μ_B per formula unit, and (iii) an enormous volume increase of $\Delta V/V \sim 10\%$ (figure 8.4(d)). Examples of technologies that could capitalize on such multifunctional phenomena include pyroelectric energy harvesting [114, 115] and solid-state refrigeration [116]. Figures 8.4(d) and (e) show, respectively, the isothermal entropy and adiabatic temperature variations induced by hydrostatic pressure in $BiFe_{0.5}Co_{0.5}O_3$ solid solutions. Those results have been obtained with the quasi-harmonic DFT method (section 8.3.1) and the Clausius–Clapeyron indirect estimation approach (section 8.4.1) [117]. First, it is noted that the $\mathcal{R} \rightarrow \mathcal{T}$ transition temperature can be enormously shifted by small pressure changes (figure 8.4(c)). For instance, in $BiFe_{0.5}Co_{0.5}O_3$ a compression decrease of only 0.06 GPa moves the transition temperature from room temperature down to about 200 K. Second, an isothermal entropy change of approximately -1.5 J $K^{-1}mol^{-1}$ (equivalent to ≈ -4.8 J K^{-1} kg^{-1}) is estimated at room temperature along with a ΔT of ≈ 5 K. These figures, although cannot be regarded as 'giant', are reasonably large and promising. And third, the size of these two BC descriptors significantly increase at lower temperatures. At $T = 200$ K, for instance, ΔS increases up to -2.8 J $K^{-1}mol^{-1}$ and ΔT to 6.6 K.

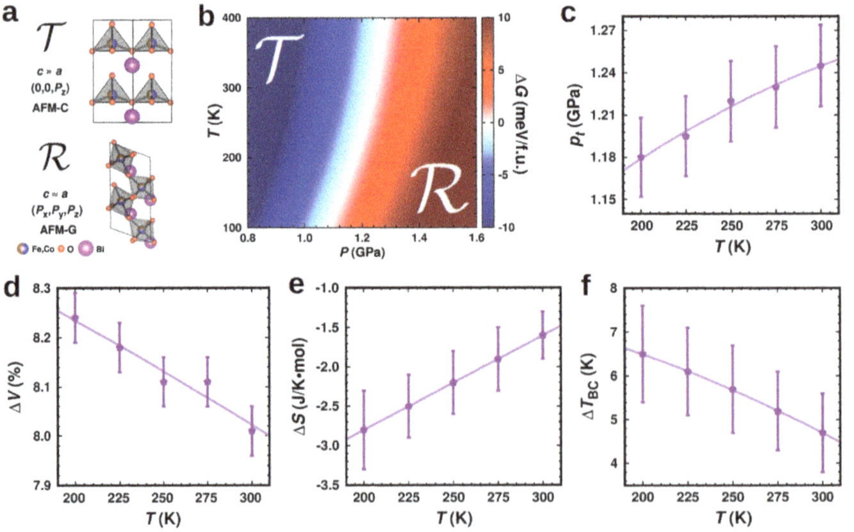

Figure 8.4. (a) Description of the two energetically competing phases in bulk BFCO under pressure. (b) Gibbs free energy difference between the $\mathcal{T}$ and $\mathcal{R}$ phases of bulk BFCO expressed as a function of temperature and pressure. (c) Phase boundary delimiting the stability region of the $\mathcal{T}$ and $\mathcal{R}$ phases for bulk BFCO. (d) Relative volume change occurring during the pressure-induced $\mathcal{T} \rightarrow \mathcal{R}$ phase transformation expressed as a function of temperature. (e) Isothermal entropy and (f) adiabatic temperature changes estimated for the pressure-induced $\mathcal{T} \rightarrow \mathcal{R}$ phase transition in BFCO expressed as a function of temperature.

The driving pressures reported in figure 8.4(d) are very large, of the order of 1 GPa; however, in practice those values can be drastically reduced (i.e., by one order of magnitude, ~0.1 GPa) by increasing the relative content of Fe ions. For instance, by using analogous first-principles computational methods a $\mathcal{R} \rightarrow \mathcal{T}$ phase transition pressure of 0.2 GPa is estimated at zero temperature for bulk $BiFe_{0.75}Co_{0.25}O_3$ [117]. This theoretical outcome is in good agreement with the available experimental data on the compositional phase diagram of BFCO solid solutions [118]. Another interesting aspect of BFCO solid solutions is that, besides hydrostatic pressure, they can react to external magnetic and electric fields since these are simultaneously magnetic and ferroelectric [113]. Such a multifunctional quality converts BFCO solid solutions into a potentially ideal playground in which to explore and take advantage of novel multicaloric effects and cycles [1, 116]. Therefore, BFCO solid solutions in particular and multiferroic materials exhibiting super-tetragonal phases in general, appear to be very promising materials for the engineering of advanced solid-state cooling applications.

8.6 Conclusions

Atomistic computational methods for the simulation of BC effects are already well established. The simulation methods reviewed in this chapter (e.g., quasi-harmonic and thermodynamic integration techniques) were originally developed for the analysis and prediction of temperature- and pressure-induced phase transitions in materials, and have already achieved great success in many scientific disciplines (e.g., high-pressure physics). Nonetheless, the number of atomistic simulation studies found in the literature on BC effects currently is quite reduced and mostly limited to indirect estimation approaches. One of the main objectives of this chapter is to improve this situation by surveying key theoretical and computational concepts in the estimation of BC effects and to promote the use of alternative and highly accurate direct and quasi-direct simulation methods.

As illustrated in this chapter, atomistic simulation approaches can be employed to reproduce with accuracy the BC performance of complex multifunctional materials like multiferroics, in which the polar, structural and magnetic order parameters may be strongly coupled, molecular crystals, in which the anionic molecules can be orientationally disordered, and FICs, in which ion transport affects most of their other physical properties. Analogous BC simulation success may be achieved also in other families of complex materials exhibiting unconventional, as well as standard, responses to hydrostatic compression.

Atomistic simulation of BC effects, therefore, can help enormously in engineering new materials and strategies for improving solid-state refrigeration based on the application of hydrostatic pressure. Thus, the present critical challenge of finding new cooling technologies that are environmentally sustainable and friendly may significantly benefit from the physically insightful atomistic computational methods surveyed in this chapter.

References

[1] Cazorla C 2019 Novel mechanocaloric materials for solid-state cooling applications *Appl. Phys. Rev.* **6** 041316

[2] Moya X, Kar-Narayan S and Mathur N D 2014 Caloric materials near ferroic phase transitions *Nat. Mater.* **13** 439

[3] Aznar A *et al* 2017 Giant barocaloric effects over a wide temperature range in superionic conductor AgI *Nat. Commun.* **8** 1851

[4] Cazorla C and Errandonea D 2016 Giant mechanocaloric effects in fluorite-Structured superionic materials *Nano Lett.* **16** 3124

[5] Scott J F 2011 Electrocaloric materials *Annu. Rev. Mater. Res.* **41** 229

[6] Pecharsky V K and Gschneidner K A Jr 1997 Giant magnetocaloric effect in $Gd_5(Si_2Ge_2)$ *Phys. Rev. Lett.* **78** 4494

[7] Bonnot E, Romero R, Mañosa L, Vives E and Planes A 2008 Elastocaloric effect associated with the martensitic transition in shape-memory alloys *Phys. Rev. Lett.* **100** 125901

[8] Shirsath S E *et al* 2020 Interface-charge induced giant electrocaloric effect in lead free rerroelectric thin-film bilayers *Nano Lett.* **20** 1262

[9] Sagotra A K, Errandonea D and Cazorla C 2017 Mechanocaloric effects in superionic thin films from atomistic simulations *Nat. Commun.* **8** 963

[10] Sagotra A K, Chu D and Cazorla C 2018 Room-temperature mechanocaloric effects in lithium-based superionic materials *Nat. Commun.* **9** 3337

[11] Min J, Sagotra A K and Cazorla C 2020 Large barocaloric effects in thermoelectric superionic materials *Phys. Rev. Mater.* **4** 015403

[12] Sau K, Ikeshoji T, Takagi S, Orimo S-I, Errandonea D, Chu D and Cazorla C 2021 Colossal barocaloric effects in the complex hydride $Li_2B_{12}H_{12}$ *Sci. Rep.* **11** 11915

[13] Menéndez C, Chu D and Cazorla C 2020 Oxygen-vacancy induced magnetic phase transitions in multiferroic thin films *npj Comput. Mater.* **6** 76

[14] Machado P *et al* 2021 Enhancement of phase stability and optoelectronic performance of $BiFeO_3$ thin films via cation co-substitution *J. Mater. Chem.* C **9** 330

[15] Kohn W and Sham L J 1965 Self-consistent equations including exchange and correlation effects *Phys. Rev.* **140** A1133

[16] Sham L J and Kohn W 1966 One-particle properties of an inhomogeneous interacting electron gas *Phys. Rev.* **145** 561

[17] Ceperley D M and Alder B J 1980 Ground state of the electron gas by a stochastic method *Phys. Rev. Lett.* **45** 566

[18] Perdew J P and Zunger A 1981 Self-interaction correction to density-functional approximations for many-electron systems *Phys. Rev.* B **23** 5048

[19] Perdew J P, Chevary J A, Vosko S H, Jackson K A, Pederson M R, Singh D J and Fiolhais C 1992 Atoms, molecules, solids, and surfaces: applications of the generalized gradient approximation for exchange and correlation *Phys. Rev.* B **46** 6671

[20] Perdew J P, Burke K and Ernzerhof M 1996 Generalized gradient approximation made simple *Phys. Rev. Lett.* **77** 3865

[21] Sun J, Ruzsinszky A and Perdew J P 2015 Strongly constrained and appropriately normed semilocal density functional *Phys. Rev. Lett.* **115** 036402

[22] Zhang Y *et al* 2018 Efficient first-principles prediction of solid stability: Towards chemical accuracy *npj Comput. Mater.* **4** 9

[23] Becke A D 1993 Density-functional thermochemistry. III. The role of exact exchange *J. Chem. Phys.* **98** 5648

[24] Adamo C and Barone V 1999 Toward reliable density functional methods without adjustable parameters: the PBE0 model *J. Chem. Phys.* **110** 6158

[25] Krukau A V, Vydrov O A, Izmaylov A F and Scuseria G E 2006 Influence of the exchange screening parameter on the performance of screened hybrid functionals *J. Chem. Phys.* **125** 224106

[26] Franchini C 2014 Hybrid functionals applied to perovskites *J. Phys.: Condens. Matter* **26** 253202

[27] Bilc D I, Orlando R, Shaltaf R, Rignanese G-M, Íñiguez J and Ghosez Ph 2008 Hybrid exchange-correlation functional for accurate prediction of the electronic and structural properties of ferroelectric oxides *Phys. Rev.* B **77** 165107

[28] Rivero P and Cazorla C 2016 Revisiting the zero-temperature phase diagram of stoichiometric $SrCoO_3$ with first-principles methods *Phys. Chem. Chem. Phys.* **18** 30686

[29] Cazorla C 2017 Lattice effects on the formation of oxygen vacancies in perovskite thin films *Phys. Rev. Appl.* **7** 044025

[30] Ercolessi F and Adams J B 1994 Interatomic potentials from first-principles calculations: the force-matching method *Europhys. Lett.* **26** 583

[31] Cazorla C and Boronat J 2015 First-principles modeling of three-body interactions in highly compressed solid helium *Phys. Rev.* B **92** 224113

[32] Mattoni A, Filipetti A, Saba M I and Delugas P 2015 Methylammonium rotational dynamics in lead halide perovskite by classical molecular dynamics: the role of temperature *J. Phys. Chem.* C **119** 17421

[33] Hata T, Giorgi G, Yamashita K, Caddeo C and Mattoni A 2017 Development of a classical interatomic potential for $MAPbBr_3$ *J. Phys. Chem.* C **121** 3724

[34] Cazorla C, Sagotra A K, King M and Errandonea D 2018 High-pressure phase diagram and superionicity of alkaline earth metal difluorides *J. Phys. Chem.* C **122** 1267

[35] Tinte S, Stachiotti M G, Sepliarsky M, Migoni R L and Rodriguez C O 1999 Atomistic modelling of $BaTiO_3$ based on first-principles calculations *J. Phys.: Condens. Matter* **11** 9679

[36] Jackson R A and Valerio M E G 2005 Structural phase transition and dynamical properties of $PbTiO_3$ simulated by molecular dynamics *J. Phys.: Condens. Matter* **17** 837

[37] Sepliarsky M, Phillpot S R, Wolf D, Stachiotti M G and Migoni R L 2001 Ferroelectric properties of $KNbO_3/KTaO_3$ superlattices by atomic-level simulation *J. Appl. Phys.* **90** 4509

[38] Hashimoto T and Moriwake H 2015 Dielectric properties of $BaTiO_3$ by molecular dynamics simulations using a shell model *Mol. Simul.* **41** 1074

[39] Gambuzzi E and Pedone A 2014 On the structure of Ce-containing silicophosphate glasses: a core-shell molecular dynamics investigation *Phys. Chem. Chem. Phys.* **16** 21645

[40] Cohen R E R E 1992 Origin of ferroelectricity in perovskite oxides *Nature* **358** 136

[41] Haomin C and Adams S 2017 Bond softness sensitive bond-valence parameters for crystal structure plausibility tests *Int. Union Crystallogr. J.* **4** 614

[42] Taioli S, Cazorla C, Gillan M J and Alfè D 2007 Melting curve of tantalum from first principles *Phys. Rev.* B **75** 214103

[43] Cazorla C, Gillan M J, Taioli S and Alfè D 2007 Ab initio melting curve of molybdenum by the phase coexistence method *J. Chem. Phys.* **126** 194502

[44] Grinberg I, Cooper V R and Rappe A M 2002 Relationship between local structure and phase transitions of a disordered solid solution *Nature* **419** 909

[45] Grinberg I, Cooper V R and Rappe A M 2004 Oxide chemistry and local structure of $PbZr_{1-x}Ti_{1-x}O_3$ studied by density-functional theory supercell calculations *Phys. Rev.* B **69** 144118

[46] Shin Y-H, Cooper V R, Grinberg I and Rappe A M 2005 Development of a bond-valence molecular-dynamics model for complex oxides *Phys. Rev.* B **71** 054104

[47] Liu S, Grinberg I, Takenaka H and Rappe A M 2013 Reinterpretation of the bond-valence model with bond-order formalism: an improved bond-valence-based interatomic potential for $PbTiO_3$ *Phys. Rev.* B **88** 104102

[48] Xu R, Liu S, Grinberg I, Karthik J, Damodaran A R, Rappe A M and Martin L W 2015 Ferroelectric polarization reversal via successive ferroelastic transitions *Nat. Mater.* **14** 79

[49] Sau K, Ikeshoji T, Kim S, Takagi S and Orimo S-I 2021 Comparative molecular dynamics study of the roles of anion–cation and cation–cation correlation in cation diffusion in $Li_2B_{12}H_{12}$ and $LiCB_{11}H_{12}$ *Chem. Mater.* **33** 2357

[50] Qi Y, Liu S, Lindenberg A M and Rappe A M 2015 Ultrafast electric field pulse control of giant temperature change in ferroelectrics *Phys. Rev. Lett.* **120** 055901

[51] Liu S and Cohen R E 2016 Response of methylammonium lead iodide to external stimuli and caloric effects from molecular dynamics simulations *J. Phys. Chem.* C **120** 17274

[52] Behler J 2010 Neural network potential-energy surfaces for atomistic simulations *Chem. Modell.* **7** 1

[53] Rupp M 2015 Machine learning for quantum mechanics in a nutshell *Int. J. Quantum Chem.* **115** 1058

[54] Batzner S *et al* 2022 E(3)-equivariant graph neural networks for data-efficient and accurate interatomic potentials *Nat. Commun.* **13** 2453

[55] Behler J 2015 Constructing high-dimensional neural network potentials: a tutorial review *Int. J. Quantum Chem.* **115** 1032

[56] Deringer V L, Caro M A and Csanyi G 2019 Machine learning interatomic potentials as emerging tools for materials science *Adv. Mater.* **31** 1902765

[57] Zhang Y and Xu X 2020 Machine learning the magnetocaloric effect in manganites from compositions and structural parameters *AIP Adv.* **10** 035220

[58] Gong J, Chu S, Mehta R K and McGaughey A J H 2022 XGBoost model for electrocaloric temperature change prediction in ceramics *npj Comput. Mater.* **8** 140

[59] Cazorla C and Boronat J 2017 Simulation and understanding of atomic and molecular quantum crystals *Rev. Mod. Phys.* **89** 035003

[60] Baroni S, Giannozzi P and Isaev E 2010 Density-functional perturbation theory for quasi-harmonic calculations *Rev. Mineral. Geochem.* **71** 39

[61] Cazorla C, Errandonea D and Sola E 2009 High-pressure phases, vibrational properties, and electronic structure of $Ne(He)_2$ and $Ar(He)_2$: a first-principles study *Phys. Rev.* B **80** 064105

[62] Cazorla C, Alfè D and Gillan M J 2012 Constraints on the phase diagram of molybdenum from first-principles free-energy calculations *Phys. Rev.* B **85** 064113

[63] Tadano T and Tsuneyuki S 2015 Self-consistent phonon calculations of lattice dynamical properties in cubic $SrTiO_3$ with first-principles anharmonic force constants *Phys. Rev.* B **92** 054301

[64] Souvatzis P, Eriksson O, Katsnelson M I and Rudin S P 2008 Entropy driven stabilization of energetically unstable crystal structures explained from first principles theory *Phys. Rev. Lett.* **100** 095901

[65] Souvatzis P 2011 Phonon lifetimes from first-principles self-consistent lattice dynamics *J. Phys.: Condens. Matter* **23** 445401

[66] Monacelli L, Bianco R, Cherubini M, Calandra M, Errea I and Mauri F 2021 The stochastic self-consistent harmonic approximation: calculating vibrational properties of materials with full quantum and anharmonic effects *J. Phys.: Condens. Matter* **33** 363001

[67] Gale J D 1997 GULP: a computer program for the symmetry-adapted simulation of solids *J. Chem. Soc. Faraday Trans.* **93** 629

[68] Plimpton S 1995 Fast parallel algorithms for short-range molecular dynamics *J. Comput. Phys.* **117** 1

[69] Frenkel D 1996 *Understanding Molecular Simulation: from algorithms to Applications* (San Diego, CA: Academic)

[70] Carrera A, Togo A and Tanaka I 2017 DynaPhoPy: a code for extracting phonon quasiparticles from molecular dynamics simulations *Comput. Phys. Commun.* **221** 221

[71] Eriksson F, Fransson E and Erhart P 2019 The hiphive package for the extraction of high-order force constants by machine learning *Adv. Theory Simul.* **2** 1800184

[72] Hellman O, Steneteg P, Abrikosov I A and Simak S I 2013 Temperature dependent effective potential method for accurate free energy calculations of solids *Phys. Rev.* B **87** 104111

[73] Nakata A *et al* 2020 Large scale and linear scaling DFT with the CONQUEST code *J. Chem. Phys.* **152** 164112

[74] Sagotra A K, Chu D and Cazorla C 2019 Influence of lattice dynamics on lithium-ion conductivity: a first-principles study *Phys. Rev. Mater.* **3** 035405

[75] Miyazaki Y, Nakano M, Krivchikov A I, Koroyuk O A, Gebbia J F, Cazorla C and Tamarit J L 2021 Low-temperature heat capacity anomalies in ordered and disordered phases of normal and deuterated thiophene *J. Phys. Chem. Lett.* **12** 2112

[76] Szewczyk D, Gebbia J F, Jezowski A, Krivchikov A I, Guldi T, Cazorla C and Tamarit J-L 2021 Heat capacity anomalies of the molecular crystal 1-fluoro-adamantane at low temperatures *Sci. Rep.* **11** 18640

[77] Polek A, Cazorla C and Kundu D 2020 Nature of alkali ion conduction and reversible Na-ion storage in hybrid formate framework materials *J. Phys. Chem.* C **124** 26714

[78] Luo X, Rawal A, Cazorla C and Aguey-Zinsou K-F 2020 Facile self-forming superionic conductors based on complex borohydrides surface oxidation *Adv. Sustain. Syst.* **4** 1900113

[79] Jinnouchi R, Lahnsteiner J, Karsai F, Kresse G and Bokdam M 2019 Phase transitions of hybrid perovskites simulated by machine-learning force fields trained on the fly with bayesian inference *Phys. Rev. Lett.* **122** 225701

[80] Togo A and Tanaka I 2015 First principles phonon calculations in materials science *Scr. Mater.* **108** 1

[81] Cazorla C, Alfé D and Gillan M J 2009 Melting properties of a simple tight-binding model of transition metals I. The region of half-filled *d*-band *J. Chem. Phys.* **130** 174707

[82] Li B *et al* 2019 Colossal barocaloric effects in plastic crystals *Nature* **567** 506

[83] Bermúdez-García J M, Sánchez-Andújar M and Señarís-Rodríguez M A 2017 A new playground for organic–inorganic hybrids: Barocaloric materials for pressure-induced solid-state cooling *J. Phys. Chem. Lett.* **8** 4419

[84] Lloveras P *et al* 2019 Colossal barocaloric effects near room temperature in plastic crystals of neopentylglycol *Nat. Commun.* **10** 1803

[85] Aznar A *et al* 2020 Reversible and irreversible colossal barocaloric effects in plastic crystals *J. Mater. Chem.* A **8** 639

[86] Mañosa Ll Planes A 2017 Materials with giant mechanocaloric effects: cooling by strength *Adv. Mater.* **29** 1603607

[87] Hull S 2004 Superionics: crystal structures and conduction processes *Rep. Prog. Phys.* **67** 1233

[88] Wang Y, Richards W D, Ong S P, Miara L J, Kim J C, Mo Y and Ceder G 2015 Design principles for solid-state lithium superionic conductors *Nat. Mater.* **14** 1026

[89] López C, Emperador A, Saucedo E, Rurali R and Cazorla C 2023 Universal ion-transport descriptors and classes of inorganic solid-state electrolytes *Mater. Horiz.* **10** 1757

[90] Cazorla C and Errandonea D 2014 Superionicity and polymorphism in calcium fluoride at high pressure *Phys. Rev. Lett.* **113** 235902

[91] Sagotra A K and Cazorla C 2017 Stress-mediated enhancement of ionic conductivity in fast-ion conductors *ACS Appl. Mater. Interfaces* **9** 38773

[92] Cazorla C 2015 In the search of new electrocaloric materials: Fast ion conductors *Res. Phys.* **5** 262

[93] Li W, Wu G, Araújo M, Scheicher R H, Blomqvist A, Ahuja R, Xiong Z, Feng Y and Chen P 2010 Li_+ ion conductivity and diffusion mechanism in α-Li_3N and β-Li_3N *Energy Environ. Sci.* **3** 1524

[94] Alpen U V, Rabenau A and Talat G H 1977 Ionic conductivity in Li_3N single crystals *Appl. Phys. Lett.* **30** 621

[95] Nazri G A, Julien C and Mavi H S 1994 Structure of Li_3X (X = N, P, As) superionic conductors: X-ray diffraction and FTIR studies *Solid State Ion.* **70** 137

[96] Cazorla C 2019 Refrigeration based on plastic crystals *Nature* **567** 470

[97] Her J-H, Yousufuddin M, Zhou W, Jalisatgi S S, Kulleck J G, Zan J A, Hwang S-J, Bowman R C and Udovic T J 2008 Crystal structure of $Li_2B_{12}H_{12}$: a possible intermediate species in the decomposition of $LiBH_4$ *Inorg. Chem.* **47** 9757

[98] Lai Q, Sun Y, Modi P, Cazorla C, Demirci U B, Ares J R, Leardini F and Aguey-Zinsou K F 2019 How to design hydrogen storage materials? Fundamentals, synthesis, and storage tanks *Adv. Sustain. Syst.* **3** 1900043

[99] Shevlin S A, Cazorla C and Guo Z X 2012 Structure and defect chemistry of low and high temperature phases of $LiBH_4$ *J. Phys. Chem.* C **116** 13488

[100] Paskevicius M, Pitt M P, Brown D H, Sheppard D A, Chumphongphan S and Buckley C E 2013 First-order phase transition in the $Li_2B_{12}H_{12}$ system *Phys. Chem. Chem. Phys.* **15** 15825

[101] Mohtadi R and Orimo S I 2016 The renaissance of hydrides as energy materials *Nat. Rev. Mater.* **2** 16091

[102] Sau K, Ikeshoji T, Kim S, Takagi S, Akgi K and Orimo S-I 2019 Reorientational motion and Li_+ ion transport in $Li_2B_{12}H_{12}$ system: molecular dynamics study *Phys. Rev. Mater.* **3** 075402

[103] Verdal N, Her J-H, Stavila V, Soloninin A V, Babanova O A, Skripov A V, Udovic T J and Rush J J 2014 Complex high-temperature phase transitions in $Li_2B_{12}H_{12}$ and $Na_2B_{12}H_{12}$ *J. Solid State Chem.* **212** 81

[104] Udovic T J, Matsuo M, Unemoto A, Verdal N, Stavila V, Skripov A V, Rush J J, Takamura H and Orimo S-I 2014 Sodium superionic conduction in $Na_2B_{12}H_{12}$ *Chem. Commun.* **50** 3750

[105] Jorgensen M *et al* 2020 Understanding superionic conductivity in lithium and sodium salts of weakly coordinating closo-hexahalocarbaborate anions *Chem. Mater.* **32** 1475

[106] Zhang L *et al* 2018 Giant polarization in super-tetragonal thin films through interphase strain *Science* **361** 494

[107] Yamada H *et al* 2013 Giant electroresistance of super-tetragonal $BiFeO_3$-based ferroelectric tunnel junctions *ACS Nano* **7** 5385

[108] Infante I C *et al* 2011 Multiferroic phase transition near room temperature in $BiFeO_3$ films *Phys. Rev. Lett.* **107** 237601

[109] Yamamoto H, Ogata T, Saki Y and Azuma M 2019 Stability of polar structure in filling-controlled giant tetragonal perovskite oxide $PbVO_3$ *Inorg. Chem.* **58** 2755

[110] Pan Z *et al* 2019 Large negative thermal expansion induced by synergistic effects of ferroelectrostriction and spin crossover in $PbTiO_3$-based perovskites *Chem. Mater.* **31** 1296

[111] Belik A A *et al* 2006 Neutron powder diffraction study on the crystal and magnetic structures of $BiCoO_3$ *Chem. Mater.* **18** 798

[112] Wang J *et al* 2003 Epitaxial $BiFeO_3$ multiferroic thin film heterostructures *Science* **299** 1719

[113] Menéndez C and Cazorla C 2020 Giant thermal enhancement of the electric polarization in ferrimagnetic $BiFe_{1-x}Co_xO_3$ solid solutions near room temperature *Phys. Rev. Lett.* **125** 117601

[114] Bowen C R, Taylor J, LeBoulbar E, Zabek D, Chauhan A and Vaish R 2014 Pyroelectric materials and devices for energy harvesting applications *Energy Environ. Sci.* **7** 3836

[115] Hoffmann M, Schroeder U, Künneth C, Kersch A, Starschich S, Böttger U and Mikolajick T 2015 Ferroelectric phase transitions in nanoscale HfO_2 films enable giant pyroelectric energy conversion and highly efficient supercapacitors *Nano Energy* **18** 154

[116] Gottschall T *et al* 2018 A multicaloric cooling cycle that exploits thermal hysteresis *Nat. Mater.* **17** 929

[117] Menéndez C and Cazorla C 2022 Colossal room-temperature electrocaloric strength aided by hydrostatic pressure in multiferroic $BiFe_xCo_{1-x}O_3$ solid solutions (To be published) arXiv:2302.07479

[118] Hojo H, Kengo O, Keisuke S, Yamamoto H, Kawabe R and Azuma M 2018 Development of bismuth ferrite as a piezoelectric and multiferroic material by cobalt substitution *Adv. Mater.* **30** 1705665

Chapter 9

The use of the barocaloric effect for refrigeration: an evaluation of the energy performances and the environmental impact

Luca Cirillo, Adriana Greco and Claudia Masselli

9.1 Introduction

Cooling, ventilating and air conditioning systems based on consolidated vapour compression (VC) have an impact upon the environment, contributing to ozone depletion and global warming. Indeed, commonly used refrigerants in VC are in vapour phase and cause harmful ozone disruption, i.e. they have an ozone depletion potential (ODP) or they contribute to global warming, i.e. they show an elevated global warming potential (GWP). Because of their ODP, CFC and HCFC refrigerants were phased out in Europe following the Montreal Protocol [1]. HFCs are the chronologically immediate substitute but they show very big GWP indexes. Consequently they are being phased out, according to the dispositions of the Kyoto Protocol [2]. The new replacing substances can be natural fluids, such as HFO. Traditional refrigeration, therefore, in addition to the problem of the environmental impact of fluids, has now reached its technological limit. The scientific community is therefore looking for alternative and eco-friendly refrigeration solutions. Alternative non-VC technologies fall into the class of non-conventional technologies or refrigeration. This class of technologies includes those based on solid-state refrigerants with caloric effect. The advantages of caloric cooling are certainly the reduction of the direct greenhouse contribution with refrigerants that are solid-state, i.e. with ODP = 0; GWP = 0. In addition, these technologies have shown a potential to achieve a lower indirect greenhouse effect linked to energy efficiencies greater than 50%–60% over VC.

The common element of caloric cooling is the observation of a temperature change in an adiabatic process (ΔT_{ad}) provoked by variation in the intensity of driving external fields [3, 4]. A magnetic field causes a magnetocaloric effect (MCE) [5–7]; an electric one, the electrocaloric effect (ECE) [8–10]; mechanical stress drives

doi:10.1088/978-0-7503-4690-0ch9

for the elastocaloric effect (eCE) [11] or barocaloric effect (BCE) [12], as a consequence of simple loading or loading with variation of the level of hydrostatic pressure. Magnetocaloric cooling has been predominant for 30 years since the beginning of the 1990s [13, 14], so interesting more than 10 000 researchers resulting in 35 000 scientific contributions presented to the scientific community and 100 proofs of concept of experimental magnetocaloric devices realized [15, 16]. More recently, all the other caloric effects have been in the spotlight. Electrocaloric and elastocaloric are branches where altogether 40–50 devices have been developed; barocaloric still lives at an embryonic level [17, 18].

The purpose of this chapter is to analyze the potentiality of barocaloric effect when applied to systems for refrigeration. To reach this goal numerical simulations have been carried out with reference to evaluating the response of a barocaloric system to different targets of cooling.

9.2 The physical phenomenon and the thermodynamic cycles

A barocaloric material (BCM) is characterized by the entropy that is a function both of the temperature and the level of hydrostatic pressure forcing it. It can be split into three aliquots as follows:

$$s = s_l(T) + s_{el}(T) + s_b(T, p) \tag{9.1}$$

the first and the second aliquots are, respectively, due to lattice vibration and electrons disorder. The latter is regulated by the barocaloric effect [19].

If the pressure changes but the temperature is constant, only the third contribution modifies (decrement with growing pressures, and vice versa), i.e. the total entropy increases or reduces (Δs_T) following s_b.

If the pressure changes adiabatically, the total entropy remains constant but s_b reduces if the pressure increases and consequently s_{el} and s_l must increase to compensate, with the result of a temperature rise in the BCM ($+\Delta T_{ad}$). Dually ($-\Delta T_{ad}$) is observed as the answer to an adiabatic reduction of pressure.

Barocaloric effect is defined according to the following relations:

$$\Delta s_T = -\int_{p_i}^{p} \left(\frac{\partial v}{\partial T}\right)_p dp, \tag{9.2}$$

$$\Delta T_{ad} = \int_{p_i}^{p} \frac{T}{c_p}\left(\frac{\partial v}{\partial T}\right)_p dp, \tag{9.3}$$

The above described is conventional barocaloric effect. There exists also the inverse barocaloric effect where a reduction of the pressure generates a positive variation of ΔT_{ad}, an increase of p provokes $\Delta T_{ad} < 0$.

A barocaloric system for cooling is composed by a cell of pressure where the BCM is inserted; two heat reservoirs must thermally communicate with the cell for heat transfer. Many are the thermodynamical cycles where barocaloric can be found.

Inverse Brayton cycle is the most exploited for caloric cooling applications because of the simplicity, since only two different field intensities are required. The Brayton cycle for barocaloric cooling is characterized by four processes shown in figure 9.1(a):

(a) Isentropic increasing of hydrostatic pressure from p_1 to p_2 where the temperature of the barocaloric material undergoes increeases of $+\Delta T_{ad,12}$;
(b) Isobaric heat rejection from the barocaloric system toward the hot heat exchanger (HHEX);
(c) Isentropic decreasing of hydrostatic pressure from p_2 to p_1 where the temperature of the barocaloric material further decreases $+\Delta T_{ad,34}$;
(d) Isobaric heat adduction by the barocaloric system from the cold heat exchanger (CHEX);

The highest coefficient of performance (COP) of the Brayton cycle is always lower than the Carnot one. In figure 9.1(b) is proposed the inverse Brayton cycle including the regeneration where the material has to pass for two more steps: (2a→3) and (4a→1), i.e. extra cooling and heating are guaranteed.

Another widely employed cycle, that easily derives from the inverse Brayton one, is called ABR cycle. ABR is the acronym of active barocaloric regenerative cooling cycle, and it is shown in figure 9.2. As visible from figure 9.2(a), there are four transformations that the system, whose evolution is visible in figure 9.2(b) must compute. There are only two required levels between which the hydrostatic pressure must adiabatically change for BCE manifesting (steps I and III). Steps II and IV are reserved for the passage of the vehiculation medium, also called heat transfer fluid (HTF), between the cold and hot heat reservoirs, whose temperatures are set at T_C and T_H, respectively. Thanks to the active regeneration, played by the BCE material itself, the maximum T jump of the system largely overtakes that of the Brayton cycle without regeneration. In the latter case it is limited by the amplitude of the adiabatic

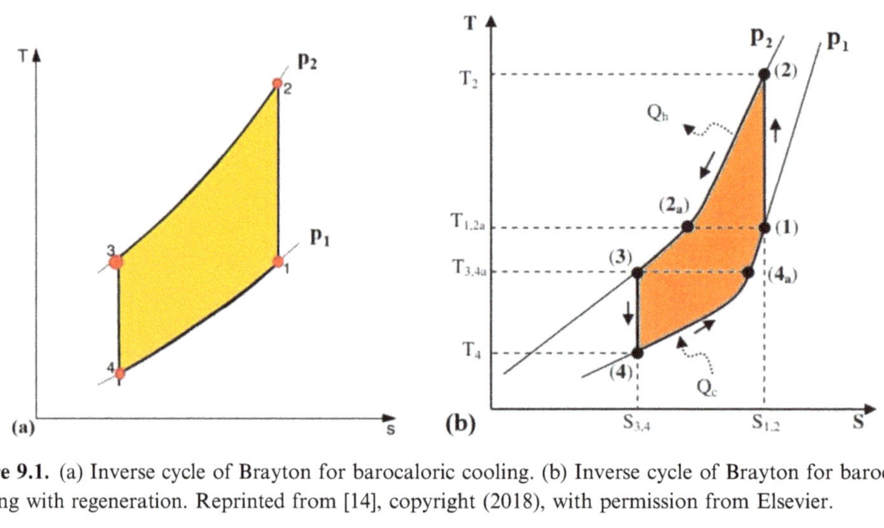

Figure 9.1. (a) Inverse cycle of Brayton for barocaloric cooling. (b) Inverse cycle of Brayton for barocaloric cooling with regeneration. Reprinted from [14], copyright (2018), with permission from Elsevier.

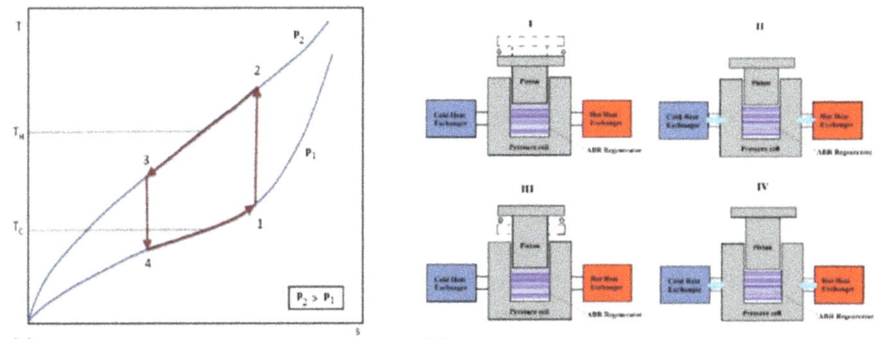

(a) (b)

Figure 9.2. The ABR cycle: (a) transformations on temperature–entropy plane; (b) the evolution of the system. Reprinted from [29], copyright (2020), with permission from Elsevier.

temperature change of the BCM. The hysteresis lying in the BCM negatively affects the adiabaticity of the process, i.e. it reduces the maximum manifestable ΔT_{ad}. The thermal hysteresis also decreases the COP of the system.

9.3 Barocaloric materials

Nature first and scientific research second have provided many barocaloric materials that can prove eligible with respect to the various usages. What is certain is that barocaloric materials must satisfy several conditions:

- remarkable barocaloric effect in terms of elevated ΔT_{ad} under small Δp;
- not toxic and chemically stable;
- competitive in terms of price and in ease in manufacturing;
- not strongly affected by thermal and mechanical hysteresis;
- elevated fatigue-life and durability with respect to the contact with the vehiculation medium.

The barocaloric materials can divided according to their phase transition that can belong to first or second order, i.e. FOT or SOT materials [20]. The difference resides in the presence of the discontinuity: for an FOT it is observable in the first derivative of the Gibbs free energy related to pressure field; in an SOT, the discontinuity is in the second order derivative. Regardless, almost all of the barocaloric materials that transition at room temperature are FOT. FOT materials show thermal and mechanical hysteresis, a change in volume and a delay in the BCE manifesting with respect to the forcing stimulus. FOT is the transition where a giant peak of BCE in terms of adiabatic temperature change can be observed [21]. Also, the entropy, as with the ΔT_{ad}, is expected to be as elevated as possible because $T_{\text{op}}\,\Delta s$ is the amount of transferable heat at constant temperature. Indeed a giant barocaloric material presents:

- a relevant latent heat:

$$\Delta H_T = \mathrm{T}\Delta S_T, \tag{9.4}$$

- an enormous volume change at the transition temperature to balance the high entropy variation coupled with the phase change soas to get a high slope in the T versus p function, i.e. larger ΔT_{ad} under slight pressure mutations.

Attention must be paid toward first order materials since they can provoke problems in the regenerator to recover the initial position at the end of all the steps of the cycle, with a consequent decrease of performance. In this chapter many applications are shown and the involved barocaloric materials belong to two groups: the magnetostructural and the elastomers. Further details are reported in table 9.1.

Among the magnetostructural BCMs, the focus of this chapter is toward the barocaloric oxyfluorides $(NH_4)_2MoO_2F_4$ [22] and the $MnCoGe_{0.99}In_{0.01}$ [23]. $(NH_4)_2MoO_2F_4$ is characterized by a peak of adiabatic temperature change at 272 K where it shows 12; 15 and 18 K as ΔT_{ad} under $\Delta p = 0.5$; 0.7; 0.9 GPa, respectively. $MnCoGe_{0.99}In_{0.01}$ exhibits [23] an inverse BCE with a peak of 18 K as adiabatic temperature change in correspondence of the magnetostructural phase transition that occurs around 298 K for a pressure change of 0.3 GPa. The elastomers are soft materials, such as rubbers and, specifically, the focus is on acetoxy silicone rubber (ASR) and vulcanized natural rubber (VNR). The former shows [24] super-enormous BCE as 41 K at 0.390 GPa; 30 K at 0.273 GPa; 22 K at 0.173 GPa. The latter (VNR) [25], for the same 0.173 GPa pressure change exhibits a $\Delta T_{ad} = 11$ K.

The above-mentioned materials were investigated as barocaloric refrigerants in an ABR cycle for a cooling system. The regenerator ($20 \times 35 \times 45$ mm^3) is made of parallel plates (0.25 mm thickness) stacked by water channels (0.125 mm thickness) and it is thermally connected to a CHEX and an HHEX. The energy performances of the system were calculated, and a relevant number of operative conditions was considered through a numerical model based on Finite Element Method (FEM). In figure 9.3 the 2D schematization of the regenerator considered in the model is reported and the supporting mathematical model is reported in equation (9.5). The CHEXs and HHEXs were considered as boundary first type conditions: $T = T_C$; $T = T_H$, respectively, at the right and left sides of the active barocaloric regenerator.

Table 9.1. Main features of the caloric materials under investigation.

Material	T_{peak} (K)	Δp (GPa)	$\Delta T_{ad,max}$ (K)	ρ (kg m^{-3})	k (W m^{-1} K^{-1})	Reference
$(NH_4)_2MoO_2F_4$	272	0.9	18	2200	1	[22]
		0.7	15			
		0.5	12			
$MnCoGe_{0.99}In_{0.01}$	298	0.3	18.3	7900	65	[23]
ASR	298	0.390	41.1	960	1.48	[24]
	273	0.273	30			
	252	0.173	22			
VNR	333	0.173	11.1	930	0.13	[25]

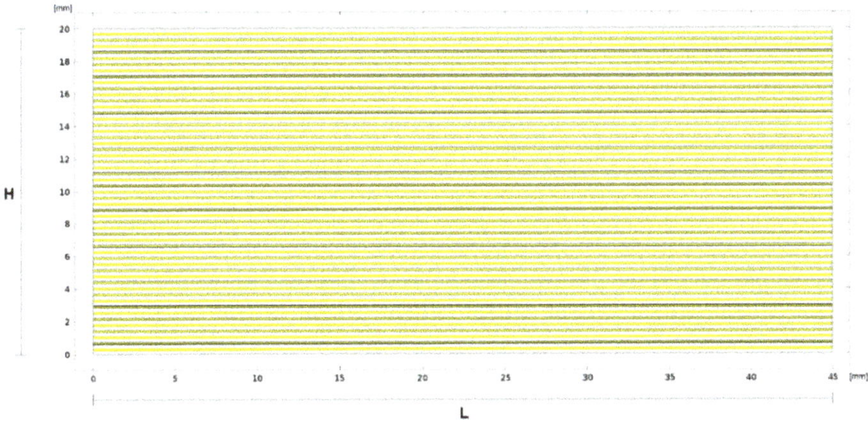

Figure 9.3. Design of the caloric regenerator. Reprinted from [4], copyright (2018), with permission from Elsevier.

$$\begin{cases} \dfrac{\partial u}{\partial x} + \dfrac{\partial v}{\partial y} = 0 \\ \dfrac{\partial u}{\partial t} + u\dfrac{\partial u}{\partial x} + v\dfrac{\partial u}{\partial y} = -\dfrac{1}{\rho_{\mathrm{f}}}\dfrac{\partial p}{\partial x} + v\left(\dfrac{\partial^2 u}{\partial x^2} + \dfrac{\partial^2 u}{\partial y^2}\right) \\ \dfrac{\partial v}{\partial t} + u\dfrac{\partial v}{\partial x} + v\dfrac{\partial v}{\partial y} = -\dfrac{1}{\rho_f}\dfrac{\partial p}{\partial y} + v\left(\dfrac{\partial^2 v}{\partial x^2} + \dfrac{\partial^2 v}{\partial y^2}\right) \\ \dfrac{\partial T_{\mathrm{f}}}{\partial t} + u\dfrac{\partial T_{\mathrm{f}}}{\partial x} + v\dfrac{\partial T_{\mathrm{f}}}{\partial y} = \dfrac{k_{\mathrm{f}}}{\rho_{\mathrm{f}} C_{\mathrm{f}}}\left(\dfrac{\partial^2 T_{\mathrm{f}}}{\partial x^2} + \dfrac{\partial^2 T_{\mathrm{f}}}{\partial y^2}\right) \\ \rho_{\mathrm{s}} C\dfrac{\partial T_{\mathrm{s}}}{\partial t} = k_{\mathrm{s}}\left(\dfrac{\partial^2 T_{\mathrm{s}}}{\partial x^2} + \dfrac{\partial^2 T_{\mathrm{s}}}{\partial y^2}\right) + Q \end{cases} \tag{9.5}$$

where u and v are the components of fluid velocity vector along x and y axes, p is the fluid pressure; T_{f} and T_{s} are the temperatures of the fluid and solid, respectively. k, ρ and C are the thermal conductivity, density and heat capacity of the fluid (if they are coupled with the f subscript) or of the solid (if they are followed by the s subscript).

In the I and III steps of the ABR cycle the fluid is not moving and the elastocaloric effect occurs through $+Q$ and $-Q$, respectively, and equation (9.5) becomes:

$$\begin{cases} \rho_{\mathrm{f}} C_{\mathrm{f}}\dfrac{\partial T_{\mathrm{f}}}{\partial t} = k_{\mathrm{f}}\left(\dfrac{\partial^2 T_{\mathrm{f}}}{\partial x^2} + \dfrac{\partial^2 T_{\mathrm{f}}}{\partial y^2}\right) \\ \rho_{\mathrm{s}} C\dfrac{\partial T_{\mathrm{s}}}{\partial t} = k_{\mathrm{s}}\left(\dfrac{\partial^2 T_{\mathrm{s}}}{\partial x^2} + \dfrac{\partial^2 T_{\mathrm{s}}}{\partial y^2}\right) + Q \end{cases} \tag{9.6}$$

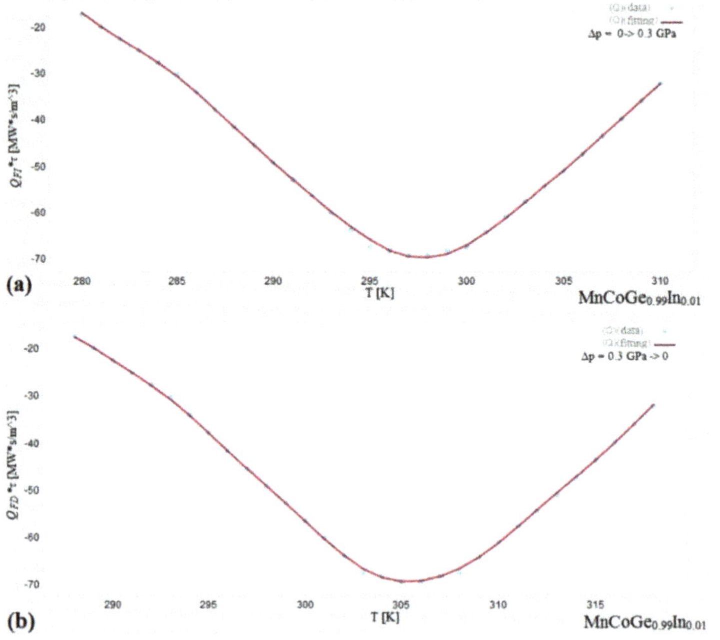

Figure 9.4. Q-functions built for the inverse barocaloric $MnCoGe_{0.99}In_{0.01}$ for field: (a) increasing; (b) decreasing in the range 0–0.3 GPa. Reprinted from [4], copyright (2018), with permission from Elsevier.

with:

$$Q = Q(\text{field}, T_S) = \frac{\rho_s C(\text{field}, T_s)\Delta T_{ad}(\text{field}, T_s)}{\tau} \tag{9.7}$$

The construction of Q functions was obtained thanks to commercial software that finds the mathematical expression closest to the table function that it receives in input. In figures 9.4 and 9.5 are reported as an example the Q functions of $MnCoGe_{0.99}In_{0.01}$ and ASR. The input data come from experimental data already published in the literature, and in table 9.2 are listed the corresponding mathematical expressions for all the investigated materials. Finite element method is used to solve the equations of the model and the regenerator was meshed through the free triangular meshing (figure 9.6).

9.4 System performance analysis

The implementation of the ABR refrigeration cycle into the model can be seen as a looped sequence of four processes. Each process is time-dependently linked to the previous and to the following since its initial condition is the final result of the previous step and the final status of the current process transforms into the initial condition of the following step. Each period has a duration of τ seconds, so the entire duration of each ABR cycle is:

$$\theta = 4\tau \tag{9.8}$$

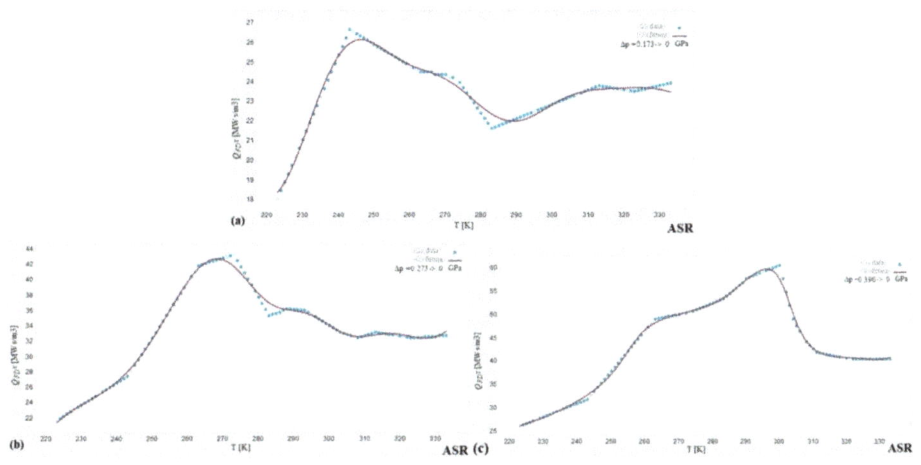

Figure 9.5. For the barocaloric elastomer ASR Q-functions built for adiabatic decompression process in the range: (a) [0.173; 0] GPa; (b) [0.273; 0] GPa; (c) [0.390; 0] GPa. Reprinted from [27], copyright (2020), with permission from Elsevier.

Table 9.2. $Q\ \tau$ functions, related to the field increasing process, obtained starting from the experimental data.

Material	Δp [GPa]	Mathematical expression
$(NH_4)_2MoO_2F_4$	0.9	$Q^*\tau = 10^6{*}[133.4+10.7{*}\cos(351.5{*}T) - 0.3822{*}T - 0.03769{*}T{*}\cos(351.5{*}T)]$
	0.7	$Q^*\tau = 10^6{*}[1.088+0.2777{*}\cos(0.3995{*}T) - 85.38-0.002365{*}T^2]$
	0.5	$Q^*\tau = 10^6{*}[96.92+0.4259{*}\sin(0.2341{*}T) - 0.2063{*}T]$
$MnCoGe_{0.99}In_{0.01}$	0.3	$Q^*\tau = 10^6{*}[-31.83 - 34.34{*}\cos(0.0002128{*}T^2)]$
ASR	0.390	$Q_{comp} * \tau = 10^6 * [(1.33 * T + 0.000211 * (T^2 * \sin(4.33 + \sin(-0.0636 * T) * \sin(4.12 + \sin(0.0726 * T) - 165 - 0.0216 * T)]$
	0.273	$Q_{comp} * \tau = 10^6 * [(0.186 * T + 1.95 * \cos(0.141{*}T) + 0.0314 * T * \sin?(0.0535 * T) + \sin(0.0545 * T) * \cos(0.144 * T) - 18.2)]$
	0.173	$Q_{comp}{*}\tau = 10^6{*}\left[23.01 + - 0.00114{*}T + 9.1{*}\cos(-0.0865{*}T) + 3.92{*}\cos(0.183{*}T) - 2980{*}\frac{\cos(-0.0865 * T)}{T} - 0.0126{*}T{*}\cos(0.183{*}T)\right]$
VNR	0.173	$Q_{comp} * \tau = 10^6 * [1.362 + 0.060\,56 * T + 0.242 * \cos(0.272 * T) - 0.2702 * \cos(0.2129 - 0.2714 * T]$

The system experiments the looped **ABR** cycle until the cycle evolves stationarily, i.e. the cyclicality criterion is addressed:

$$T(x, y, 0 + n\theta) = T(x, y, 4\tau + n\theta) \tag{9.9}$$

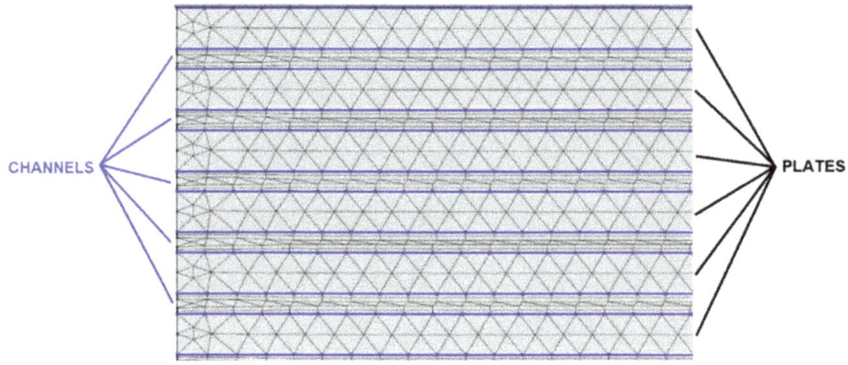

Figure 9.6. A focus on regenerator triangular mesh built for solving the model.

To do this a limit quantity δ must be fixed and it represents the maximum tolerances allowed on the deviance between the two solutions:

$$\delta = \max\{T(x, y, 0 + n\theta) - T(x, y, 4\tau + n\theta)\} < \bar{\varepsilon} \tag{9.10}$$

The map of the energy performances was drawn basing on the following energy parameters:

$$\Delta T_{\mathrm{span}} = \frac{1}{\tau} \int_{\tau+n\theta}^{2\tau+n\theta} T_{\mathrm{f}}(L, y, t)dt - T_{\mathrm{c}} \tag{9.11}$$

$$\dot{Q}_{\mathrm{H}} = \frac{1}{\theta} \int_{\tau+n\theta}^{2\tau+n\theta} \dot{m}_{\mathrm{f}} C_{\mathrm{f}}(T_{\mathrm{f}}(L, y, t) - T_{\mathrm{H}})dt \tag{9.12}$$

$$\dot{Q}_{\mathrm{ref}} = \frac{1}{\theta} \int_{3\tau+n\theta}^{4\tau+n\theta} \dot{m}_{\mathrm{f}} C_{\mathrm{f}}(T_{\mathrm{C}} - T_{\mathrm{f}}(0, y, t))dt \tag{9.13}$$

$$\mathrm{COP} = \frac{\dot{Q}}{\dot{W}_{\mathrm{TOT}}} \frac{\dot{Q}}{\dot{W}_{\mathrm{field}} + \dot{W}_{\mathrm{pump}}} = \frac{\dot{Q}}{\dot{W}_{\mathrm{field}} + \dot{W}_{\mathrm{pump}}} \tag{9.14}$$

Equation (9.11) reports the temperature span across the regenerator that operates in heat pumping mode; dually one can easily derive the temperature span of the system working in cooling mode. Equations (9.12) and (9.13) are the useful thermal powers of the system working in heat pump and cooling mode, respectively. Equation (9.14) represents the COP, i.e. the useful thermal power and total power ($\dot{W}_{\mathrm{TOT}}$) needed to make the system achieve the useful effect. Two are the contributions accountable as power expense: $\dot{W}_{\mathrm{field}}$ that measures the power required to make variations in the driving field, i.e. to observe BCE; $\dot{W}_{pump}$ is the contribution associated with the mechanical power needed to move the water inside the ABR system:

$$\dot{W}_{field} = \dot{Q}_H - \dot{Q}_{ref} \tag{9.15}$$

$\dot{W}_{pump}$ was calculated according to the following equation:

$$\dot{W}_{pump} = \frac{\dot{m}_f(\Delta p_{CF} + \Delta p_{HF})}{\eta_{pump}\rho_f} \tag{9.16}$$

Specifically, if the system operates in heating power or cooling modes, the COP is particularized as:

$$COP = \frac{\dot{Q}_H}{\dot{W}_{TOT}} \tag{9.17}$$

$$COP = \frac{\dot{Q}_c}{\dot{W}_{TOT}} \tag{9.18}$$

The environmental impact has also been evaluated through the calculation of the total equivalent warming impact (TEWI). The latter is an index that considers the greenhouse effect of a system by dividing it into a direct and an indirect contribution, as:

$$TEWI = CO_{2,\,dir} + CO_{2,\,indir} \quad \left[\text{kg } CO_2\right] \tag{9.19}$$

$$CO_{2,\,dir} = RC\left[P_L + \left(\frac{1 - P_R}{V}\right)\right]V \cdot GWP \tag{9.20}$$

$$CO_{2,\,indir} = \alpha \cdot \frac{\dot{Q}_{ref}}{COP} \cdot H \cdot V \tag{9.21}$$

Equation (9.20) explains the direct contribution that is related to the GWP of the employed refrigerant. Specifically, the contribution is proportional to the amount of refrigerant leaked during the normal operation of the systems (P_L) and the one due to inadvertently accidents or due to a wrong emission ($1 - P_R$).

The estimations of the analysis perpetuated in this chapter focus on 5% per year of leakage during regular operations (P_L), whereas it is considered that there are no accidents over the lifetime of the plant and that it is ecologically dismissed ($1 - P_R) = 0$. The literature provides some indicative, average levels of CO_2 released per kWh of electrical energy for various countries. For Italy, the value is 0.6 kg CO_2/kWh. Specifically the TEWI is evaluated as percentage deviancy with respect to the TEWI index of a VC plant, working under the same operative conditions:

$$\Delta TEWI = \frac{(TEWI_{ABR} - TEWI_{VC})}{TEWI_{VC}} \tag{9.22}$$

Specifically it has to be considered that the $CO_{2,\ dir}$ of the barocaloric system is zero since BCMs are solid-state, i.e. their GWP is zero. Indicatively, considering also the GWP of the currently employed HFC refrigerants, the direct contribution provided by the VC system is 10% of the total TEWI.

9.5 Results and discussion

This section introduces a panorama of results obtained in barocaloric systems devoted to the most variable applications.

9.5.1 Wine cooler

Aprea *et al* investigated a barocaloric system developed to operate as a wine cooler [4] with the following operating conditions: 1.25 Hz as frequency, 292 K and 300K as temperatures of the CHEXs and HHEXs. The HTF is water and the flow rate was varied 150–250 kg s^{-1} m^{-2}, whereas the system is tested with $(NH_4)_2MoO_2F_4$ and $MnCoGe_{0.99}In_{0.01}$.

The energy performances were evaluated: in figure 9.7 the temperature span versus normalized flow rate (G_f) is reported. One can observe that the temperature span and the velocity of the fluid are related through a function of inverse proportionality; values are confined in the 12–15.5 K range with a peak of 15.3 K touched employing $(NH_4)_2MoO_2F_4$ under $\Delta p = 0.9$ GPa for $G_f = 150$ kg m^{-2} s^{-1}.

Figure 9.8 plots the cooling power versus normalized warer flow rate and in this case the relation between these two parameters is direct proportionality. The growing trends with velocity, for all the materials, reveal a peak of 400 W for ($\Delta p = 0.9$ GPa and $G_f = 250$ kg m^{-2} s^{-1}).

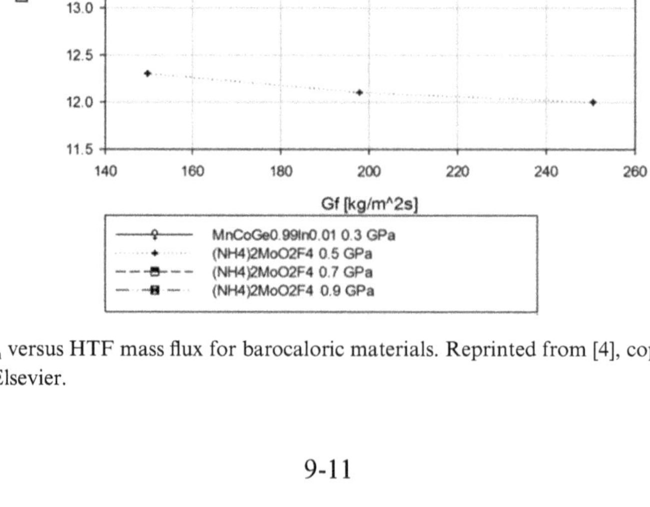

Figure 9.7. ΔT_{span} versus HTF mass flux for barocaloric materials. Reprinted from [4], copyright (2018), with permission from Elsevier.

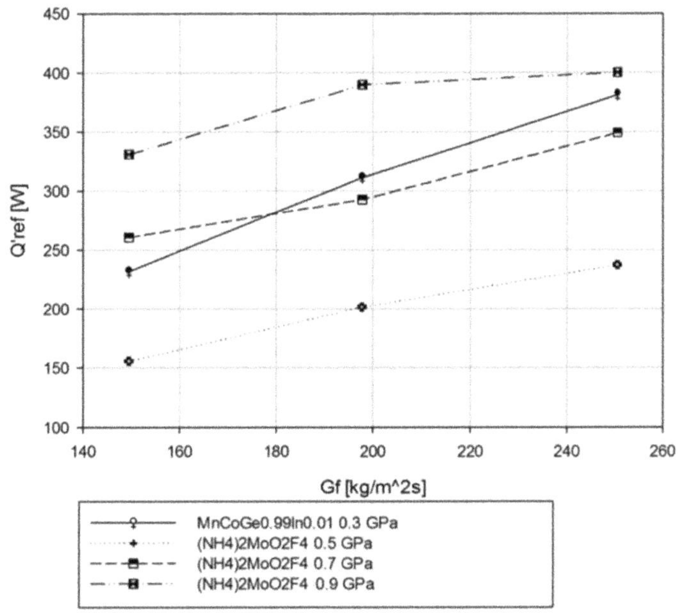

Figure 9.8. $\dot{Q}_{ref}$ versus HTF mass flux for barocaloric materials. Reprinted from [4], copyright (2018), with permission from Elsevier.

As is visible in figure 9.9, direct proportionality is observed also between COP versus G_f, for all the materials and the peak is 10.3, calculated through the employment of $(NH_4)_2MoO_2F_4$ ($\Delta p = 0.9$ GPa; $G_f = 250$ kg s^{-1} m^{-2}).

The environmental impact of the system was also calculated [26] and compared with a VC-based wine cooler operating with HFC134 (GWP = 1430) and with an annual power consumption of 290 kWh yr^{-1}. The percentage ΔTEWI is plotted in figure 9.10 and the trends reveal that the employment of $(NH_4)_2MoO_2F_4$ at pressure variation of 0.9 GPa, guarantees the highest reduction in terms of environmental impact of the barocaloric wine cooler with respect to the VC one: the peak value is −70%; the mean value is −64%. Real affordability and comfort derives from the utilization of $MnCoGe_{0.99}In_{0.01}$ with an average −60% as reduction of environmental impact in terms of TEWI.

9.5.2 Heat pump

Aprea *et al* also investigated [27] the employment of barocaloric material in a solid-state heat pump based on the ACR cycle: 278–298 K is the outdoor–indoor temperature range; 1.25 H is the operational frequency; water is the HTF and the normalized flow rate was varied in the [150; 250] kg s^{-1} m^{-2}. The system was tested employing the following barocaloric materials: $(NH_4)_2MoO_2F_4$; $MnCoGe_{0.99}In_{0.01}$; ASR and VNR.

Temperature spans reported in figure 9.11 reveal that all the materials provide performances greater than 20 K. ASR ensures 28.7 K rate 150 kg m^{-2} s^{-1}. The lowest spans are measured using VNR since they do not overcome 25K.

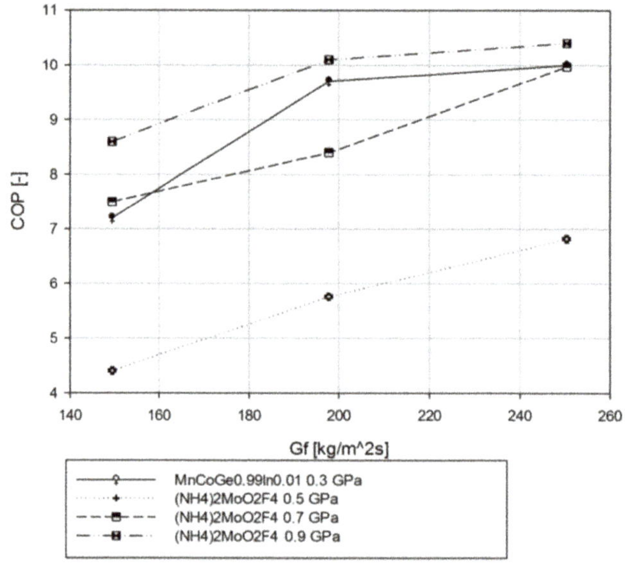

Figure 9.9. COP versus HTF mass flux for barocaloric materials. Reprinted from [4], copyright (2018), with permission from Elsevier.

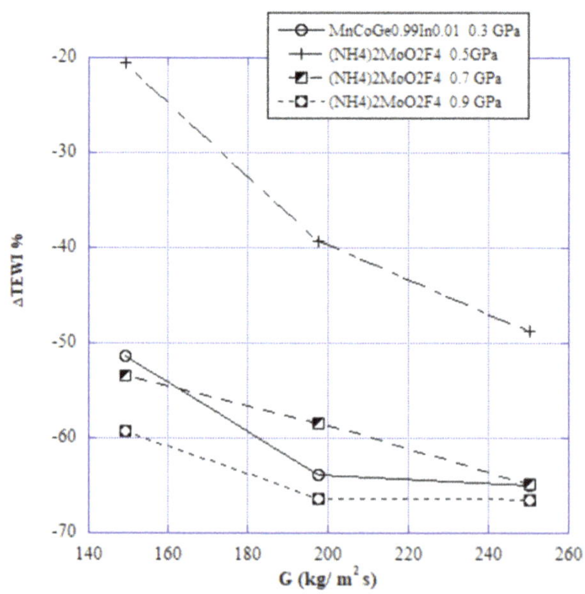

Figure 9.10. ΔTEWI as a function of water mass flux for the different BCMs.

In figure 9.12 the heating power versus G_f are displayed with a maximum of 430 W at 250 kg m^{-2} s^{-1} given by ASR. Direct is the proportionality between the two plotted terms. Considering the scalability of the mass of the barocaloric refrigerant with respect to the power target of the application, the data are very

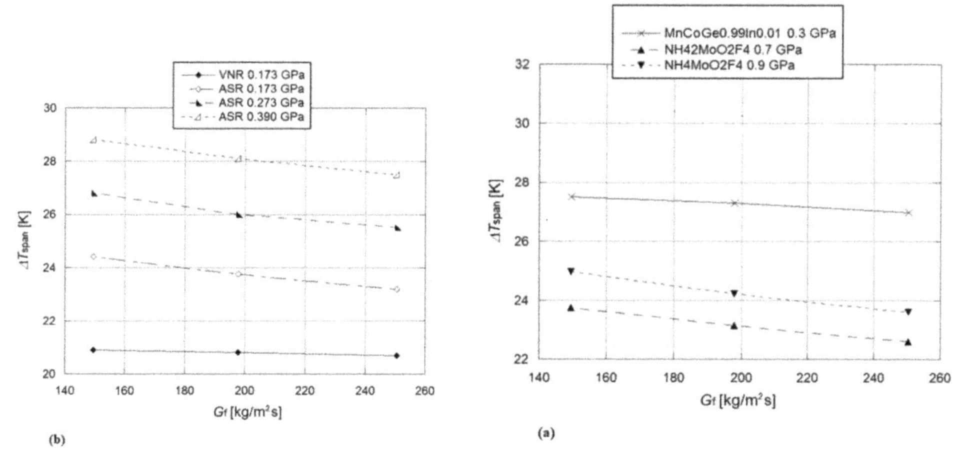

Figure 9.11. ΔT_{span} versus mass flux for: (a) magnetostructural compounds, (b) elastometer BCMs. Reprinted from [27], copyright (2020), with permission from Elsevier.

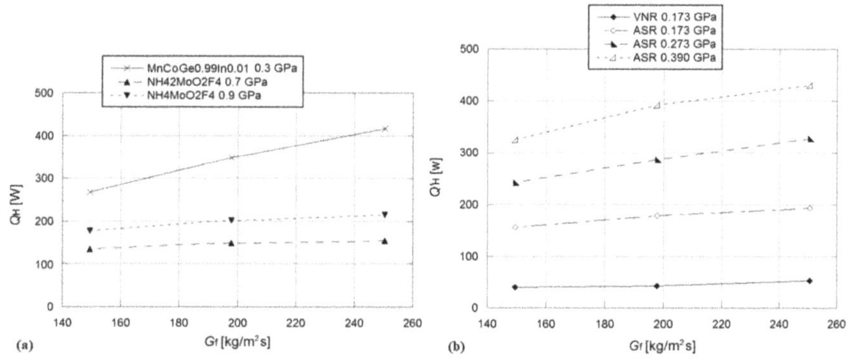

Figure 9.12. $\dot{Q}_{ref}$ versus mass flux for: (a) magnetostructural compounds, (b) elastometer BCMs. Reprinted from [27], copyright (2020), with permission from Elsevier.

promising since multiple ABRs can be implemented to cover greater demands in terms of thermal load.

In figure 9.13 COPs as a function of mass flux are reported for the different BCMs. The best values of the COP can be obtained with $MnCoGe_{0.99}In_{0.01}$ at 0.3 GPa, reaching the maximum value of around 10 at G 250 kg m^{-2} s^{-1} (corresponding to 68% of Carnot COP). More than satisfactory values can be obtained with ASR.

The environmental impact, calculated in terms of TEWI and compared with a VC heat pump [28] working with HFC134a reveals, as shown in figure 9.14, the barocaloric heat pump having a TEWI impact lower than the VC system, except if VNR is employed. $MnCoGe_{0.99}In_{0.01}$ at 0.3 GPa, ensures a maximum saving (−70%) in terms of equivalent emissions of carbon dioxide. ASR, on its own, ensures −50% as reduction.

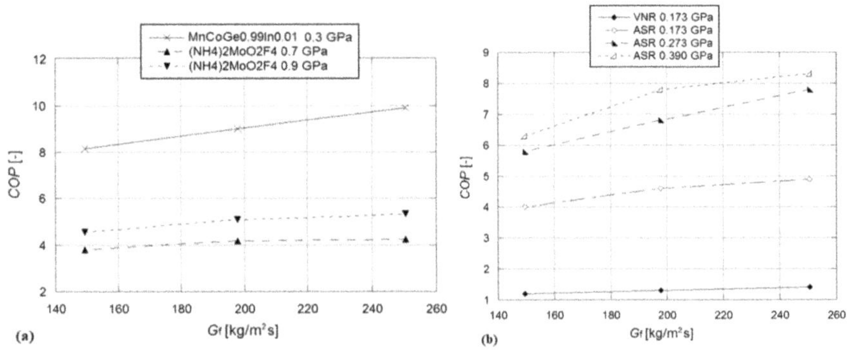

Figure 9.13. COP versus mass flux for: (a) magnetostructural compounds, (b) elastometer BCMs. Reprinted from [27], copyright (2020), with permission from Elsevier.

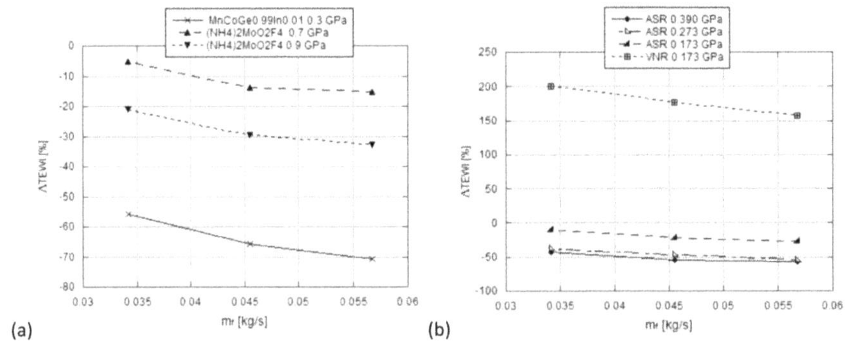

Figure 9.14. ΔTEWI as a function of water mass flow rate for the different BCMs. Reprinted from [40], with the permission of AIP Publishing.

9.5.3 Domestic refrigerator

If the application of the barocaloric system is the cooling one, e.g. a domestic refrigerator, as presented in the investigations [29, 30], the working temperature range is 255–290 K.

The cooling system, mounting an ABR regenerator made of ASR, is crossed by 50% ethylene–glycol (EG)–50% water mixture as auxiliary fluid to operate up to the freezing point (236K). Additionally, improved HTFs were considered through the addition of Cu nanoparticles (properties at 293 K: $\rho = 8933$ kg m^{-3}; $C_p = 383$ J kg^{-1} K^{-1}; $k = 401$ W m^{-1} K^{-1}) to increase the convective heat transfer between solid and fluid, in a variable concentration between 0% and 10%. The resultant thermodynamical properties of the nanofluid are evaluated through the following relations provided by scientific open literature [31–33]:

$$\rho_{nf} = (1 - \varphi)\rho_{bf} + \varphi\rho_{np} \tag{9.23}$$

$$\rho_{nf} C_{p,\, nf} = (1 - \varphi)\left(\rho C_p\right)_{bf} + \varphi\left(\rho C_p\right)_{np} \tag{9.24}$$

$$\mu_{nf} = \frac{1}{(1-\varphi)^{2.5}}\mu_{bf} \tag{9.25}$$

$$k_{nf} = \frac{k_{np} + (n-1)k_{bf} - (n-1)\varphi\left(k_{bf} - k_{np}\right)}{k_{np} + (n-1)k_{bf} + \varphi\left(k_{bf} - k_{np}\right)} \tag{9.26}$$

where, supposing spherical nanoparticles, n is equal to 3.

The period of the ABR cycle is 0.8 s, whereas the nanaofluid flows with variable velocity [0.04; 0.20] m s^{-1}. The energy performances have been evaluated and the following considerations are derived:

- Temperature span that in figure 9.15 is plotted versus the velocity of the HTF for growing nanoparticle concentrations, presents a maximum that is located at 0.06 m s^{-1} for all the concentrations trends and whose absolute value is increasing with the nanoparticle concentration (peak of 40.7 K with Φ = 10%). The lowest ΔT_{span} values are evaluated for 0.20 m s^{-1}: 38.6 K with Φ = 0%, 38.9 K with Φ = 10%. Furthermore, one can observe that on equal Φ, the temperature span improvement, with setting the optimal velocity (0.06 m s^{-1}), touches a maximum of +4%.
- Cooling powers, plotted in figure 9.16 are increasing functions of fluid velocity and nanoparticle concentration. The percentage enhancement of the cooling power brought by the increment of nanofluid concentration does not follow the same trend for each fluid velocity. Indeed, for v = 0.20 m s^{-1} the highest $\frac{\dot{Q}_{\text{ref}}}{W}$ is 4340 W m^{-1} (at Φ = 10%) and the percentage enhancement of cooling power given by increasing nanofluid concentration up to 10%, is just 2% with respect of working with the base fluid as HTF. On the other side, when the HRF moves with 0.10 m s^{-1}, even if the maximum $\frac{\dot{Q}_{\text{ref}}}{W}$ is 3141 W m^{-1} (for Φ = 10%), the increment of $\frac{\dot{Q}_{\text{ref}}}{W}$ with increasing Φ from 0% to 10% is

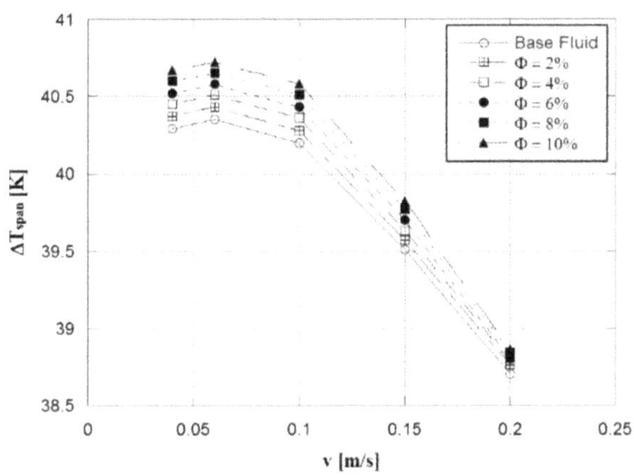

Figure 9.15. Temperature span of the barocaloric refrigerator as a function of the velocity of the heat-transfer nanofluid, for the base fluid and different concentrations of Cu nanoparticles. Reprinted from [29], copyright (2020), with permission from Elsevier.

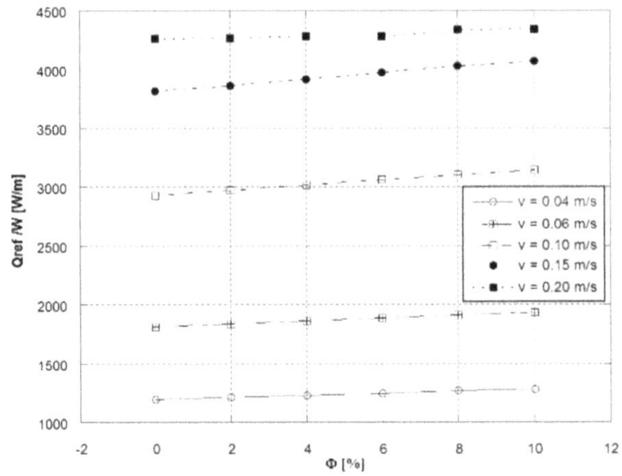

Figure 9.16. $\frac{\dot{Q}_{\mathrm{ref}}}{W}$ versus of the concentration of Cu nanoparticles in the heat-transfer nanofluid, parametrized for nanofluid velocity. Reprinted from [29], copyright (2020), with permission from Elsevier.

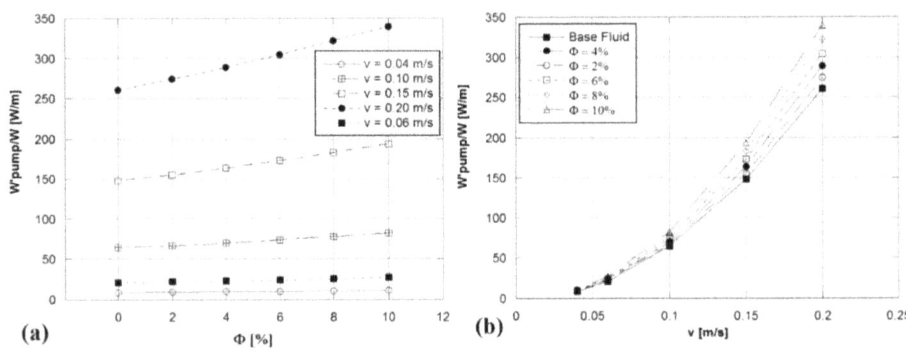

Figure 9.17. Mechanical power on width associated to the work of the pump for fluid motion as a function of: (a) concentration of Cu nanoparticles in the nanofluid; (b) nanofluid velocity. Reprinted from [29], copyright (2020), with permission from Elsevier.

+7.3% (larger than the one calculated for $v = 0.20$ m s^{-1}). For $v = 0.04$ m s^{-1} and $v = 0.06$ m s^{-1} such augmentations are 7.0% and 6.7%, respectively. This attests nevertheless to the positive contribution on cooling power given by working with nanofluids.

- The mechanical power needed to pump the fluid was plotted in figure 9.17 verus (a) concentration of Cu nanoparticles in the nanofluid; (b) nanofluid velocity. The quantity increases strongly with the velocity of the fluid; slightly with the concentration of nanoparticles. Anyhow the values do not grow up to 30 W m^{-1} for velocities lower than 0.06 m s^{-1} that represent a threshold value to get acceptable values of mechanical power of the pump.
- The trend of COP, plotted in figure 9.18, is strongly influenced by the latter term. Figure 9.18(a) displays COP versus nanoparticle concentration; figure 9.18(b) COP versus fluid velocity. Largest COP are calculated for

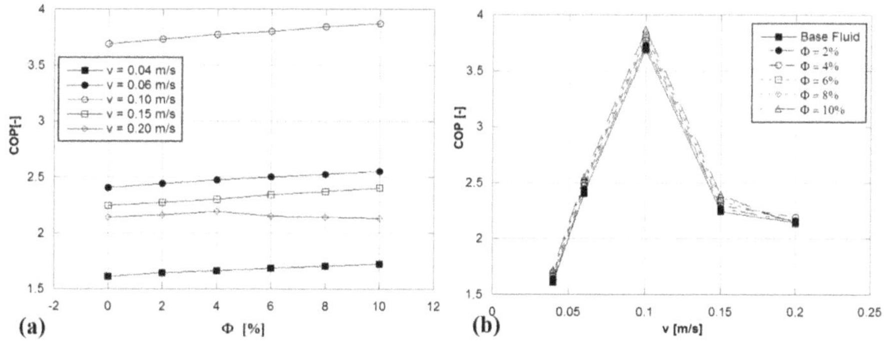

Figure 9.18. COP of the ABR as a function of: (a) concentration of Cu nanoparticles in the nanofluid; (b) nanofluid velocity. Reprinted from [29], copyright (2020), with permission from Elsevier.

0.10 m s^{-1}. The simultaneous growing of nanoparticle concentration provides a +8.5% as further increment of COP with a maximum of 3.9 ($\Phi = 10\%$). The work of the pump determines a decisive falling of the COP after the peak of 0.10 m s^{-1}. For $v = 0.20$ m s^{-1} the influence of $\frac{\dot{W}_{\text{pump}}}{W}$ with increasing Φ is so strong that COP increases up to working with $\Phi = 4\%$, then it falls since $\frac{\dot{W}_{\text{pump}}}{W}$ is more influential than the benefits carried by cooling power growing.

The comparison with a VC cooler was made considering the deviance of the COP of the ABR system with respect to the VC as visible in equation (9.27):

$$\Delta\text{COP} = \frac{\text{COP}_{\text{ABR}} - \text{COP}_{\text{VC}}}{\text{COP}_{\text{VC}}} \tag{9.27}$$

The COP of the VC systems are calculated by means of experimental measurements using an experimental apparatus introduced in previous works [34–39]. HFC134a is the refrigerant used in the VC system. The temperature of the cell was set at 255 K, whereas the temperature of the environment where the measures were perpetuated was 290 K. The percentage deviance of COP is reported in figure 9.19 with respect to the nanoparticle concentration of the nanofluid of the ABR system. The results underline an improvement of the COPs of ABR with respect to VC ranging between +5% and +150%, with a maximum at 0.10 m s^{-1} and 10% as Cu-concentration.

In figure 9.20 is shown the environmental impact of the ABR cycle in terms of ΔTEWI with respect to the VC system. Trends are shown as a function of nanoparticles concentration and the addition of the latter carries to a further reduction of the environmental impact reaching also −60% at 10% as Cu-concentration and 0.10 m s^{-1} as nanofluid velocity.

9.6 Conclusions

This chapter provides a panorama on the energy performances of barocaloric systems determined through numerical 2D models. Temperature span, useful

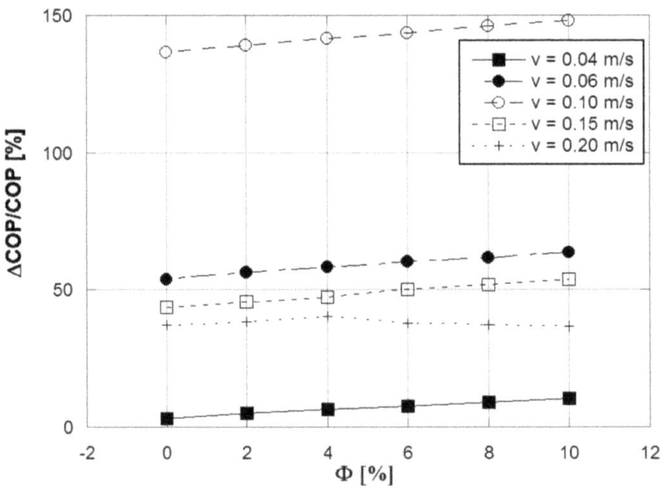

Figure 9.19. The energy performances of the Active Barocaloric refrigerator compared to a vapor compression cooler ones, in terms of ΔCOP %. Reprinted from [29], copyright (2020), with permission from Elsevier.

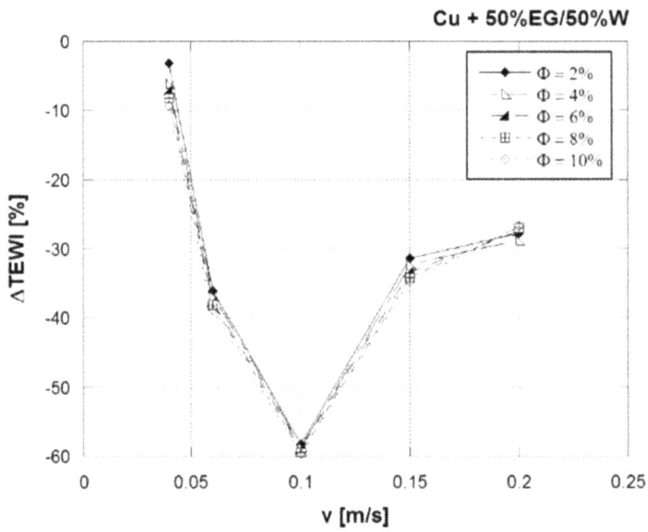

Figure 9.20. ΔTEWI percentage versus fluid flow rate. Reproduced from [28]. CC BY 4.0.

thermal power, COP are calculated. Additionally, the environmental impact is measured in terms of TEWI and compared with VC systems.

The barocaloric materials investigated are: $(NH_4)_2MoO_2F_4$; $MnCoGe_{0.99}In_{0.01}$; ASR; VNR. The auxiliary fluids employed are: pure water (over 273 K); 50% ethylene-glycol–50% water (under 273 K); Cu-based nanofluids derived by the previously-mentioned anti-freezing mixture.

The working temperature ranges are those in the examples of: (1) a wine cooler; (2) a heat pump; (3) a cooler, and the ABR frequency is 1.25 Hz.

The following considerations can be drawn:

- In the wine cooler application, the barocaloric refrigeration can be used instead of traditional technologies based on VC with encouraging results (ΔT_{span} greater than 12 K, COP greater than 4.5).
- The best BCM in the wine cooler application is $(NH_4)_2MoO_2F_4$ with a Δp of 0.9 GPa that ensures a ΔT_{span} of 15.3 K, a cooling power of 400 W with a COP of 10.3.
- The barocaloric wine cooler has a TEWI impact lower than the VC one. The best case is working with $(NH_4)_2MoO_2F_4$ at 0.9 GPa, which carries a maximum 67% reduction of TEWI. Similar results can be obtained with $MnCoGe_{0.99}In_{0.01}$ (with a mean value of –64%).
- Barocaloric refrigeration in the application related to heat pumps gives very satisfactory results both in terms of ΔT_{span} (always greater than 20 K) and of the heating powers.
- The temperature spans are promising in the heat pumps temperature range given by the barocaloric elastomers ASR (from 27.3 K to 28.8 K) if subjected to a pressure field of 0.390 GPa, as well as the ones proper of $MnCoGe_{0.99}In_{0.01}$ (with a middle value of 27.3 K under $\Delta p = 0.3$ GPa).
- The highest powers of the investigated caloric heating pump are associated with ASR ($\Delta p = 0.390$ GPa) once again. The maximum heating power is 430 W at 250 kg m^{-2} s^{-1}. Also with $MnCoGe_{0.99}In_{0.01}$ very suitable heating power values are obtained.
- The best values of the COP in the heat pumps temperature range can be obtained with $MnCoGe_{0.99}In_{0.01}$ at 0.3 GPa, reaching the maximum value of around 10 at G 250 kg m^{-2} s^{-1} (corresponding to 68% of Carnot COP). More than satisfactory values can be obtained with ASR.
- In almost all the cases (except the barocaloric VNR), the barocaloric heat pump has a TEWI impact lower than the VC system. The best case is working with $MnCoGe_{0.99}In_{0.01}$ at 0.3 GPa, which carries a maximum 70% reduction of TEWI.
- The energy performances of the barocaloric domestic refrigerator are always better than those of the VC system (maximum value of +150%).
- Adding nanoparticles to the HTF of the barocaloric refrigerator results in an effect favorable to heat transfer and, consequently, to temperature span (+4% maximum) and cooling power (+7.3% maximum). The nanoparticles volume fraction has a detrimental effect on mechanical power coupled with the pump for nanofluids motion (maximum increase of +30%). Nevertheless, the positive effect prevails, and the COP slightly increases using nanofluids (mean value of +5%) until reaching an optimal working-point.
- The environmental impact of the barocaloric refrigerator is always lower than that of the traditional vapour compression one, touching a minimum of –60% at $\varphi = 10\%$ with 0.10 m s^{-1} fluid velocity.

Therefore, it is possible to assess that barocaloric refrigeration is a very promising non-convectional technique since in all the applications considered it has always allowed obtaining better energy performance than VC and lower environmental impact in terms of greenhouse effect.

References

[1] Office of Environmental Quality 1987 *Montreal Protocol on Substances That Deplete the Ozone Layer* (New York: United Nations Environment Program (UN))

[2] United Nations 1997 *Kyoto Protocol to the United Nation Framework Convention on Climate Change* (Kyoto: United Nations Environment Program (UN))

[3] Kitanovski A, Plaznik U, Tomc U and Poredoš A 2015 *Int. J. Refrig.* **57** 288–98

[4] Aprea C, Greco A, Maiorino A and Masselli C 2018 *Energy* **165** 439–55

[5] Pecharsky V K and Gschneidner K A 1999 *J. Magn. Magn. Mater.* **200** 44–56

[6] Ohnishi T, Soejima K, Yamashita K and Wada H 2017 *Magnetochemistry* **3** 6

[7] Thang N, Dijk N and Brück E 2016 *Materials* **10** 14

[8] Scott J F 2011 *Annu. Rev. Mater. Res.* **41** 229–40

[9] Sun X, Huang H, Jafri H M, Wang J, Wen Y and Dang Z M 2019 *Appl. Sci.* **9** 1672

[10] Jiang Z Y, Zheng G P, Zheng X C and Wang H 2017 *Polymers* **9** 315

[11] Wu Y, Ertekin E and Sehitoglu H 2017 *Acta Mater.* **135** 158–76

[12] Strässle T and Furrer A 2000 *Int. J. High Pressure Res.* **17** 325–33

[13] Brück E, Tegus O, Thanh D C and Buschow K H J 2007 *J. Magn. Magn. Mater.* **310** 2793–9

[14] Aprea C, Greco A, Maiorino A and Masselli C 2018 *Therm. Sci. Eng. Progr.* **6** 370–9

[15] Greco A, Aprea C, Maiorino A and Masselli C 2019 *Int. J. Refrig.* **106** 66–88

[16] Kitanovski A 2020 *Adv. Energy Mater.* **10** 1903741

[17] Kabirifar P, Žerovnik A, Ahčin Ž, Porenta L, Brojan M and Tušek J 2019 *J. Mech. Eng.* **65** 615–30

[18] Cirillo L, Greco A and Masselli C 2022 *Therm. Sci. Eng. Prog.* **33** 101380

[19] Gorev M V, Bogdanov E V and Flerov I N 2017 *J. Phys. D: Appl. Phys.* **50** 384002

[20] Lei T, Engelbrecht K, Nielsen K K, Bez H N and Bahl C R H 2016 *J. Phys. D: Appl. Phys.* **49** 345001

[21] Gottschall T, Skokov K P, Fries M, Taubel A, Radulov I, Scheibel F and Gutfleisch O 2019 *Adv. Eng. Mater.* **9** 1901322

[22] Gorev M V, Bogdanov E V, Flerov I N, Kocharova A G and Laptash N M 2010 *Phys. Solid State* **52** 167–75

[23] Wu R R *et alet al* 2015 *Sci. Rep.* **5** 18027.1

[24] Imamura W, Usuda E O, Paixão L S, Bom N M, Gomes A M and Carvalho A M G 2020 *Chin. J. Pol. Sci.* **38** 999–1005

[25] Usuda E O, Bom N M and Carvalho A M G 2017 *Eur. Pol.* J **92** 287–93

[26] Aprea C, Greco A, Maiorino A and Masselli C 2018 *Int. J. Heat Technol* **36** 1155–62

[27] Aprea C, Greco A, Maiorino A and Masselli C 2020 *Int. J. Refrig.* **109** 1–11

[28] Aprea C, Greco A, Maiorino A and Masselli C 2019 *Climate* **7** 115

[29] Aprea C, Greco A, Maiorino A and Masselli C 2020 *Energy* **190** 116404

[30] Aprea C, Greco A, Maiorino A and Masselli C 2019 *Energies* **12** 2902

[31] Bianco V, Manca O, Nardini S and Vafai K (ed.) 2015 *Heat Transfer Enhancement with Nanofluids* (Boca Raton, FL: CRC Press)

[32] Brinkman H C 1952 *J. Chem. Phys.* **20** 571–1

[33] Hamilton R L and Crosser O K 1962 *I & EC Fundamentals* **1** 187–91

[34] Aprea C, Greco A and Maiorino A 2016 *Appl. Therm. Eng.* **106** 959–67

[35] Aprea C, Greco A, Maiorino A, Masselli C and Metallo A 2016 *Int. J. Heat Technol.* **34** S212–8

[36] Aprea C, Greco A and Maiorino A 2017 *Int. J. Refrig.* **76** 109–17

[37] Aprea C, Greco A and Maiorino A 2017 *Int. J. Refrig.* **82** 71–82

[38] Aprea C, Greco A, Maiorino A and Masselli C 2018 *Int. J. Therm. Sci.* **127** 117–25

[39] Aprea C, Greco A and Maiorino A 2018 *Appl. Therm. Eng.* **141** 226–33

[40] Greco A, Aprea C, Maiorino A and Masselli C 2019 The environmental impact of a caloric heat pump working with solid-state materials based on TEWI analysis *AIP Conf. Proc.* **2191** 020091

www.ingramcontent.com/pod-product-compliance
Ingram Content Group UK Ltd.
Pitfield, Milton Keynes, MK11 3LW, UK
UKHW051859100926
8556IPUK00005B/701

9 780750 346887